Phenomics

Phenomics

Editor

John M. Hancock

Department of Physiology
Development and Neuroscience
Cambridge University
Cambridge, UK

CRC Press
Taylor & Francis Group
Boca Raton London New York

CRC Press is an imprint of the
Taylor & Francis Group, an **informa** business
A SCIENCE PUBLISHERS BOOK

CRC Press
Taylor & Francis Group
6000 Broken Sound Parkway NW, Suite 300
Boca Raton, FL 33487-2742

First issued in paperback 2019

ISBN-13: 978-1-4665-9095-3 (hbk)
ISBN-13: 978-0-367-37892-9 (pbk)

**Visit the Taylor & Francis Web site at
http://www.taylorandfrancis.com**

**and the CRC Press Web site at
http://www.crcpress.com**

I would like to dedicate this book
to my father, Ronald Hancock
who passed away on 29 September 2013

Preface

The idea for this book arose when I was deeply involved with the development of mouse phenotyping projects during the first decade of this century. It was clear that we were developing a new approach in biology, but that similar efforts were taking place using other model and non-model species. Although the term phenomics had been invented, and we occasionally used it, it was not widely used or recognized and, as a consequence, themes common to phenotyping efforts across this wide range of species were not being explored. The aim of this book, then, is to lay out the approaches used for systematic phenotyping in different species and to pull out their commonalities as well as their differences.

I would like to thank all of the contributing authors in making this volume what is, I hope, a fascinating introduction to a relatively new field of biology.

Wantage   **John Hancock**
October 2013

Contents

Dedication   v

Preface   vii

1. **Introduction to "Phenomics"**   1
   John M. Hancock

2. **Computational Phenotype Analysis in Human Medicine**   8
   Peter N. Robinson

3. **Phenomics of the Laboratory Mouse**   24
   John M. Hancock and *Michael S. Dobbie*

4. **Phenotyping Zebrafish**   65
   Elisabeth M. Busch-Nentwich

5. **Systematic Cell Phenotyping**   86
   Jean-Karim Hériché

6. **Systematic Phenotyping of Plant Development in**   111
   Arabidopsis thaliana
   Christine Granier, Vincent Nègre and *Fabio Fiorani*

7. **Challenges of Crop Phenomics in the Post-genomic Era**   142
   Vasilis C. Gegas, Alan Gay, Anyela Camargo and *John H. Doonan*

8. **Yeast Phenomics—Large-scale Mapping of the Genetic**   172
   Basis for Organismal Traits
   Jonas Warringer and *Anders Blomberg*

9. **Phenomics in Bacteria**   208
   Robert J. Nichols and *Carol A. Gross*

10. **Phenotype Databases**   237
    Philip Groth and *Bertram Weiss*

Index   263

Color Plate Section   269

1

Introduction to "Phenomics"

John M. Hancock

Aims of This Volume

Now that we have complete or draft genome sequences for many organisms, attention has moved on to their detailed characterization. Genome sequencing has stimulated massive progress in computational approaches to the detection of genes and other genomic features, although these are still far from perfect. Computational function prediction lags far behind, however, especially for previously uncharacterised genes. This has led to large-scale experimental projects, such as the 1,000 genomes project (1000 Genomes Project Consortium et al. 2010) and ENCODE (Encode Project Consortium et al. 2012), which aim to gather cross-genome polymorphism and molecular function data. At the whole organism level, experimental approaches are developing to relate genomic features to phenotypes with the aim of linking genes to phenotypic traits directly and these are coming to fall under the general rubric of "phenomics". Although the term phenome first appears in the literature in 1989 (Conrad and Rizki 1989), its use in the sense of the measurable set of all phenotypes of an organism really only emerged in a 1995 publication by Strohman (Strohman 1995) and was given major impetus by the proposal of a human phenome project in 2003 (Freimer and Sabatti 2003). The term has become increasingly common in the literature during the 2000s and 2010s with applications in a widening variety of organisms.

Department of Physiology, Development and Neuroscience, University of Cambridge, Downing Street, Cambridge CB2 3EG, UK.
Email: jmhancock@gmail.com

The aim of this volume is to bring together descriptions of phenomics applications in a range of organisms. The motivation for doing this is twofold. Firstly it will provide the wider scientific community with a broad practical definition of the meaning of the term phenomics. This is instructive because, although the term can be loosely taken to mean "the systematic, reproducible measurement of phenotypic characteristics of an organism," in practice this can have differing meanings. In many species, phenomics is an experimental approach whereby specific questions are asked (for example, specific environmental or chemical challenges are applied) and phenomic responses are assayed. In other species, most notably the mouse, yeast (*Saccharomyces cerevisiae*) and increasingly zebrafish, the phenomics approach is being or has been applied to systematically assay the effects of genetic changes, such as gene knockouts. Secondly, it will provide an overview of the current state-of-the-art in the field both for those already involved and for those wishing to gain more knowledge of a new and developing subject.

Scope of This Volume

Although the general principle of phenomics is to link gene sequence to function, this volume illustrates three main application areas. The first of these is the identification of genes involved in human health and diseases. Chapter 2 outlines what are still early stages in linking human phenotypes to disease. The relatively limited progress in phenotype collection in humans has two primary causes—there are major concerns about data security but also there has been far more focus on identifying genome alterations associated with disease than on gene function *per se*. Chapters 3 and 4 outline the current state of the art in two major vertebrate models that are currently developing phenomics approaches to underpin studies of human health: the lab mouse and zebrafish. Because zebrafish is particularly suited to the study of developmental defects while mice are currently better suited to adult phenomics studies, the two systems are often seen as complementary and, when combined, are likely to provide a treasure trove of information. The mouse research community is currently unique amongst communities studying complex organisms in starting to undertake a comprehensive, gene knockout-based phenotyping programme and it will be interesting to see how cost-effective this turns out to be in generating new knowledge. As well as the organism level phenotyping being carried out in mouse and zebrafish, phenotyping of vertebrate organisms at the cellular level can provide complementary information that is, in principle, relatively cheap and easy to acquire. Approaches to cellular phenotyping are covered in Chapter 5.

Another major area in which phenomics applications are developing is in the analysis of plant systems, where the emphasis is on the improvement of crop plants to aid long-term food security. In this area, progress is being made both on the model organism of choice, *Arabidopsis thaliana*, and on crop plants themselves. The two approaches are covered in Chapters 6 and 7.

The third application area for phenomics is the understanding of fundamental biology. Yeast was the first free-living organism with a sequenced genome and serves as a model organism for understanding eukaryotic biology. Because of its small size and ease of genetic manipulation, yeast serves as an exemplar of what is possible when linking genomics with phenomics. Yeast phenomics is covered in Chapter 8. Chapter 9 describes the application of phenomics approaches to understanding genome function in bacteria, of relevance both for understanding pathogenesis in humans and other species and for understanding genome-phenome relationships in another domain of life.

The final chapter in this volume considers the informatics of phenomics. Informatics is critical to the subject at many levels. Data needs to be made available via databases and web sites, but it also needs to be captured systematically and analysed. A particularly important aspect in this regard is the development of ontologies that will enable integration of phenotype data with other kinds of data within and, critically, across species. This brings us full circle as Chapter 2 discusses in depth the importance of the Human Phenotype Ontology for human phenomics.

Reproducibility, Data Accessibility and Coverage

Although the experimental approaches described in this volume are very different in the different species, there are underlying similarities across organisms. The key issues are reproducibility, data accessibility and coverage.

Reproducibility

Measuring phenotypes in a way that is reproducible between labs, so that results can be compared quantitatively and integrated, is a major issue which remains to be comprehensively tackled for most species. It has been addressed to some extent in the mouse in the pilot high-throughput phenotyping projects, notably EUMODIC (Ayadi et al. 2012), but reproducibility remains an issue for all the systems covered in this volume.

Sources of Error

Phenotypes (P) are classically regarded as the results of the interactions of genotype (G) and environment (E):

$$P = G + E$$

Although genotype is relatively easy to control, environments are not as it is not always clear which aspects of the environment have an effect on a particular phenotype (and, for systematic approaches, variables need to be controlled over a long timescale). Additionally, what is often forgotten is that not only can experimental error be substantial, with numerous and some unexplained causes (as seen, for example, in attempts to produce reproducible phenotyping tests in the mouse case) but there is likely to be stochastic variation in the expression of phenotypes even when genotype and environment are constant because of the complex relationship between genotype and phenotype. Both kinds of additional stochasticity need to be borne in mind when designing and interpreting phenomic studies. We can therefore express phenotypic measurement more generally as:

$$\hat{P} = G + E + \Delta_{exp} + \Delta_{bio}$$

where Δ_{exp} is the experimental error, Δ_{bio} is the biological variation and $\hat{P}$ is the measured phenotype.

Data Accessibility

Whereas vertebrate phenomics is developing unified approaches to data, in other species data are available in a more fragmentary manner and in some cases may not be available from databases at all. As biology develops into more of a data-rich discipline, efforts will need to be made to make data available and to use consistent formalisms—including but not restricted to interoperable ontologies—to make the data as useful as possible. The genomics community has led the way in this respect, with a series of declarations on access to genome sequencing data dating back to the early days of genome sequencing (HUGO 1996, The Wellcome Trust 2003, Toronto International Data Release Workshop Authors et al. 2009) but calls have recently been made to extend these principles to systematically gathered phenotype data and physical genetic resources (Bedu et al. 2010, Schofield et al. 2009).

Coverage

Whereas covering the entire genome with annotations is a realistic goal in bacterial systems or in yeast, in slower-growing organisms with larger genomes, this is more of a challenge. In plants, phenomics approaches have generally been restricted to traits of interest, so that much of the genome is only marginally covered or not at all. In vertebrates, there is more motivation to annotate the whole genome but time and cost constraints restrict the possibilities. For example, the International Mouse Phenotyping Consortium aims to generate phenotype data for 20,000 genes over ten years but on current plans this will be restricted to gene knockouts in one genetic background, so that studies of gene-gene interactions will not be possible. Individual communities will, no doubt, address these issues over time as their needs evolve.

Phenomics is a relatively new field, but techniques for acquiring phenomic data are becoming increasingly mature in a number of organisms. Hopefully this volume will support and stimulate further developments in the field.

References Cited

1000 Genomes Project Consortium, G.R. Abecasis, D. Altshuler, A. Auton, L.D. Brooks, R.M. Durbin, R.A. Gibbs, M.E. Hurles and G.A. McVean. 2010. A map of human genome variation from population-scale sequencing. Nature 467(7319): 1061–1073.

Ayadi, A., M.C. Birling, J. Bottomley, J. Bussell, H. Fuchs, M. Fray, V. Gailus-Durner, S. Greenaway, R. Houghton, N. Karp, S. Leblanc, C. Lengger, H. Maier, A.M. Mallon, S. Marschall, D. Melvin, H. Morgan, G. Pavlovic, E. Ryder, W.C. Skarnes, M. Selloum, R. Ramirez-Solis, T. Sorg, L. Teboul, L. Vasseur, A. Walling, T. Weaver, S. Wells, J.K. White, A. Bradley, D.J. Adams, K.P. Steel, M. Hrabe de Angelis, S.D. Brown and Y. Herault. 2012. Mouse large-scale phenotyping initiatives: overview of the European Mouse Disease Clinic (EUMODIC) and of the Wellcome Trust Sanger Institute Mouse Genetics Project. Mamm Genome. 23(9-10): 600–610.

Bedu, E., J. Blake, S. Brown, T. Bubela, C. Bult, D. Church, D. Einhorn, J. Eppig, C. Fletcher, J. Hancock, J. Kim, A.-M. Mallon, H. Masuya, A. McKenzie, C. McKerlie, M. Moore, T. Nguyen, R. Ramirez-Solis, P. Schofield, D. Smedley, T. Suzuki, N. Tanaka, S. Wakana and J. Yen. 2010. Legal Issues on Phenotype Data Accessibility. http://www.casimir.org.uk/fullstory.php?storyid=76 Date accessed: 13/3/13.

Conrad, M. and M.M. Rizki. 1989. The artificial worlds approach to emergent evolution. Biosystems 23(2-3): 247–258.

Encode Project Consortium, I. Dunham, A. Kundaje, S.F. Aldred, P.J. Collins, C.A. Davis, F. Doyle, C.B. Epstein, S. Frietze, J. Harrow, R. Kaul, J. Khatun, B.R. Lajoie, S.G. Landt, B.K. Lee, F. Pauli, K.R. Rosenbloom, P. Sabo, A. Safi, A. Sanyal, N. Shoresh, J.M. Simon, L. Song, N.D. Trinklein, R.C. Altshuler, E. Birney, J.B. Brown, C. Cheng, S. Djebali, X. Dong, J. Ernst, T.S. Furey, M. Gerstein, B. Giardine, M. Greven, R.C. Hardison, R.S. Harris, J. Herrero, M.M. Hoffman, S. Iyer, M. Kellis, P. Kheradpour, T. Lassman, Q. Li, X. Lin, G.K. Marinov, A. Merkel, A. Mortazavi, S.C. Parker, T.E. Reddy, J. Rozowsky, F. Schlesinger, R.E. Thurman, J. Wang, L.D. Ward, T.W. Whitfield, S.P. Wilder, W. Wu, H.S. Xi, K.Y. Yip, J. Zhuang, B.E. Bernstein, E.D. Green, C. Gunter, M. Snyder, M.J. Pazin, R.F. Lowdon, L.A. Dillon, L.B. Adams, C.J. Kelly, J. Zhang, J.R. Wexler, P.J. Good, E.A. Feingold, G.E. Crawford, J. Dekker, L. Elnitski, P.J. Farnham, M.C. Giddings, T.R. Gingeras, R. Guigo, T.J. Hubbard, M. Kellis, W.J. Kent, J.D. Lieb, E.H. Margulies, R.M. Myers, J.A. Starnatoyannopoulos, S.A. Tennebaum, Z. Weng, K.P. White, B. Wold, Y. Yu, J. Wrobel, B.A. Risk, H.P. Gunawardena, H.C. Kuiper, C.W. Maier, L. Xie, X. Chen, T.S. Mikkelsen,

S. Gillespie, A. Goren, O. Ram, X. Zhang, L. Wang, R. Issner, M.J. Coyne, T. Durham, M. Ku, T. Truong, M.L. Eaton, A. Dobin, T. Lassmann, A. Tanzer, J. Lagarde, W. Lin, C. Xue, B.A. Williams, C. Zaleski, M. Roder, F. Kokocinski, R.F. Abdelhamid, T. Alioto, I. Antoshechkin, M.T. Baer, P. Batut, I. Bell, K. Bell, S. Chakrabortty, J. Chrast, J. Curado, T. Derrien, J. Drenkow, E. Dumais, J. Dumais, R. Duttagupta, M. Fastuca, K. Fejes-Toth, P. Ferreira, S. Foissac, M.J. Fullwood, H. Gao, D. Gonzalez, A. Gordon, C. Howald, S. Jha, R. Johnson, P. Kapranov, B. King, C. Kingswood, G. Li, O.J. Luo, E. Park, J.B. Preall, K. Presaud, P. Ribeca, D. Robyr, X. Ruan, M. Sammeth, K.S. Sandu, L. Schaeffer, L.H. See, A. Shahab, J. Skancke, A.M. Suzuki, H. Takahashi, H. Tilgner, D. Trout, N. Walters, H. Wang, Y. Hayashizaki, A. Reymond, S.E. Antonarakis, G.J. Hannon, Y. Ruan, P. Carninci, C.A. Sloan, K. Learned, V.S. Malladi, M.C. Wong, G.P. Barber, M.S. Cline, T.R. Dreszer, S.G. Heitner, D. Karolchik, V.M. Kirkup, L.R. Meyer, J.C. Long, M. Maddren, B.J. Raney, L.L. Grasfeder, P.G. Giresi, A. Battenhouse, N.C. Sheffield, K.A. Showers, D. London, A.A. Bhinge, C. Shestak, M.R. Schaner, S.K. Kim, Z.Z. Zhang, P.A. Mieczkowski, J.O. Mieczkowska, Z. Liu, R.M. McDaniell, Y. Ni, N.U. Rashid, M.J. Kim, S. Adar, Z. Zhang, T. Wang, D. Winter, D. Keefe, V.R. Iyer, K.S. Sandhu, M. Zheng, P. Wang, J. Gertz, J. Vielmetter, E.C. Partridge, K.E. Varley, C. Gasper, A. Bansal, S. Pepke, P. Jain, H. Amrhein, K.M. Bowling, M. Anaya, M.K. Cross, M.A. Muratet, K.M. Newberry, K. McCue, A.S. Nesmith, K.I. Fisher-Aylor, B. Pusey, G. DeSalvo, S.L. Parker, S. Balasubramanian, N.S. Davis, S.K. Meadows, T. Eggleston, J.S. Newberry, S.E. Levy, D.M. Absher, W.H. Wong, M.J. Blow, A. Visel, L.A. Pennachio, L. Elnitski, H.M. Petrykowska, A. Abyzov, B. Aken, D. Barrell, G. Barson, A. Berry, A. Bignell, V. Boychenko, G. Bussotti, C. Davidson, G. Despacio-Reyes, M. Diekhans, I. Ezkurdia, A. Frankish, J. Gilbert, J.M. Gonzalez, E. Griffiths, R. Harte, D.A. Hendrix, T. Hunt, I. Jungreis, M. Kay, E. Khurana, J. Leng, M.F. Lin, J. Loveland, Z. Lu, D. Manthravadi, M. Mariotti, J. Mudge, G. Mukherjee, C. Notredame, B. Pei, J.M. Rodriguez, G. Saunders, A. Sboner, S. Searle, C. Sisu, C. Snow, C. Steward, E. Tapanari, M.L. Tress, M.J. van Baren, S. Washieti, L. Wilming, A. Zadissa, Z. Zhengdong, M. Brent, D. Haussler, A. Valencia, A. Raymond, N. Addleman, R.P. Alexander, R.K. Auerbach, K. Bettinger, N. Bhardwaj, A.P. Boyle, A.R. Cao, P. Cayting, A. Charos, Y. Cheng, C. Eastman, G. Euskirchen, J.D. Fleming, F. Grubert, L. Habegger, M. Hariharan, A. Harmanci, S. Iyenger, V.X. Jin, K.J. Karczewski, M. Kasowski, P. Lacroute, H. Lam, N. Larnarre-Vincent, J. Lian, M. Lindahl-Allen, R. Min, B. Miotto, H. Monahan, Z. Moqtaderi, X.J. Mu, H. O'Geen, Z. Ouyang, D. Patacsil, D. Raha, L. Ramirez, B. Reed, M. Shi, T. Slifer, H. Witt, L. Wu, X. Xu, K.K. Yan, X. Yang, K. Struhl, S.M. Weissman, S.A. Tenebaum, L.O. Penalva, S. Karmakar, R.R. Bhanvadia, A. Choudhury, M. Domanus, L. Ma, J. Moran, A. Victorsen, T. Auer, L. Centarin, M. Eichenlaub, F. Gruhl, S. Heerman, B. Hoeckendorf, D. Inoue, T. Kellner, S. Kirchmaier, C. Mueller, R. Reinhardt, L. Schertel, S. Schneider, R. Sinn, B. Wittbrodt, J. Wittbrodt, G. Jain, G. Balasundaram, D.L. Bates, R. Byron, T.K. Canfield, M.J. Diegel, D. Dunn, A.K. Ebersol, T. Frum, K. Garg, E. Gist, R.S. Hansen, L. Boatman, E. Haugen, R. Humbert, A.K. Johnson, E.M. Johnson, T.M. Kutyavin, K. Lee, D. Lotakis, M.T. Maurano, S.J. Neph, F.V. Neri, E.D. Nguyen, H. Qu, A.P. Reynolds, V. Roach, E. Rynes, M.E. Sanchez, R.S. Sandstrom, A.O. Shafer, A.B. Stergachis, S. Thomas, B. Vernot, J. Vierstra, S. Vong, M.A. Weaver, Y. Yan, M. Zhang, J.A. Akey, M. Bender, M.O. Dorschner, M. Groudine, M.J. MacCoss, P. Navas, G. Stamatoyannopoulos, J.A. Stamatoyannopoulos, K. Beal, A. Brazma, P. Flicek, N. Johnson, M. Lukk, N.M. Luscombe, D. Sobral, J.M. Vaquerizas, S. Batzoglou, A. Sidow, N. Hussami, S. Kyriazopoulou-Panagiotopoulou, M.W. Libbrecht, M.A. Schaub, W. Miller, P.J. Bickel, B. Banfai, N.P. Boley, H. Huang, J.J. Li, W.S. Noble, J.A. Bilmes, O.J. Buske, A.O. Sahu, P.V. Kharchenko, P.J. Park, D. Baker, J. Taylor and L. Lochovsky. 2012. An integrated encyclopedia of DNA elements in the human genome. Nature 489(7414): 57–74.

Freimer, N. and C. Sabatti. 2003. The human phenome project. Nat. Genet. 34(1): 15–21.

HUGO. 1996. Summary of Principles Agreed at the First International Strategy Meeting on Human Genome Sequencing. http://www.ornl.gov/sci/techresources/Human_Genome/research/bermuda.shtml Date accessed: 13/3/13.

Schofield, P.N., T. Bubela, T. Weaver, L. Portilla, S.D. Brown, J.M. Hancock, D. Einhorn, G. Tocchini-Valentini, M. Hrabe de Angelis, N. Rosenthal and C.R.M. participants. 2009. Post-publication sharing of data and tools. Nature 461(7261): 171–173.

Strohman, R.C. 1995. Linear genetics, non-linear epigenetics: complementary approaches to understanding complex diseases. Integr. Physiol. Behav. Sci. 30(4): 273–282.

The Wellcome Trust. 2003. Sharing Data from Large-scale Biological Research Projects: A System of Tripartite Responsibility. http://www.wellcome.ac.uk/stellent/groups/corporatesite/@policy_communications/documents/web_document/wtd003207.pdf Date accessed: 13/3/13.

Toronto International Data Release Workshop Authors, E. Birney, T.J. Hudson, E.D. Green, C. Gunter, S. Eddy, J. Rogers, J.R. Harris, S.D. Ehrlich, R. Apweiler, C.P. Austin, L. Berglund, M. Bobrow, C. Bountra, A.J. Brookes, A. Cambon-Thomsen, N.P. Carter, R.L. Chisholm, J.L. Contreras, R.M. Cooke, W.L. Crosby, K. Dewar, R. Durbin, S.O. Dyke, J.R. Ecker, K. El Emam, L. Feuk, S.B. Gabriel, J. Gallacher, W.M. Gelbart, A. Granell, F. Guarner, T. Hubbard, S.A. Jackson, J.L. Jennings, Y. Joly, S.M. Jones, J. Kaye, K.L. Kennedy, B.M. Knoppers, N.C. Kyrpides, W.W. Lowrance, J. Luo, J.J. MacKay, L. Martin-Rivera, W.R. McCombie, J.D. McPherson, L. Miller, W. Miller, D. Moerman, V. Mooser, C.C. Morton, J.M. Ostell, B.F. Ouellette, J. Parkhill, P.S. Raina, C. Rawlings, S.E. Scherer, S.W. Scherer, P.N. Schofield, C.W. Sensen, V.C. Stodden, M.R. Sussman, T. Tanaka, J. Thornton, T. Tsunoda, D. Valle, E.I. Vuorio, N.M. Walker, S. Wallace, G. Weinstock, W.B. Whitman, K.C. Worley, C. Wu, J. Wu and J. Yu. 2009. Prepublication data sharing. Nature 461(7261): 168–170.

2

Computational Phenotype Analysis in Human Medicine

Peter N. Robinson

What is a Phenotype?

The word **phenotype** is used with many different meanings. In biology, the most widely accepted definition of phenotype is "the observable traits of an organism." The concept of phenotype is usually contrasted with the concept of genotype, which again has been used with a number of different meanings in the literature, but generally can be taken to mean the totality of the genetic information of an individual.

Although there are at least five definitions of phenotype used in the literature on biology (Table 1), it can be seen that they are used to refer either to a single trait (a phenotypic feature) or to the entirety of traits of an organism. Correspondingly, genotype is generally defined as the alleles that are associated with two or more different variants of a single trait (e.g., the color of pea pods is an inherited trait wherely green is a recessive trait [aa], and yellow a dominant trait [Aa or AA]), or with the entire genome. We will here take the definition of phenotype in biology to mean the collection of observable traits of an organism, comprising its morphology, its physiology at the level of the cell, the organ, and the body, and its behavior, comprising even characteristics such as the gene expression profiles in response to environmental cues (Nachtomy et al. 2007).

Institut für Medizinische Genetik und Humangenetik, Charité –Universitätsmedizin Berlin, Augustenburger Platz 1, 13353 Berlin, Germany.
Email: peter.robinson@charite.de

Table 1. Various definitions of the concepts of phenotype and genotype in the biological literature.

Phenotype	Genotype
(1) The physical appearance of a trait	Alleles written in pairs (DD, Dd, dd)
(2) The observable properties (structural and functional) of an organism	The sum total of the genetic information (genes) contained in the [...] (chromosomes)
(3) The specific allelic composition of a cell—either of the entire cell or, more commonly, for a certain gene or a set of genes	The form taken by some character (or group of characters) in a specific individual; [...or] The detectable outward manifestation of a specific genotype
(4) The observable properties of an organism	The genotype is the genetic constitution of an organism; AA, Aa, and aa are examples
(5) The manifestation [...] of the interaction of [the genetic] information with the physical and chemical factors—the environment in the broadest sense—that enable the blueprint to be realized	A blueprint for an organism, the set of instructions for development
(6) [The] appearance [of an organism], its morphology, physiology, and ways of life—what we can observe	The sum total of hereditary materials of the organism

The definitions are by Klug and Cummings (1991), Rieger et al. (1991), Suzuki et al. (1989), Hartl et al. (1987), Futuyma (1986), and Dobzhansky (1977). See Mahner and Karys (1997) for original citations and an in-depth analysis of the concepts phenotype and genotype as used in biology.

In medical contexts, however, the word "phenotype" is more often used to refer to some deviation from normal morphology, physiology, or behavior, and this is the definition that we will use here. Thus, physicians characterize the phenotype of their patients (although they rarely speak of it in this way) by taking a medical history or by means of a physical examination, diagnostic imaging, blood tests, psychological testing, and so on, in order to make the diagnosis. This can be one of the most challenging tasks for the physician, and this is especially the case for rare diseases, with currently over 8,000 named diseases and presumably many thousands more waiting to be discovered and classified. This has been recognized as an important challenge for bioinformatics, and numerous algorithms and applications have been published for prioritizing candidate genes for disease-gene discovery [reviewed in Bauer et al. (2012)], and increasingly, various types of phenotype data are being used to inform the analysis.

As with the definitions of phenotype in biology, usage varies also in medical contexts. One sees the word used to describe both an individual phenotypic abnormality (for instance, hypertelorism) and to refer to all of the phenotypic characteristics of some disease (for instance, Stevens–Johnson syndrome). We find it more useful to reserve the use of the word phenotype to describe individual phenotypic abnormalities rather than diseases, and this is the sense in which we will use the word in the rest of this chapter.

Although clearly phenotypic analysis has always been at the core of clinical medicine, in recent years interest has emerged in the study of disease phenotypes as a way of understanding biological networks as a way of gaining insight into the molecular mechanisms underlying different diseases, which, among other things, might generate novel hypotheses for therapeutic strategies (Linghu and DeLisi 2010). This chapter will review some of the most important vocabularies and ontologies in current use to describe human diseases and human disease phenotypes, and will conclude by reviewing how phenotype analysis has been used to improve our understanding of management of selected human diseases and a roadmap for integrating detailed phenotype analysis ("Deep phenotyping") into medical genomic research.

Describing Phenotypes in Human Medicine

Many phenotypic descriptions in medical publications describe the phenotype in sloppy or imprecise ways. For instance, a description such as "myopathic EMG" is used instead of describing the reasons for this diagnosis (which can include reduced duration and reduced amplitude of the action potentials, increased spontaneous activity with fibrillations, positive sharp waves, or a reduced number of motor units in the muscle). There is no way of knowing what exactly the authors observed in their patients, and it may be extremely difficult or impossible to compare studies from different centers on the basis of such descriptions. Likewise, descriptions such as "still walking 25 years after onset" to describe a neurological phenotype may be evocative in a certain sense, but are likely to evoke a different picture in different readers depending on their experiences, knowledge, and imagination.

Similarly, many current gene mutation databases record little or no phenotype information beyond the fact that a disease was diagnosed in the person carrying the mutation. While this sort of information is useful for a diagnostician who is writing a report on a mutation, and finds a report in the database stating that some mutation has been previously found in an unrelated patient, and thus has at least some suggestive evidence for pathogenicity, it does not help much in understanding the natural history of the disease, the spectrum of complications of the disease, or genotype-phenotype correlations—all very useful clinical information.

The analysis of the medical phenotype as defined above is quite simply the daily work of practicing physicians: get the medical history, perform a physical examination, blood and laboratory tests. Routine, you might say. So what is to be gained by a deeper and more precise analysis of the phenotype? One of the goals of personalized medicine is to classify patients into subpopulations that differ with respect to disease susceptibility, phenotypic or molecular subclass of a disease, or to the likelihood of a positive or adverse response to a specific therapy. The related concept of "precision medicine," whose goal is to provide the best available care for each individual, refers to the stratification of patients into subsets with a common biological basis of disease, such that stratified medical management is most likely to benefit the patients (Committee on a Framework for Development of a New Taxonomy of Disease; National Research Council 2011). All medically relevant disease subclassifications can be said to have a distinct phenotype, with the understanding that a medical phenotype comprises not only the abnormalities described above but also the response of a patient to a certain type of treatment (e.g., responsiveness of seizures to valproic acid can be considered to be a phenotype of certain forms of epilepsy). However, responsivity to treatment is by no means the only or even the most important characteristic by which clinically relevant subclassifications can be identified. There are countless examples of clinically actionable items that were discovered by a precise phenotypic analysis (Doelken et al. 2012); for instance, clinical analysis of persons with a particular type of familial thoracic aortic aneurysms and dissections caused by mutations in the *ACTA2* gene were shown to have a substantially increased risk for coronary artery disease and premature ischemic strokes, which has obvious implications for clinical management (Guo et al. 2009). Therefore, comprehensive and precise phenotype data, combined with ever increasing amounts of genomic data, appear to have an enormous potential to accelerate the identification of clinical actionable complications, of disease subtypes with prognostic or therapeutic implications, and in general to improve our understanding of human health and disease.

Computational Analysis of the Human Phenotype

DNA sequence, protein structure, and later gene expression were arguably the most important areas in the early days of bioinformatics in the 1980s and 1990s. Although there are of course many challenges and difficulties in storing and analyzing this data, there are relatively obvious ways of representing data of this form in a computer: biological sequences can be represented as strings, protein structures can be represented by the three-dimensional coordinates of a subset of its atoms, and gene expression data by a matrix of numbers. In comparison to that, the human phenotype is immensely more complex and difficult to observe in a precise and comprehensive fashion. The great majority of specialists for the human phenotype (i.e., physicians) do not have computational training and are primarily or even exclusively involved in patient care rather than research. Ethical and privacy issues preclude putting many kinds of human phenotype data into open-access databases such as the Mouse Genome Informatics group has done for the mouse phenotype (Bult et al. 2010), and the ethical norms in use in medical research have not quite caught up to the technical possibilities of the World Wide Web for exchanging information and catalyzing research. This has meant that the great majority of documentation of the human phenotype in hospitals, medical practice, and even medical research is recorded in free text or with the use of *ad hoc* lists of terms rather than standards. This has come to be recognized as a major roadblock to further research in human medicine, which increasingly relies on international databases of genetic, genomic, and phenotypic data.

The rise to prominence of the Gene Ontology (GO) following the landmark publication in the year 2000 (Ashburner et al. 2000) brought a new kind of integrative bioinformatics to center stage, in which ontologies were used to connect experimental data to notions of biological function. The use of GO term over-representation analysis became a standard way of interpreting long lists of genes identified as differentially expressed (Robinson and Bauer 2011), and the original article, by now cited by many thousands of other papers, has become one of the most important publications in the history of the field of bioinformatics. The success of the GO inspired other groups to develop ontologies for other fields, many of which have united under the banner of the OBO Foundry, which has developed standards and tools to promote best practices in ontology development as well as interoperability between ontologies in related fields of science and medicine (Smith et al. 2007). This chapter will not attempt to provide an overview of all the resources within and beyond the OBO Foundry that are being used for medical research, but instead will provide a very selective overview of resources that are most likely to be of use for bioinformaticians, ontology specialists, and researchers in settings where mainly open-access and open-source resources are *de rigeur*. A few important resources with more restrictive usage licenses will be presented. The chapter will conclude with a presentation of the Human Phenotype Ontology, which was developed by the author's group.

Medical Terminologies

We will now introduce a selection of important terminological resources and ontologies needed for the computational study of the human phenotype. The coverage is far from complete, and no attempt has been made to cover specialist terminologies for specific areas of medicine.

UMLS

The United Medical Language System (UMLS) (Bodenreider 2004) comprises databases and software designed to facilitate the analysis of information such as patient records, scientific literature, guidelines, and public health data. The UMLS Metathesaurus combines data on biomedical concepts from multiple sources, including synonyms and some relationships between concepts. All concepts in the Metathesaurus are assigned to at least one semantic type. The UMLS Semantic Network defines 133 semantic types (categories) as well as 54 relationship types. In essence, the semantic types represent nodes in a network, and the relationships represent links between the nodes. The third major component of the UMLS is the Specialist Lexicon with its lexical tools. These are designed to recognize inflectional variants of the same word (e.g., "examine," "examined", and "examining"), to recognize spelling variants (e.g., British and American English), as well as alphabetic case variants, and are useful in text mining operations.

The UMLS does not intend to provide a comprehensive ontology of medicine with all relevant concepts and relations, but rather attempts to reflect the relations embedded into its source terminologies, even if these are inconsistent. A wide variety of terminologies have been incorporated in the UMLS, including all clinical standards designated as target U.S. government-wide standards (e.g., LOINC, SNOMED CT, RxNorm).

The Metathesaurus is organized by concepts, which are used to connect different names for the same entity from many different vocabularies. Each Metathesaurus concept has a unique and permanent concept identifier (CUI). Additionally, each unique name (string) has a unique and permanent string identifier (SUI). Finally, the Metathesaurus keeps tracks of instances of concept names (strings) from each of the source vocabularies by means of unique atom identifiers (AUIs). For instance, if the same string occurs in two different vocabularies within the UMLS, each string will receive a different AUI but be assigned to the same SUI, which in turn is assigned to the corresponding CUI.

SNOMED CT

The *Systematized Nomenclature of Medicine—Clinical Terms* (SNOMED CT) is a clinical reference terminology designed to enable electronic clinical decision support, disease screening, and enhanced patient safety. A major goal is to provide a concept-based terminology, which means that each medical concept is uniquely identified and can have multiple descriptions. SNOMED CT contains more than three hundred thousand active concepts with unique meanings and formal logic-based definitions organized into hierarchies. SNOMED CT is intended to make electronic health records

(EHRs) maintainable, sharable, and accessible to semantic interpretation and to provide a foundation for clinical decision support programs. In contrast to classification schemes, such as the ICD, that are often used to specify a single primary diagnosis for morbidity/mortality statistics or billing purposes, SNOMED CT is designed to document information about patient and medical encounters in EHRs and health information systems.

SNOMED CT comprises hundreds of thousands of terms and relations for describing findings and diseases, medical procedures and other events of relevance for the medical history, and a number of other things. In addition to providing a subsumption hierarchy for the terms (e.g., pneumonia is a subclass of lung disease), other kinds of relations are used to define the logical relationships between concepts (e.g., virus is a causative agent of viral pneumonia).

SNOMED CT has not been widely used in academic bioinformatics research, which is probably related to the fact that SNOMED CT is not available under an open source license, the fact that much of the data annotated with SNOMED CT is private patient data in health care IT systems, and perhaps because of the very complexity of SNOMED CT, with its comprehensive coverage of multiple aspects of the health care process. Nonetheless, EHR-driven research is rapidly growing in importance, and it would be highly desirable to improve interoperability of SNOMED CT with ontologies typically used in the research community such as the ontologies of the OBO Foundry.

Disease Ontologies and Classifications

The first modern medical classifications recognizable as an ontology of diseases was developed by Carl Linnaeus (1707–1778), who divided diseases into 11 classes, 37 orders, and 325 species (Robinson et al. 2011). For instance, the class *Exanthematici* contained an order *contagiosi* that contains six diseases including variola, rubeola, and syphilis (Fig. 1). Although this classification contained the occasional error, for instance the notion that leprosy can be caught by eating herring worms, it was enormously influential, paving the way for the work of William Farr (1807–1883) and Jacques Bertillon (1851–1922) that culminated in the first edition of the international classification of diseases (ICD) in 1891 (Robinson et al. 2011).

The ICD has become the standard diagnostic tool for epidemiology, health management, and clinical purposes, and is used to monitor the incidence and prevalence of diseases and other health problems. However, the ICD was not designed to be used as a scientific tool for identifying the molecular causes of diseases or understanding their pathophysiology. Possibly because of the fact that the ICD is used in national health systems for statistics and reimbursement purposes, it cannot be easily changed in

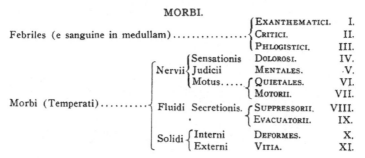

			MORBI.	

			EXANTHEMATICI.	I.
Febriles (e sanguine in medullam)			CRITICI.	II.
			PHLOGISTICI.	III.
	Nervii	Sensationis	DOLOROSI.	IV.
		Judicii	MENTALES.	.V.
		Motus	QUIETALES.	VI.
Morbi (Temperati)			MOTORII.	VII.
	Fluidi	Secretionis.	SUPPRESSORII.	VIII.
		.	EVACUATORII.	IX.
	Solidi	Interni	DEFORMES.	X.
		Externi	VITIA.	XI.

Figure 1. Excerpt of the *Genera morborum* (1759), the classification of diseases developed by Carl Linnaeus. Illustration from Schulz et al. (2011).

response to new scientific findings (for instance, molecular subtypes of a certain kind of cancer). In fact, the United States is still using ICD-9, whereas Europe and much of the rest of the world have adopted ICD-10 (often with national amendments), and the ICD-11 is currently being prepared. This makes it relatively difficult to use data encoded with the ICD for scientific research.

In addition to the classifications in the ICD and SNOMED CT, there are several important resources of high relevance to computational research on human diseases with use licenses that are compatible with academic research settings. The Disease Ontology (DO) currently comprises classes for over 8000 human diseases (Srour et al. 2006). One of the goals of DO is to provide a single structure for the classification of disease, which unifies the representation of disease among the many and varied terminologies and vocabularies, with cross-references to other disease nomenclatures such as MeSH, OMIM, ICD, and SNOMED CT. By placing diseases into a relational ontology, the DO permits inference and reasoning of the relationships between disease terms and concepts. While the DO includes some representation of Mendelian diseases, most of its concepts refer to other classes of disease.

In the field of rare diseases and human genetics, two nomenclatures have achieved a widely used status in the field. Orphanet is a multilingual information portal on rare diseases and orphan drugs (Schriml et al. 2012). It takes a clinical approach to the classification of rare diseases, and follows the organization of medical specialties according to organs or body systems. Other criteria (e.g., histopathology) may be followed at deeper levels of the ontology. Each disease entry is classified into one or more sections, with pleiotropic diseases having multiple parentage. For instance, Bardet–Biedl syndrome is a hereditary disease characterized clinically by a wide range of findings. Table 2 displays the multiple parents of the term for Bardet–Biedl syndrome.

Table 2. Parents of the Orphanet rare disease ontology "Bardet–Biedl syndrome" (Orpha number ORPHA110).

Parent	Orpha Number
Retinal ciliopathy due to mutation in Bardet–Biedl gene	156183
Nephronophthisis-associated ciliopathy	156162
Syndromic obesity	240371
Syndromic renal or urinary tract malformation	93547
Syndromic retinitis pigmentosa	98661
Syndrome with hypogonadotropic hypogonadism	181387
Multiple congenital anomalies/dysmorphic syndrome—variable intellectual deficit	102284
Syndromic developmental defect of the eye	108987
Unclassified primitive or secondary maculopathy	98666
Congenital hypogonadotropic hypogonadism	174590
Syndromic intestinal malformation	108969

This term has cross-references to corresponding ICD codes, OMIM entries, as well as information about age of onset, prevalence, and mode of inheritance. The multiple parents of this entry are shown in the table and reflect clinical aspects that might be important in management (e.g., syndromic retinitis pigmentosa) as well as etiological classifications (e.g., Retinal ciliopathy due to mutation in Bardet–Biedl gene).

Deeper in the different sections of the Orphanet classification of rare diseases, other criteria are followed, according to the intrinsic logic of the medical management in a given specialty. For instance, in the neuromuscular section of neurological diseases, a histopathological approach is adopted for muscular diseases, distinguishing myopathies from dystrophies, because muscle biopsy is a first step in the diagnostic workup. Etiopathological classifications of particular groups of diseases are also produced, in order to have disorders classified by mechanisms or causative genes (Table 2).

OMIM, the Online Mendelian Inheritance in Man, has cataloged human Mendelian disease, has focused on the relationship between genes and their molecular variants and associated phenotypes for over four decades, and is widely regarded as the gold standard for disease nomenclature in genetics and as a central resource for research and clinical care in human genetics. The OMIM team classifies Mendelian disease based on recognizable patterns of phenotypic features, including especially those that allow one condition to be distinguished from another (Amberger et al. 2011). More recently, a set of phenotypic features taken from OMIM's clinical synopsis has been developed for quick data entry in whole-exome sequencing projects (http://phenodb.net).

Phenotype Ontologies

It is important to distinguish between diseases and phenotypic features. A disease may have multiple features, e.g., the disease "common cold" can have the features "sneezing", "runny nose", "fatigue", and "fever". On the other hand, a feature can occur with multiple different diseases. For instance, "fever" occurs not only with the common cold but also with a wide range of other infectious diseases, as well as hyperthyroidism, leukemia, rheumatic arthritis, and many other noninfectious diseases. Thus, there is an m-to-n relationship between diseases and phenotypic features, which are interrelated in complex ways that may partially reflect disruption of overlapping biochemical pathways and cellular networks.

Human Phenotype Ontology

The Human Phenotype Ontology (HPO) (Robinson et al. 2008) is being developed to provide a standard for describing the manifestations of human disease that can be used in databases, registries, and publications, as well as for sophisticated bioinformatics analysis. As discussed above, the word phenotype is used with many different meanings. In biology, the most widely accepted definition of phenotype is "the observable traits of an organism," or perhaps the collection of observable traits of an organism, which result from the interaction of the genetic constitution of the organism and the environment. In medical contexts, the "phenotype" more often refers to some deviation from "normal", and this is the definition taken by the HPO. Many different types of phenotypic abnormality are represented in the HPO, including not only morphological signs, but also cellular, physiological, and behavioral abnormalities (see Table 3). In general, HPO terms describe the abnormality and not the diagnostic procedure used to identify it. Thus, the term *Agenesis of corpus callosum* (HP:0001274) does not state whether the abnormality was found by computer tomography or magnetic resonance imaging. On the other hand, there are abnormalities that are only observable with one kind of diagnostic modality (e.g., EEG with hyperventilation-induced focal epileptiform discharges, HP:0011183), and, in these cases, the diagnostic modality is generally included in the name of the term.

The HPO is available under an open-source license and it currently contains over 10,000 terms, each describing a phenotypic abnormality. The terms are related to one another by subclass ("is_a") relations, such that the ontology can be represented as a so-called directed acyclic graph in which any term except the root can have one or more parent terms and cycles are forbidden.

The HPO terms themselves do not describe any specific disease. To describe a disease, annotations to HPO terms can be used. For instance, to assert that patients with sialidosis type I have cherry red spot of the macula, we annotate the disease sialidosis type I with the corresponding HPO term. The network of diseases, associated HPO features, and genes can

Table 3. Types of phenotypic abnormality covered by the HPO.

Type of phenotypic feature	Example HPO term
Morphological abnormality	Broad nose (HP:0000445), Arachnodactyly (HP:0001166)
Abnormal process (organ)	Ileus (HP:0002595)
Abnormal process (cellular)	Abnormality of Krebs cycle metabolism (HP:0000816)
Abnormal laboratory finding	Hyperlipidemia (HP:0003077)
Electrophysiological abnormality	Decreased nerve conduction velocity (HP:0000762)
Abnormality by medical imaging	Choroid plexus cyst (HP:0002190)
Behavioral abnormality	Head nodding (HP:0001361), Self-mutilation (HP:0000742)

now be used for a number of purposes including differential diagnostics, prioritization of candidate genes, and research in the relationships between genotype and phenotype. Some examples follow.

The Phenomizer: Clinical diagnostics in human genetics with semantic similarity searches in ontologies

Combinations of phenotypic abnormalities (signs, symptoms, laboratory or imaging findings, etc.) are often used to inform differential diagnosis, which is one of the most important tasks of physicians. It can be extremely challenging to identify all diseases that should be considered in the differential diagnosis, and to develop a plan for efficiently evaluating each potential diagnosis, especially in the field of human genetics with the thousands of syndromes, many of which are characterized by complex and highly variable phenotypic manifestations. On the other hand, timely and correct diagnosis is essential to avoid unnecessary diagnostic procedures and to initiate therapeutic measures and clinical management plans, and to ensure that genetic counseling is correct and complete. Although next-generation sequencing and especially exome sequencing is opening up new possibilities for molecular diagnostics in human genetics, currently we are far from being able to offer all patients a precise, etiological diagnosis. For instance, an etiological diagnosis cannot be made in up to about half of all children presenting with dysmorphism or mental retardation.

An individual patient with a hereditary disease may not necessarily have all of the manifestations that can characterize that disease, and may have additional signs or symptoms that are not primarily related to the disease. Individual phenotypic features may have different degrees of specificity. For instance, a symptom such as fever may not narrow down a differential diagnosis as much as the finding of a blood culture positive for, say, *Streptococcus pneumoniae*. For an individual patient, clinical findings may be available at different levels of granularity depending on the expertise of the investigating physician or the availability of specialized laboratory or imaging procedures. Thus, phenotypic abnormalities can be described at either a specific or general level, and computational algorithms for differential diagnosis should be able to exploit data at any level of specificity, with specific features having a higher weight than general ones. Algorithms should not be overly sensitive to the lack of particular phenotypic features or to the presence of unrelated ones.

We have developed the Phenomizer, which exploits the semantic structure of the HPO to weight clinical features according to their specificity in order to improve the differential diagnosis. Each of the terms of the HPO is assigned an information content based on the frequency of annotations to the term. Users of the software can enter multiple HPO terms corresponding to the phenotypic abnormalities found in a patient. The Phenomizer searches amongst all entries in its database and returns a list of differential diagnoses ranked according to their match to the query entered by the physician. A full description of the algorithm was given in the original publication, but here we will provide an intuitive explanation. We begin with a database of diseases (syndromes), each of which is annotated to one or more terms of the HPO. For instance, the disease Marfan syndrome is annotated to the HPO term for *scoliosis*, the term for *ectopia lentis*, and many more. The specificity of a term is measured by its information content, which in turn is determined by the frequency with which the term is used for annotation. For the sake of explanation, let us assume we have 1000 diseases in our database. If only one disease is annotated to the term *ectopia lentis*, then its frequency is $1/1000$ and its information content is then calculated[1] as $-\log \dfrac{1}{1000} = 3$. On the other hand, if scoliosis is used to annotate ten diseases in the database, its frequency is $1/100$, and its information content is $-\log \dfrac{1}{100} = 2$. Thus, the higher specificity of the term *ectopia lentis* is reflected in its higher information content. There are many methods that have been used to calculate the similarity between two ontology terms, but perhaps the best-known method is the one introduced by Resnik (Robinson et al. 2011) that defines the similarity between two terms as the information content of their most specific common ancestor. Thus, two terms that describe a specific abnormality of the hypothalamus might have *Abnormality of the hypothalamus* as their most specific common ancestor. Assuming that *Abnormality of the hypothalamus* is a relatively rare disease manifestation, it will have a high information content. On the other hand, the most specific ancestor of,

---

[1] In this example, we are using the decadic (base 10) logarithm for simplicity. It does not matter which base is used for the calculations, but usually base 2 or natural logarithms are used in practice.

say, *Hypoplasia of the thalamus* and *Peripheral nerve demyelination* might be *Abnormality of the nervous system*, which is a much more general term with a much lower information content. Furthermore, the only common ancestor of *Hypoplasia of the thalamus* and *Liver cirrhosis* is the root of the ontology, which as the root of the ontology has no information content at all (thus, *Hypoplasia of the thalamus* and *Liver cirrhosis* have no similarity). Assuming now that a physician enters three search terms, the Phenomizer iterates over all diseases in the databases, and for each query term it finds the best match amongst all terms used to annotate the disease in question, defined again as the information content of the most specific common ancestor of the query term and a term used to annotate the disease. The semantic similarity of the query to the disease is then defined as the average match score, and the diseases are ranked according to this score. A number of improvements to this basic algorithm further improve the performance of semantic similarity searches using ontologies. Recent work has led to computer-interpretable logical definitions for the terms of the HPO using PATO, the ontology of phenotypic qualities, to link terms of the HPO to the anatomic and other entities that are affected by abnormal phenotypic qualities. These definitions enable quality control of the HPO via semantic reasoning (Resnik 1995) and allow interspecies phenotypic mapping.

Other Phenotype Terminologies and Ontologies

Even in the world of DNA sequences, there are multiple standards, including the formats FASTA, genbank, and more recently at least three varieties of the FASTQ format. Given the enormous complexity of the human disease phenotype, it is not to be expected that a single ontology or terminology will satisfy all needs in the field. The HPO was designed especially for computational analysis and scientific research on genotype phenotype correlations in human genetics and other fields of medicine. Many other resources are available with different focus areas. The Elements of Morphology group has created state-of-the-art consensus definitions of specific phenotypic abnormalities commonly encountered in human genetics, with a focus on dysmorphology (Gkoutos et al. 2009). OMIM has long been the most important knowledge base in human genetics, and in addition to the occasionally inconsistent terminology used in the Clinical Synopsis section of the disease descriptions, a new ontology has been developed with a view towards efficient and consistent data entry for exome sequencing projects at the Centers for Mendelian Genomics (see www.mendelian.org). Orphanet is a multilingual information portal on rare diseases and orphan drugs with links to many other resources including the International Classification of Diseases (10th version), SNOMED CT, MeSH, MedDRA, and UMLS. Orphanet uses a thesaurus of clinical signs and symptoms that is useful to access the information in the portal. Each of these knowledge bases, as well as the commercial diagnosis support programs of the London Dysmorphology Database (LDDB) (Köhler et al. 2011), and the Pictures of Standard Syndromes and Undiagnosed Malformations database (POSSUM; http://www.possum.net.au) have used their own vocabularies to describe phenotypic abnormalities. The multiplicity of mutually incompatible vocabularies in databases and publications on human genetics has hindered integrative research in clinical genetics (Doelken et al. 2012). The coordinators of many of the phenotype ontologies and terminologies for human genetics have recently (2012) planned to establish cross-links between the various resources in a way that will improve interoperability.

Outlook

While large-scale phenotyping initiatives have become almost routine for model organisms such as the mouse, rat, and zebrafish (Allanson et al. 2009, Cheng et al. 2011, Fryns and de Ravel 2002, Lindblom and Robinson 2011), and even for organisms such as *Toxoplasma gondii* (Maddatu et al. 2012), similar large-scale efforts for human disease have been lacking. Recently, however, it is widely acknowledged that a precise analysis of the phenotype will be needed to take full advantage of findings from whole-exome and whole-genome sequencing in order to understand human biology and disease. With the current revolutions in DNA sequencing technologies likely to bring us previously unimaginable amounts of genotype data for patients with rare and common disease in the next decade, the need for a Human Phenotype Project has become all the more pressing, and in fact, adequate methods for capturing and analyzing human phenotype data now appear as one of the major bottlenecks towards progress in our understanding of human genome biology. Ontologies represent important computational tools that are in the position to provide a foundation for a Human Phenotype Project.

References Cited

Allanson, J.E., L.G. Biesecker, J.C. Carey and R.C. Hennekam. 2009. Elements of morphology: introduction. Am. J. Med. Genet. A 149A(1): 2–5.
Amberger, J., C. Bocchini and A. Hamosh. 2011. A new face and new challenges for Online Mendelian Inheritance in Man (OMIM(R)). Hum. Mutat. 32(5): 564–567.
Ashburner, M., C.A. Ball, J.A. Blake, D. Botstein, H. Butler, J.M. Cherry, A.P. Davis, K. Dolinski, S.S. Dwight, J.T. Eppig, M.A. Harris, D.P. Hill, L. Issel-Tarver, A. Kasarskis, S. Lewis, J.C. Matese, J.E. Richardson, M. Ringwald, G.M. Rubin and G. Sherlock. 2000. Gene ontology: tool for the unification of biology. The Gene Ontology Consortium. Nat. Genet. 25(1): 25–29.

Bauer, S., S. Köhler, M.H. Schulz and P.N. Robinson. 2012. Bayesian ontology querying for accurate and noise-tolerant semantic searches. Bioinformatics 28(19): 2502–2508.

Bodenreider, O. 2004. The Unified Medical Language System (UMLS): integrating biomedical terminology. Nucleic Acids Res. 32(Database issue): D267–270.

Bult, C.J., J.A. Kadin, J.E. Richardson, J.A. Blake and J.T. Eppig. 2010. The Mouse Genome Database: enhancements and updates. Nucleic Acids Res. 38(Database issue): D586–592.

Cheng, K.C., X. Xin, D.P. Clark and P. La Riviere. 2011. Whole-animal imaging, gene function, and the Zebrafish Phenome Project. Curr. Opin. Genet. Dev. 21(5): 620–629.

Committee on a Framework for Development a New Taxonomy of Disease; National Research Council. 2011. Toward Precision Medicine: Building a Knowledge Network for Biomedical Research and a New Taxonomy of Disease. The National Academies Press, Washington, D.C.

Doelken, S.C., S. Kohler, C.J. Mungall, G.V. Gkoutos, B.J. Ruef, C. Smith, D. Smedley, S. Bauer, E. Klopocki, P.N. Schofield, M. Westerfield, P.N. Robinson and S.E. Lewis. 2013. Phenotypic overlap in the contribution of individual genes to CNV pathogenicity. Dis. Model Mech. 6(2): 358–72.

Fryns, J.P. and T.J. de Ravel. 2002. London Dysmorphology Database, London Neurogenetics Database and Dysmorphology Photo Library on CD-ROM [Version 3] 2001R. M. Winter, M. Baraitser, Oxford University Press, ISBN 019851-780, pound sterling 1595. Hum. Genet. 111(1): 113.

Gkoutos, G.V., C. Mungall, S. Dolken, M. Ashburner, S. Lewis, J. Hancock, P. Schofield, S. Kohler and P.N. Robinson. 2009. Entity/quality-based logical definitions for the human skeletal phenome using PATO. Conf. Proc. IEEE Eng. Med. Biol. Soc. 2009: 7069–7072.

Guo, D.C., C.L. Papke, V. Tran-Fadulu, E.S. Regalado, N. Avidan, R.J. Johnson, D.H. Kim, H. Pannu, M.C. Willing, E. Sparks, R.E. Pyeritz, M.N. Singh, R.L. Dalman, J.C. Grotta, A.J. Marian, E.A. Boerwinkle, L.Q. Frazier, S.A. LeMaire, J.S. Coselli, A.L. Estrera, H.J. Safi, S. Veeraraghavan, D.M. Muzny, D.A. Wheeler, J.T. Willerson, R.K. Yu, S.S. Shete, S.E. Scherer, C.S. Raman, L.M. Buja and D.M. Milewicz. 2009. Mutations in smooth muscle alpha-actin (ACTA2) cause coronary artery disease, stroke, and Moyamoya disease, along with thoracic aortic disease. Am. J. Hum. Genet. 84(5): 617–627.

Köhler, S., S. Bauer, C.J. Mungall, G. Carletti, C.L. Smith, P. Schofield, G.V. Gkoutos and P.N. Robinson. 2011. Improving ontologies by automatic reasoning and evaluation of logical definitions. BMC Bioinformatics 12: 418.

Lindblom, A. and P.N. Robinson. 2011. Bioinformatics for human genetics: promises and challenges. Hum. Mutat. 32(5): 495–500.

Linghu, B. and C. DeLisi. 2010. Phenotypic connections in surprising places. Genome Biology 11: 116.

Maddatu, T.P., S.C. Grubb, C.J. Bult and M.A. Bogue. 2012. Mouse Phenome Database (MPD). Nucleic Acids Res. 40(Database issue): D887–894.

Mahner, M. and M. Karys. 1997. What exactly are genomes/ genotypes and phenotypes? And what about phenomes? Journal of Theoretical Biology 186: 55–63.

Nachtomy, O., A. Shavit and Z. Yakhini. 2007. Gene expression and the concept of the phenotype. Studies in History and Philosophy of Biological and Biomedical Sciences 38: 238–254.

Rath, A., A. Olry, F. Dhombres, M.M. Brandt, B. Urbero and S. Ayme. 2012. Representation of rare diseases in health information systems: the Orphanet approach to serve a wide range of end users. Hum. Mutat. 33(5): 803–808.

Resnik, P. 1995. Using information content to evaluate semantic similarity in a taxonomy. Proceedings of the 14th IJCAI 1: 448–453.

Robinson, P.N. and S. Bauer. 2011. Introduction to Bio-Ontologies. Chapman & Hall/CRC Mathematical & Computational Biology, Boca Raton.

Robinson, P.N., S. Kohler, S. Bauer, D. Seelow, D. Horn and S. Mundlos. 2008. The Human Phenotype Ontology: a tool for annotating and analyzing human hereditary disease. Am. J. Hum. Genet. 83(5): 610–615.

Robinson, P.N., P. Krawitz and S. Mundlos. 2011. Strategies for exome and genome sequence data analysis in disease-gene discovery projects. Clin. Genet. 80(2): 127–132.

Schriml, L.M., C. Arze, S. Nadendla, Y.W. Chang, M. Mazaitis, V. Felix, G. Feng and W.A. Kibbe. 2012. Disease Ontology: a backbone for disease semantic integration. Nucleic Acids Res. 40(Database issue): D940–946.

Schulz, M.H., S. Köhler, S. Bauer and P.N. Robinson. 2011. Exact score distribution computation for ontological similarity searches. BMC Bioinformatics 12: 441.

Smith, B., M. Ashburner, C. Rosse, J. Bard, W. Bug, W. Ceusters, L.J. Goldberg, K. Eilbeck, A. Ireland, C.J. Mungall, N. Leontis, P. Rocca-Serra, A. Ruttenberg, S.A. Sansone, R.H. Scheuermann, N. Shah, P.L. Whetzel and S. Lewis. 2007. The OBO Foundry: coordinated evolution of ontologies to support biomedical data integration. Nat. Biotechnol. 25(11): 1251–1255.

Srour, M., B. Mazer and M.I. Shevell. 2006. Analysis of clinical features predicting etiologic yield in the assessment of global developmental delay. Pediatrics 118(1): 139–145.

3

Phenomics of the Laboratory Mouse

John M. Hancock[1], and *Michael S. Dobbie[2]*

Introduction—Why the Mouse?

Using model organisms to study human disease has many benefits for the experimental scientist. Primary amongst these is the ability to carry out experimental manipulations, an approach not possible in humans. Because of their relatively close phylogenetic relationships to man and consequently their more similar physiology and development, mammals are the most popular group used as models of human disease. Other organisms, most notably zebrafish (*Danio rerio*; see Chapter 4 of this book), fruit fly (*Drosophila melanogaster*) and nematode worm (*Caenorhabditis elegans*), are also used and can provide useful insights.

Traditionally a wide range of mammals has been used for physiological and developmental study including rodents (rat, mouse, guinea pig), rabbits, dogs and pigs. Rats in particular have been used heavily and have the advantage over mice of being larger and therefore more easily worked

---

[1] Department of Physiology, Development & Neuroscience, University of Cambridge, Downing Street, Cambridge CB2 3EG, UK.
  Email: jmhancock@gmail.com
[2] Australian Phenomics Facility, The Australian National University, Hugh Ennor Building 117, Garran Road, Canberra ACT 0200, Australia.
  Email: michael.dobbie@anu.edu.au
* Corresponding author

on. However, in recent years the laboratory mouse has increasingly become the experimental mammal of choice for a number of reasons:

- Their small size is an advantage as it allows them to be kept at a high density. Modern mouse houses can efficiently accommodate tens of thousands of individuals.
- Their short generation time (about 12 weeks) means that the creation, breeding and expansion of new strains can take place rapidly.
- Over 100 years of inbreeding has generated over 450 different strains, which provide a wide range of genotypes and phenotypes.
- Many inbred mouse strains have fully sequenced genomes providing an exquisitely defined and maintained genetic background. Ninety-nine percent of mouse genes have a human homologue (Waterston et al. 2002), 75% having a 1:1 orthologous relationship with a human gene (Church et al. 2009). Mouse and human orthologous proteins show 88.2% sequence identity (Church et al. 2009).
- There is an extensive and expanding genetic toolkit allowing a range of manipulations including gene deletion, conditional inactivation and modification (see Rosenthal and Brown (2007) for a good overview).
- Increasingly sophisticated small-scale measurement technologies (e.g., *in vivo* imaging such as microCT, fluorescence microscopy) are being developed, allowing ever-more detailed phenotype description of mouse models.

Because of these advantages, the use of the mouse has been increasing in recent years, even in domains of experimentation in which the rat has, for many years, been the dominant model organism. This is illustrated in Fig. 1, which shows increasing use of mouse versus rat overall and the mouse overtaking rat, or almost so, in areas in which rat was previously dominant such as diabetes and heart disease.

The mouse genome was one of the first to be sequenced (Waterston et al. 2002) and this initial sequencing was followed up by the generation of representative genome sequences for 17 inbred mouse strains (Keane et al. 2011). The existence of genome sequence information has massively eased the problem of identifying mutations causing novel phenotypes, particularly in mouse strains generated by chemical mutagenesis where no sequence tags for the causative mutation or its location are available (Andrews et al. 2012). In this respect, advances in massively parallel sequencing (High-Throughput Sequencing/Next Generation Sequencing) have revolutionised the ability to link phenotypic data with specific genomic sequences and mutations quickly and cost effectively.

Despite this invaluable resource, however, many mouse genes remain functionally uncharacterised (even if information from other species is used to supplement data collected directly in the mouse): the Mouse Genome

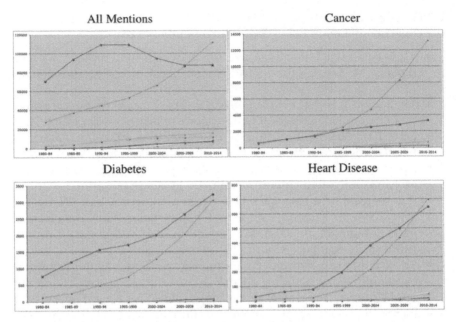

Figure 1. Citations to common model organisms in PubMed by era. The four panels represent the raw numbers of hits to searches using different search terms or combinations thereof to model organisms. X-axis: year range; Y-axis: number of hits (counts for 2010–2014 normalised from counts for 2010–2012). **All Mentions**: searches for mouse (blue), rat (red), *Drosophila* (green), yeast (orange; search for "cerevisiae"), zebrafish (yellow), *C. elegans* (purple; search for "elegans"); **Cancer**: search for keyword "cancer" plus organism term; **Diabetes**: search for keyword "diabetes" plus organism term; **Heart Disease**: search for keyword "Heart Disease" or "Cardiovascular Disease" plus organism term.

Color image of this figure appears in the color plate section at the end of the book.

Database (Eppig et al. 2011), which provides the most comprehensive information on genes and their phenotypes in the laboratory mouse, currently contains functional or phenotype information on 17,576 genes out of 24,530 with protein sequence data (data taken from the Mouse Genome Informatics (MGI) web site http://www.informatics.jax.org/genes.shtml on 2 October 2013).

Furthermore, for most and perhaps all genes, the functional information is incomplete. This is because, even for the most well-studied genes, experimental laboratories have tended to focus on particular aspects of gene function and may not have paid attention to other phenotypes because of lack of expertise or interest in an apparently unrelated area of biology. This is now being somewhat addressed through systematic phenotyping of selected mouse strains by the International Mouse Phenotyping Consortium (IMPC, see below for further details). Nonetheless, the vast majority of mouse strains will remain partially phenotyped and the challenge will be to improve the way that data on a specific strain can be compiled in a meaningful and comprehensive way as it is generated.

In summary, the mouse has become the preeminent organism for modelling mammalian biology, both to assign gene function and model human disease. The scale required to efficiently achieve these twin goals is increasingly being built through the development of national and international consortia enabling cooperation on an unprecedented level. These challenges are being met through systematic approaches using the mouse, not just in phenotyping (phenomics) but also in genotyping (genomics) and the nomenclature and bioinformatics required to filter and link the data. Therefore, in the mouse we have powerful ways to create new mutant strains, archive them and, importantly, to systematically characterise the new strains. The projects underway and the resources available to create, cryopreserve and characterise these research tools are described below.

Mouse Mutagenesis

One of the greatest advantages of using the mouse for phenomics approaches is the variety of ways to manipulate its genome, both in targeted and random (unbiased) manners.

ENU Mutagenesis

N-Ethyl-N-nitrosourea (ENU) is the most potent germ-line mutagen in the mouse and continues to be an effective tool in creating new functional mutations in the mouse since its use was pioneered at Oak Ridge National Laboratory, Tennessee, USA, in the 1950s. ENU induces single nucleotide variations (SNVs, point mutations) in a random, relatively unbiased and genome-wide manner. In the past 20–30 years, many laboratories throughout the world have used ENU to produce a steady production line of mutants that can be screened for a large variety of phenotypic abnormalities (phenodeviants) in mice of any age, including embryos. Genome-wide ENU mutagenesis projects include those being undertaken by the Medical Research Council, Harwell, UK (http://www.har.mrc.ac.uk/); the Institute of Experimental Genetics, Munich, Germany (http://www.helmholtz-muenchen.de/en/ieg/); Mouse Mutagenesis for Developmental Defects, Baylor College of Medicine, USA (http://www.nih.gov/science/models/mouse/resources/fmmpdd.html); Mutagenetix Project at the University of Texas Southwestern Medical Center, Dallas, USA (http://mutagenetix.utsouthwestern.edu/); a number of members of the Australian Phenomics

Network (http://www.australianphenomics.org.au/) (including the Australian Phenomics Facility, Canberra (http://apf.anu.edu.au/); the Walter and Eliza Hall Institute of Medical Research, Melbourne (http://www.wehi.edu.au/faculty_members/research_projects/mouse_genetics); Macquarie University, Sydney (http://mq.edu.au/)); the Jackson Laboratory, Maine, USA (http://www.jax.org/); the NIH Neurogenomics Project, Northwestern University, Illinois, USA (http://genomics.northwestern.edu/neuro/); RIKEN, Japan (http://www.brc.riken.jp/lab/gsc/mouse/); and the Toronto Centre for Modelling Human Disease, Canada (http://www.cmhd.ca/). Together these and other ENU mutagenesis projects would have likely saturated the genome with point mutations, albeit on a variety of inbred strain genetic backgrounds. Although only a minority of these mutations have been annotated with functional phenotypic data, they potentially represent an allelic series at each mouse gene.

Targeted Mutation and Gene Traps

The ability to engineer the mouse genome in a targeted manner using mouse embryonic stem cells (mESCs) has led the species to become the preeminent mammalian model. Genome engineering and gene expression strategies are sophisticated and include conditional mutagenesis so that expression of the mutant allele can be temporally and spatially controlled. This, for example, can enable the propagation and study of mutations that would normally cause embryonic lethality but which can be silenced until adulthood. The full gamut of mutations can be engineered, including deletions, insertions, translocations, inversions and gene replacements. Alternatively, a gene can be disrupted by gene-trapping with the advantage of being able to tag the location of the disruption with a reporter. Point mutations can also be engineered through a knock-in. The International Knockout Mouse Consortium (IKMC) aims to produce mutations in every gene for distribution in the form of mESCs; further details are provided below.

In recent years, new mutagenesis techniques for mouse functional genetics have been developed. Transposon insertional mutagenesis (e.g., using "Sleeping Beauty" or "piggyBac" systems) has been achieved using mobile DNA segments to disrupt gene function and can include an activated reporter and tracker (Landrette et al. 2011). Another method is to engineer mice to express short hairpin RNAs, which trigger RNA interference (RNAi) causing the knock-down of gene function in a conditional and reversible manner (Premsrirut et al. 2011). Finally, the most recent developments in genome editing are the adaptation of transcription activator-like (TAL) effector nucleases (TALENs) (Sung et al. 2013) and the CRISPR/Cas system to facilitate RNA-guided DNA cleavage (Cong et al. 2013) to induce mutations in specific genomic loci. The latter technique can produce new mouse strains carrying multiple targeted mutations in a single generation, and is transforming the genome engineering field.

Physical Infrastructure Underpinning Mouse Phenotyping

Because of the cost and logistical difficulty of undertaking reliable phenotypic characterisation of mouse lines, significant underpinning infrastructure is needed to support such projects. Two main kinds of physical infrastructure are needed for community-wide phenotyping projects in the mouse: archives providing access to frozen embryos or sperm for the mouse lines to be studied, and well-equipped mouse facilities (mouse clinics) capable of generating the numbers of mice required to produce reliable datasets and carry out a range of phenotyping tests under controlled conditions.

Mutant Mouse Archives

Mutant mouse archives (MMAs) contain frozen germ-line material that can be used to reconstitute mutant mouse lines, which may be spontaneous variants of various kinds, ENU or other chemical mutants, or other kinds of mutants such as gene traps, targeted mutants or transposon-induced mutations. Large collections of variants are archived along with relevant metadata in a systematic manner to permit searching and retrieval based on genotype or phenotype. For example, there are established and growing archives of DNA from chemically (ENU) mutagenised mice in the UK (MRC Harwell; http://www.har.mrc.ac.uk/services/enu-dna-archive), Australia (Missense Mutation Library at the Australian Phenomics Facility; http://databases.apf.edu.au/mutations), Japan (RIKEN; http://www.brc. riken.jp/lab/mutants/RGDMSavailability.html), and USA (Mutagenetix, at University of Texas Southwestern Medical Center; http://mutagenetix. utsouthwestern.edu).

Material is typically stored as frozen sperm, ES cells or induced pluripotent stem (iPS) cell lines, embryos or ovaries. Material is stored in liquid nitrogen for long-term preservation. Some popular lines are also made available as live mice although transporting live mice, especially across international borders, is becoming increasingly problematic. MMAs serve not only as local repositories but as centres for the distribution of archived material, with distribution typically being at cost price from academic institutions (there are also some commercial suppliers of commonly used lines, such as Taconic (http://www.taconic.com) and Charles River Laboratories (http://www.criver.com)). An important aspect of MMAs is that by carrying out quality control tests and adhering to international guidelines they guarantee the quality of the material they make available.

MMAs also carry out specialist services such as reanimation (germplasm to live mice) and microinjection or aggregation (ES cells to live mice). MMAs are generally located in large mouse research facilities: most major European countries host one such archive, for example. For more detail on the background and activities of MMAs, see the review by Donahue et al. (2012).

As well as these general MMAs, specialised MMAs dedicated to particular types of mutant material have started to emerge. From the perspective of mouse phenotyping the most relevant of these is the KOMP (Knockout Mouse Project) repository, which makes ES cell lines from the KOMP project available for academic use (http://www.komp.org). Another recent development is the CreZOO database, which is a virtual repository of Cre and other targeted conditional driver strains (http://bioit.fleming. gr/crezoo/; http://www.creline.org/search_cre_mice).

The International Mouse Strain Resource (IMSR) web site (http:// www.findmice.org) (Eppig and Strivens 1999) provides an overview of the availability of the material held in the large repositories worldwide (it currently has information from 22 repositories in nine countries) allowing researchers to identify the nearest or most convenient source of any publicly available strain. MMAs may also provide more localised search engines providing more detailed information (e.g., Blake et al. (2010)). In recent years, there has been increasing coordination of archiving and supply of archived material between MMAs. In Europe, the European Mutant Mouse Archive (EMMA) (Wilkinson et al. 2010) coordinates the activities of 13 MMAs in 11 countries (http://www.emmanet.org). The Mutant Mouse Regional Resource Centers (MMRRC) in the USA and the Australian Phenome Bank (hhtp://www.pb.apf.edu.au) perform similar coordinating, data and biobanking functions in other regions. At the global level, the Federation of International Mouse Resources (FIMRe) (http:// www.fimre.org) (FIMRe Board of Directors 2006) has acted as a coordinating body. FIMRe aims to "ensure availability, assure quality, promote sharing and preservation of genetically defined mice, and disseminate information, affiliated resources, and expertise in their use to the global biomedical research community."

In summary, these various MMAs offer the resources to facilitate the archiving and distribution of mouse strains from all sources and each of the major repositories collaborates with the broader mouse genetic community regarding the dissemination of technological advances and enables the exchange of biological material and data.

Mouse Clinics and Infrafrontier

The second key element of infrastructure for carrying out the large-scale phenotyping experiments in the mouse is the so-called mouse clinic—a large facility capable of breeding and maintaining the necessary stocks and carrying out the range of phenotyping tests needed for exhaustive systematic phenotyping. The concept of the mouse clinic was originally developed as a means to fully characterise individual mouse mutants at the German Mouse Clinic in Munich, Germany (Gailus-Durner et al. 2005), founded in 2001. Subsequently the concept of the mouse clinic became increasingly widely recognised and became embedded in a number of institutions including the Mary Lyon Centre at Harwell, UK, the Wellcome Trust Sanger Institute at Hinxton, UK, the Institut Clinique de la Souris (ICS) in Strasbourg, France, the Toronto Centre for Phenogenomics, Canada, the Australian Phenomics Network, based in Canberra, Australia, and the Japan Mouse Clinic at Tsukuba, Japan. A list of currently existing mouse clinics is provided in Table 1.

Table 1. Existing and nascent mouse clinics.

Name	Location	URL
German Mouse Clinic	Munich, Germany	http://www.helmholtz-muenchen.de/en/ieg/
Mary Lyon Centre	Harwell, UK	http://www.har.mrc.ac.uk/
Wellcome Trust Sanger Institute	Hinxton, UK	http://www.sanger.ac.uk/
Institut Clinique de la Souis	Strasbourg, France	http://www.ics-mci.fr/
Toronto Centre for Phenogenomics	Toronto, Canada	http://www.phenogenomics.ca/
Mutagenetix	University of Texas Southwestern Medical Center	http://mutagenetix.utsouthwestern.edu
Australian Phenomics Network	Canberra, Australia	http://www.australianphenomics.org.au/
Asian Mouse Phenotyping Consortium	China, Korea, Taiwan, Japan	http://www.ampc.asia
Japan Mouse Clinic	Tsukuba, Japan	http://www.brc.riken.jp/lab/jmc/mouse_clinic/en/
Czech Centre for Phenogenomics	Prague, Czech Republic	http://www.biocev.eu/en/
Center of Animal Biotechnology and Gene Therapy	Barcelona, Spain	http://www.infrafrontier.eu/partners/UAB.php
CNR EMMA-International Monterotondo mouse clinic	Monterotondo, Italy	http://www.emma.cnr.it/

In view of the need for an increasing number of such facilities to enable both large-scale systematic phenotyping and more detailed phenotyping of individual mouse mutant lines, the European Commission established the Infrafrontier project (http://www.infrafrontier.eu) in 2008 as part of its ESFRI (European Strategy Forum on Research Infrastructures) project. ESFRI Infrastructures are intended to be large-scale, Europe-wide infrastructures that are essential for the future development of European science. Infrafrontier aims to support the construction of an enlarged network of mouse clinics in Europe by encouraging the individual EU member countries to invest in local mouse clinics. It has had success so far in stimulating investment in new mouse clinics in the Czech Republic (Prague), Spain (Barcelona) and Italy (Monterotondo).

An important aspect of Infrafrontier was to review the design and capabilities of existing mouse clinics both architecturally and from an informatics point of view. Kollmus et al. (2012) considered the layouts of a number of large mouse facilities in both academic and industrial organisations and identified a small number of common features despite considerable variation reflecting different local needs and regulatory regimes. They concluded that core breeding and holding, phenotyping, archiving, transgenics and rederivation, quarantine, and supporting infrastructure units were common to all facilities. Infrafrontier's survey on Animal and Laboratory Information Management Systems (AMS and LIMS) (Lengger et al. 2012) concluded that there was a wide range of systems in use but that large facilities tended to use complex in-house systems that were closely adapted to local needs. Although the development and use of these systems was a significant overhead both in terms of software development and to users, these systems provided significant benefits in terms of consistent and easily searchable storage of data and data security.

Mouse Phenomics Projects

Mouse Phenome Project and Database

The Mouse Phenome Project (MPP) (Paigen and Eppig 2000) was established in 1999 to provide phenotype information on inbred mouse lines. The objective was to collect phenotype data on a wide range of mouse lines (see http://phenome.jax.org/db/q?rtn=docs/workshop1999). Thirty-six of these lines were designated as priority strains, i.e. strains for which there was a strong community interest in collecting comprehensive data. The list was revised in 2007 (see http://jaxmice.jax.org/jaxnotes/507/507m.html). The current list of core strains is given in Table 2.

Table 2. The 36 core strains of the Mouse Phenome Project.

Strain[1]	Type[2]	Measurements[3]
129S1/SvImJ	IN	1701
A/J	IN	2040
AKR/J	IN	1526
BALB/cByJ	IN	1661
BPL/1J	IN	73
BTBR T+ tf/J	IN	1248
BUB/BnJ	IN	816
C3H/HeJ	IN	2134
C57BL/6J	IN	2696
C57BLKS/J	IN	1011
C57BR/cdJ	IN	840
C57L/J	IN	902
C58/J	IN	642
CAST/EiJ	WD	997
CBA/J	IN	1375
CZECHII/EiJ	WD	238
DBA/2J	IN	2140
DDY/JclSidSeyFrkJ	IN	84
FVB/NJ	IN	1722
KK/HlJ	IN	883
LP/J	IN	1052
MA/MyJ	IN	390
MOLF/EiJ	WD	597
MRL/MpJ	IN	575
MSM/Ms	WD	275
NOD/ShiLtJ	IN	1085
NON/ShiLtJ	IN	744
NZL/LtJ	IN	80
NZW/LacJ	IN	1014
PL/J	IN	1106
PWD/PhJ	WD	500
RIIIS/J	IN	794
SJL/J	IN	1490
SM/J	IN	1305
SWR/J	IN	1212
WSB/EiJ	WD	649

[1]Strain names are according to the standard mouse nomenclature; [2]IN = inbred line; WD = wild derived line; [3]number of phenotype measurements in the mouse phenome database on 15 March 2012. Information downloaded on 1 February 2013 from http://phenome.jax.org/db/q?rtn=strains/search&reqpanel=mpd_priority.

In the MPP, individual laboratories sign up to provide a particular type of phenotype data, using a particular protocol. Ideally data are gathered from two laboratories to provide confirmation. By using a standard protocol in a single laboratory, the project would ensure comparability of data sets (in the nomenclature of the MPP, a "project" refers to the set of data generated in a particular laboratory using a particular protocol; projects may generate more than one "measurement" on a set of strains). A number of standards and quality controls for testing have been established, including: the number of mice to be tested (10 males and 10 females), controlling for age (10–12 weeks of age) and sex, standardising environment (i.e., temperature and photoperiod) and housing (type of bedding, number of animals per cage), using consistent feed, and keeping health status reports. Many of these standards and quality controls were carried forward into subsequent mouse phenomics projects such as EUMODIC (see below). A key point to note is that the project is confined to inbred and a few wild-derived strains of mice and does not include information on any mutants. However, as genome sequence information becomes available for a wider range of mouse strains, the potential to relate phenotypic differences to their genetic composition will grow.

A key advance of the MPP was the development of a central database to collect the data from the participating laboratories: the Mouse Phenome Database (MPD; http://phenome.jax.org) (Bogue et al. 2007, Maddatu et al. 2012), established in 2001. The use of a central database provided significant advantages over simply presenting data on a web page, which would limit data mining and reporting. Data are typically submitted to the MPD from individual laboratories carrying out projects using a standardised format. Data are then quality checked and loaded into the database. Because the back end is a database, tools can be built to allow analysis and comparison of data sets, allowing the comparison of strains and searches for strains with particular phenotypic profiles. Data sets, including selections of data, can also be downloaded by users for further analysis.

Tools available at the MPD site include the ability to:

- Compare pairs of strains for their phenotypic attributes
- Identify sex differences in phenotype
- Find strains that are phenotypic outliers
- See the distribution of a given phenotype across strains
- See correlations of phenotypes between strains
- Search for correlated phenotypes.

The Knockout Mouse Project and Systematic Mouse Phenotyping

Although the MPP is providing an invaluable resource for mouse geneticists, new developments in mouse genetics mean that another approach to unscrambling the genotype-phenotype relationship is possible by manipulating the genome of a single strain. To pursue this, a discussion meeting was held at the Banbury Center at Cold Spring Harbor Laboratories in 2003 to discuss how best to exploit the mouse genome sequence. The outcome of this meeting, published in 2004 (Austin et al. 2004), outlined a strategy of firstly generating a targeted null allele at every gene in the mouse genome and then comprehensively phenotyping each line to characterise the broad phenotypes associated with each mouse gene.

A project such as this can realistically only be carried out on a global level as neither funding nor experimental capacity can be provided from a single source. The project that resulted was the Knockout Mouse Project (KOMP) and the consortium that ran it the International Mouse Knockout Consortium (IMKC; the consortium subsequently became known as the IKMC—see for example their web site at http://www.knockoutmouse. org) (Collins et al. 2007).

As this article focusses on phenotype, we will not address the details of the generation of KOMP mutations in detail. However, an outline is worthwhile so that the implications of the project for subsequent phenotyping can be understood. Firstly, it needs to be understood that the IKMC is an umbrella organisation and as such subsumes more than one approach to producing gene knockouts. These approaches are summarised in Collins et al. (2007). Two main, complementary approaches are being used to generate targeting constructs to produce the IKMC knockout alleles in C57BL/6 mouse embryonic stem (ES) cells. The first is gene trapping to delete the entire protein-coding sequence (null mutation) or to produce conditional alleles (the approach used by VelociGene and Regeneron). The second approach is gene targeting by homologous recombination using promoter-driven and promoterless cassettes to produce deleted or conditional alleles (the approach used by EUCOMM and KOMP-CSD).

High-Throughput Phenotyping Projects: EUMORPHIA

The first moves towards developing a systematic pipeline for producing phenotyping data for the laboratory mouse were two Integrated Projects funded by the European Union under Framework Programmes 5 and 6: EUMORPHIA (2002–2006; EUMORPHIA stands for European Union Mouse Research for Public Health and Industrial Applications; see http://www.

eumorphia.org) and EUMODIC (2007–2012; European Mouse Disease Clinic; see http://www.eumodic.org). The aim of these two projects was to establish the basic requirements for a larger, international project to carry out systematic phenotyping in the laboratory mouse. These requirements were:

- A set of easy to execute, cheap and reproducible phenotyping tests covering the body systems relevant to disease in humans
- Efficient means to generate sufficiently large cohorts of mice for phenotyping
- Assessment of the necessary cohort (sample) size to detect significantly abnormal phenotypes of moderate effect size
- Assessment of the best statistical approaches to detect abnormal phenotypes in a systematic phenotyping pipeline
- Development of exemplar bioinformatics infrastructure for capturing, validating and disseminating the data

EUMORPHIA primarily addressed the first of these aims by collecting together a group of mouse biologists from 49 centres across Europe to develop panels of phenotyping tests for 14 body systems. These are:

- Clinical Chemistry & Haematology
- Hormonal & Metabolic Systems
- Cardiovascular
- Allergy & Infectious Diseases
- Renal
- Sensory Systems (Vision & Auditory)
- Peripheral/Central Nervous System & Muscle
- Behaviour, Cognition & Neurological
- Cancer Phenotyping
- Bone & Cartilage
- Gene Expression
- Whole-body Imaging
- Necropsy, Histology & Pathology
- Respiratory

Groups met a number of times over a three-year period to develop panels of tests in each area. Once preliminary panels had been drawn up, the tests were validated using inbred mouse lines to test their reproducibility and a second round of test definition was carried out to provide a final, validated set of protocols (or SOPs). This set of SOPs was named EMPReSS (European Mouse Phenotyping Resource of Standardised Screens) and hosted in an online database at http://empress.har.mrc.ac.uk/ (Brown et al. 2005, Green et al. 2005). The set of SOPs developed by EUMORPHIA is listed in Table 3.

In addition to the EMPReSS set of phenotyping tests, EUMORPHIA laid the basis for the informatics infrastructure needed for systematic phenotyping in two ways. Firstly, it considered the best way to represent abnormal and normal phenotypes detected by the phenotyping screens. A standardised description is essential for computational sharing and analysis of data and the favoured way of representing biological annotations is to use biological ontologies such as the Gene Ontology (Ashburner et al. 2000). For mouse phenotypes the equivalent ontology is the Mammalian Phenotype Ontology (MP) (Smith et al. 2005). However, the MP suffers

Table 3. Summary table of SOPs developed by EUMORPHiA (taken from EUMORPHIA Consortium (2006)).

SOP
Clinical chemistry and haematology
Differential blood count
Blood collection by retro-orbital puncture
Blood collection by intra-cardiac puncture
Blood sample handling—Clinical chemistry
Blood sample handling—Haematology
Blood sample handling—Coagulation
Clinical chemistry parameters
Haematology tests
Gas anaesthesia
Coagulation tests
Annex 1: Reagents for blood clinical chemistry on AU400
Annex 2: Calibrators for clinical chemistry on AU400
Annex 3: Controls for clinical chemistry on AU400
Hormonal and metabolic systems
Simplified metabolic cages
Metabolic cages
Dexa-scan analysis
Simplified Intra-Peritoneal Glucose Tolerance Test (I.P.G.T.T.)
Intra-Peritoneal Glucose Tolerance Test (I.P.G.T.T.)
Cold test
Meal Tolerance Test (M.T.T.)
Oral Glucose Tolerance Test (O.G.T.T.)
Cardiovascular
Non invasive blood pressure and heart rate
Invasive blood pressure
Invasive left ventricular haemodynamics
Blood pressure by Telemetry
Surface electrocardiography (ECG)
Echocardiography (ECG)

Table 3. contd....

Table 3. contd.

Allergy and infectious diseases
Titration of antibody solutions for FACS analysis
Cross linking phycoerythrin to antibodies
Coupling fluorescein to antibodies
FACS analysis of peripheral blood cells
Determination of immunoglobulin concentrations in the serum of mice
Isolation of murine bone marrow-derived macrophages
Isolation and culturing of proteose peptone-elicited peritoneal macrophages
Isolation and culturing of thioglycolat-elicited peritoneal macrophages
Quantification of TNFα production by PAMP stimulated macrophages
Infection of proteose peptone macrophages with Listeria monocytogenes
Quantitative measurement of iNOS activity after macrophage stimulation
Staining protocols for Listeria monocytogenes-infected macrophages
Counting cells using a Thoma chamber
Annex: Characterisation of macrophage functions
Inoculation of Listeria monocytogenes EGD for infection*
Infection of mice with Listeria monocytogenes EGD*
Tissue monitoring (Listeria monocytogenes)*
Inoculation of Streptococcus pyogenes A20 for infection*
Infection of mice with Streptococcus pyogenes A20*
Blood monitoring (Streptococcus pyogenes A20)*
Health monitoring of mice in infection experiments*
Sensory Systems
Vision
Optokinetic response test
Fundus and angiography
Using an indirect ophthalmoscope
Using a slit lamp
Auditory and vestibular
Modified SHIRPA protocol specific to sensory systems
Elevated platform and reaching response test
Acoustic startle and pre-pulse inhibition
Swim ability test
Auditory brainstem response (ABR)
Behaviour and cognition
Open Field
Modified SHIRPA
Grip strength
Y-maze
Acoustic startle and pre-pulse inhibition
Tail flick

Table 3. contd....

Table 3. contd.

Tail suspension

Swim ability test

Rotarod test

Cancer phenotyping

The macroscopic description of a tumor process

Freezing of murine tumor tissue

Fixation and processing of murine tumor tissues enabling diagnostic, immunohistochemical and molecular analysis

Annex 1: Staining with Harrison's fixative

Annex 2: Optimisation of specificity and sensitivity in immunohistochemical investigations on formalin- fixed, paraffin-embedded murine (tumor) tissue

Immunostaining using streptavidin-biotin-horseradish peroxidase enhancement

Manual production of Tissue Micro Arrays

Isolation of DNA from frozen tissue, immunohistochemical and molecular analysis

Isolation of DNA from formalin-fixed, paraffin-embedded tissue

Isolation of total RNA from frozen tissue

Monitoring the quality of total RNA by agarose gel electrophoresis

Reverse transcription of mRNA. First strand cDNA synthesis

Isolation of total RNA from formalin-fixed, paraffin-embedded tissue

Bone, cartilage, arthritis, osteoporosis

Dysmorphology

Ionic fraction of Ca^{2+} in whole blood

Bone densitometry

X-ray

Micro CT imaging

Skeletal preparation of a mouse

Gene expression

In vitro transcription of digoxigenin-labelled riboprobes

In situ hybridisation of cryosections with digogixenin-labelled probes

Whole-mount *in situ* hybridisation of mouse embryos

Automated whole-mount *in situ* hybridisation of mouse embryos

In vitro transcription of 35S-labelled riboprobes

In situ hybridisation of cryosections with 35S-labelled probes

Annex 1: *In situ* hybridisation; optimisation of specificity and sensitivity-troubleshooting notes

Annex 2: *In situ* hybridisation; working with RNA

Necropsy examination, pathology, histology

Annex 1: A systematic histopathology flow scheme

First line phenotyping necropsy

Trimming fixed tissues from necropsy

Table 3. contd....

Table 3. contd.

Fixation

Tissue fixation with Bouin's solution

Tissue fixation with Davidson's solution

Tissue fixation with 10% buffered neutral formalin

Tissue fixation with gluteraldehyde

Tissue fixation with 4% buffered paraformaldehyde

Tissue fixation by perfusion

Tissue fixation with periodate-lysine-2% paraformaldehyde (PLP)

Embedding

Demineralisation of long bones using EDTA

Fixation decalcification processing and paraffin embedding of whole adult mouse head

Freezing tissues for histopathological analyses

Tissue processing and embedding in paraffin

Sectioning

Sectioning from paraffin-embedded tissues

Cryosectioning

Proliferation Apoptosis

Immunodetection of the cell proliferation marker Ki67

Detection of 5-bromo-2-deoxyuridine (BrdU) incorporation

Immunodetection of cells in mitosis

In situ detection of cell death

Staining

Haematoxylin and eosin staining of histological sections

Modified Mallory's trichrome staining of histological sections

Luxol fast blue and cresyl violet staining of brain and spinal cord

Orcein staining for elastic fibres

Von Koss's silver nitrate method for detection of calcified tissue deposits

Periodic Acid Schiff (PAS) staining of histological sections

Oil red O staining of histological sections

Detection of cholesterol esters in histological stains

Mammary gland whole mount preparation

Alizarin red and alcian blue method for staining bones and cartilage in the adult mouse

Sirius red staining for collagen

Congo red for amyloidosis

Oral

Phenotyping mouse oral cavity—primary first line

Phenotyping mouse oral cavity—primary extended

(*SOPs validated ready for inclusion in the release of the new version of EMPReSS).

from disadvantages that make it unsuitable for annotating systematic phenotyping data, most notably that it is only useful for annotating abnormal phenotypes. The EUMORPHIA informatics work package therefore developed an alternative approach to annotating phenotype data known as the EQ approach (Gkoutos et al. 2004, Gkoutos et al. 2005) and described in Chapter 10 of this volume.

High-Throughput Phenotyping Projects: EUMODIC and the Wellcome Trust Sanger Institute Mouse Genetics Project

EUMODIC and the Wellcome Trust Sanger Institute Mouse Genetics Project (WTSI-MGP), which both commenced work in 2006, built on the groundwork carried out in the EUMORPHIA and KOMP projects to pilot full scale systematic phenotyping programmes (Ayadi et al. 2012).

Three key innovations were needed for the execution of these projects:

1. A well-structured pipeline for the production of appropriate numbers of homozygous or heterozygous mice for phenotyping,
2. Phenotyping pipelines designed to optimise the amount of phenotype data that could be obtained from the minimum number of mice without interference between phenotyping tests carried out in series,
3. A well-founded basis for calling abnormal phenotypes in mouse lines based on appropriate controls.

Mutant Mouse Production

In both programmes, mutant mice were produced from germ-line transmission (GLT) mice derived from targeted ES cells. Both projects used targeted C57BL/6N ES cells and hosts. The C57BL/6N strain is a close relative of C57BL/6J which is a commonly used strain in transgenic experiments, is a good breeder, and was the first mouse inbred line for which a genome sequence was available. GLT mice were crossed to "wild type" C57BL/6N mice and the resulting litter used to found a heterozygous colony which was then used to derive the desired number of homozygous mice for phenotyping (14 males and 14 females). When mutations caused homozygous lethality (i.e., no or very few homozygous mice were born), heterozygous mice were instead used for phenotyping.

Phenotyping Pipelines

EUMODIC and WTSI-GMP used different approaches to constructing their phenotyping pipelines. The EUMODIC screen, which made use of EMPReSS Slim, a subset of the EMPReSS set of protocols developed by EUMORPHIA, comprised three pipelines (represented in Fig. 2). Two of these comprised the majority of the phenotyping tests and the third was a "virtual" pipeline that reported on homozygote viability and fertility. The aim of splitting the EUMODIC screen into two main pipelines, despite its higher cost in

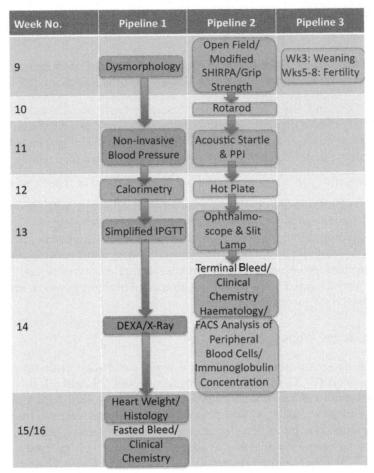

Week No.	Pipeline 1	Pipeline 2	Pipeline 3
9	Dysmorphology	Open Field/ Modified SHIRPA/Grip Strength	Wk3: Weaning Wks5-8: Fertility
10		Rotarod	
11	Non-invasive Blood Pressure	Acoustic Startle & PPI	
12	Calorimetry	Hot Plate	
13	Simplified IPGTT	Ophthalmo-scope & Slit Lamp	
14	DEXA/X-Ray	Terminal Bleed/ Clinical Chemistry Haematology/ FACS Analysis of Peripheral Blood Cells/ Immunoglobulin Concentration	
15/16	Heart Weight/ Histology Fasted Bleed/ Clinical Chemistry		

Figure 2. Overview of the EUMODIC phenotyping pipelines. Phenotyping protocols are arranged by week vertically and by pipeline horizontally. Colouring reflects the class of test carried out: Blue = Morphology & Metabolism; Red = Cardiovascular; Grey = Bone; Green = Neurobehavioural & Sensory; Purple = Haematology & Clinical Chemistry; Orange = Allergy & Immune. Redrawn from http://empress.har.mrc.ac.uk/viewempress/index. php?map=EUMODIC+Pipeline.

Color image of this figure appears in the color plate section at the end of the book.

terms of the number of mice used and consequent more complex logistics, was to avoid interference between the different tests. For example, as the screen contained a number of behavioural tests and a blood pressure test, it was decided to separate these two classes of test as behavioural tests could affect subsequent blood pressure. The WTSI-GMP project took a different approach by using fewer phenotyping tests in a single pipeline to minimise the number of mice used and therefore overall cost.

The phenotyping tests used by the two projects covered many of the same body systems. This reflects the aim of both projects to produce phenotype data that reflect all major body systems so that the projects should maximise the discovery of new mouse models of human disease, although the individual tests used were in some cases significantly different. In particular, the EUMODIC screen contained systematic phenotyping for abnormalities in fasted blood chemistry and the rotarod performance test (which assesses motor behaviour), while the WTSI-GMP screen added a number of more sophisticated tests (Auditory brainstem response, MicroCT and quantitative Faxitron, Citrobacter challenge, Salmonella challenge, presence of micronuclei, stress-induced hyperthermia, tail epidermis whole mount, and eye, skin and brain histopathology; see Table 5 of Ayadi et al. (2012)).

Statistical Analysis

Assessment of the outcomes of these two phenotypic screens was also carried out differently. EUMODIC developed a statistical pipeline which processed experimental and matched control sets, measured at nearly the same time, and calculated P-values for the experimental and control values being different (Morgan et al. 2010). The statistical tests used depended on whether quantitative values were being investigated, in which case t-tests were used, or qualitative values, in which case frequencies of abnormal results in cohorts were tested for abnormality using chi-squared or Fisher's exact test. Phenotyping parameters reaching significance at a predefined level (the default was $P < 0.0001$) were then automatically annotated with appropriate Mammalian Phenotype ontology terms which were then made available via the EuroPhenome web site (see below). The WTSI-MGP took a different approach to calling abnormal phenotypes. For quantitative data it made use of a reference range test whereby if more than 60% of mice lay outside the 5–95% distribution range of the baseline, a call was made. For qualitative data, a Fisher's exact test was used with $P < 0.05$ cutoff and a requirement for a 60% change in frequency. These calls were supplemented by expert input from investigators (Mallon et al. 2012). Analysis of the false positive and negative calls generated by these approaches suggests a need for improved statistical analysis (Smedley et al. 2013). A first step towards improved analysis has been taken with the development of an improved, hybrid statistical approach (Karp et al. 2012) but it is likely that further improvement will be needed.

Data Capture and Dissemination

Data capture in EUMODIC was carried out by the individual mouse clinics via their local LIMS. In order to collect the data from these LIMS into a central database, the EUMODIC Bioinformatics work package developed an XML format which not only acted as a repository for the data captured in each phenotyping test but also specified the metadata required for each of them. Data were then transferred to the data centre at MRC Harwell where they underwent some quality checks before being deposited in the EuroPhenome database (http://www.europhenome.org) (Mallon et al. 2008, Morgan et al. 2010). Data analysis was carried out as described in the previous section and MP annotations associated with the relevant lines.

The EuroPhenome site presents a number of options for downloading and browsing the EUMODIC data. Lines with abnormalities in particular phenotyping tests can be identified by traversing the Mammalian Phenotype ontology hierarchy (only MP terms used to annotate EUMODIC tests are shown). Alternatively the range of normal and abnormal phenotypes for one or a set of lines can be visualised using a heatmap-like interface. Data sets for individual lines can be downloaded and visualised graphically. Summary phenotype data is also provided to the Mouse Genome Informatics group at the Jackson Laboratory for inclusion in the Mouse Genome Database. However, EuroPhenome does not provide the level of sophistication in data analysis provided by the Mouse Phenome Database.

Associated Activities

As a precursor to the eventual development of the International Mouse Phenotyping Consortium (see below), the EUMODIC bioinformatics work package led a number of international efforts relating to data openness and exchange. Interphenome (http://www.interphenome.org) was an international group that met twice yearly from 2006 to 2009 to identify and coordinate the development of standards and approaches to facilitate the exchange of mouse phenotype data (Mouse Phenotype Database Integration Consortium 2007). As part of this initiative a minimum standard for the description of mouse phenotyping data—MIMPP—was developed (Taylor et al. 2008).

CASIMIR (Coordination and Sustainability of International Mouse Informatics Resources; http://www.casimir.org.uk) (Hancock et al. 2008) was funded separately from EUMORPHIA and EUMODIC to develop policies for the open exchange of mouse data and resources and their long-term sustainability. This project held a number of high profile international conferences and generated a number of publications including developing the Rome Agenda for post-publication data sharing (Schofield et al. 2009), proposing mechanisms to support the long-term sustainability of resources for the mouse community (Schofield et al. 2010), and carrying out some pilot projects on data integration (Gruenberger et al. 2010, Smedley et al. 2008). CASIMIR and Interphenome also produced a briefing paper "Legal Issues on Phenotype Data Accessibility" (http://www.casimir.org.uk/fullstory. php?storyid=76) which proposed that mouse phenotype data be shared without restriction and, if a formal licence is needed for administrative reasons, the CC0 licence be used. CC0 is a legal code developed by Creative Commons (http://creativecommons.org/about/cc0) which allows data producers to waive intellectual property rights, allowing data to be used freely.

The International Mouse Phenotyping Consortium (IMPC)

Starting with initial discussions in late 2007, the aim of the International Mouse Phenotyping Consortium (Brown and Moore 2012a,b) is to build on the success of EUMODIC to construct a global, virtual mouse clinic and to generate systematic phenotype data on as complete a set of mouse genes as possible (nominally 20,000 genes over a ten-year period—5,000 from 2011 to 2016 and a further 15,000 from 2016 to 2021). The IMPC will build on the groundwork carried out by EUMODIC to link together a global network of laboratories to carry out a core set of mandatory phenotyping tests which may be supplemented by additional tests at individual laboratories. Reflecting experience during EUMODIC that there is a necessity to ensure that all the data going onto the final analysis pipeline and database are of the highest possible quality, the IMPC data will initially be collected by a Data Coordination Centre (DCC) which will be hosted at MRC Harwell and ensure data quality and carry out annotation. The statistical model to be used for calling abnormal phenotypes is currently under development. Data will then be deposited into a Core Data Archive (CDA) to be hosted at EMBL-EBI which will provide the public face of IMPC data dissemination (Brown and Moore 2012a, Mallon et al. 2012). The set of phenotyping tests and protocols to be used, which are still in development, can be seen at the IMPRESS database (https://www.mousephenotype.org/impress) while project progress and, in future, access to data will be through the main web portal (https://www.mousephenotype.org/).

Other Phenotyping Efforts

The Asian Mouse Phenotyping Consortium (AMPC; http://www.ampc. asia) has been driving major developments in the phenomics capability in the Asia-Pacific region. Several institutions in China, Korea, Taiwan and Japan are cooperating to promote the coordination and development of advanced strategies and service platforms for phenotyping and informatics, and the AMPC is a member of the IMPC.

Commercial Mouse Phenotyping Efforts

In the early 2000s, two commercial organisations, Deltagen and Lexicon Genetics, started to generate mutant mouse lines and accompanying systematic phenotype data as a means to attract sales of the individual lines. In 2005, NIH purchased rights to distribute 256 of these lines via the Jackson Laboratory (with other distributors stepping in subsequently). Genetic and phenotype data on the strains were deposited and are accessible via the Mouse Genome Informatics web site (http://www.informatics.jax. org/external/ko/) as part of the broader MGI dataset. Phenotype data are available via individual factsheets rather than via a database. As can be seen from Table 4, many of the phenotypic features assayed by the two companies were similar. Lexicon generally produced more detailed descriptions, particularly on oncology and ophthalmology, but Deltagen provided additional information on embryonic development and necropsy as well as detailed phenotyping protocols. It is noteworthy that the same broad areas of phenotyping carried over into subsequent public high-throughput phenotyping projects; adjustments in those subsequent projects reflected a number of factors including cost, time consumption and reproducibility. Deltagen continued to produce mutant mouse lines up to a total of 750 with information available via their web site at https://deltaone.deltagen.com/ (registration required).

Related Academic Projects

Some systematic phenotyping projects are being carried out by individual institutions or consortia focussing on particular aspects of phenotype or phenotypic profiles that are different from that covered by the IMPC.

The Nathan Shock Center of Excellence in the Basic Biology of Aging at the Jackson Laboratory is carrying out studies of specific, ageing-related phenotypes on 32 commonly used inbred strains (http://agingmice.jax. org/index.php/research/center-research/). Phenotypes characterised include median lifespan and insulin-like growth factor levels (Leduc et al. 2010, Yuan et al. 2009), albumin:creatinine ratios (Tsaih et al. 2010), relative percentages of peripheral blood leukocyte subtypes (Petkova et al. 2008), cardiac conduction activity (Xing et al. 2009), serum sodium concentration (Sinke et al. 2011), disease-associated histopathology (Sundberg et al. 2011) and DNA damage. Data from this project are available via the Mouse Phenome Database (see http://phenome.jax.org/db/q?rtn=docs/ collab&group=Shock).

Table 4. Overview of phenotyping tests carried out by Lexicon and Deltagen.

Lexicon		Deltagen	
Assay type	Feature assayed	Assay type	Feature assayed
Blood chemistry	Alkaline phosphatase	Clinical chemistry	Sodium
	Albumin		Potassium
	Total Cholesterol		Chloride
	Triglyceride		Bicarbonate
	Blood urea nitrogen		Blood urea nitrogen
	Glucose		Creatinine
	Sodium		Osmolality
	Chloride		Calcium
	Alanine Aminotransferase		Phosphorus
	Bilirubin Total		Glucose
	Phosphorus		Creatine kinase
	Creatinine		Alkaline phosphatase
	Calcium		Alanine aminotransferase
	Uric acid		Aspartate transferase
	Potassium		Lactate dehydrogenase
	HgbA1C		Protein
	Insulin		Albumin
	Oral glucose tolerance		Globulin
			Bilirubin
			Cholesterol
			HDL
			LDL
			Triglyceride
Urinalysis	Protein		
	Glucose		
	Ketones		
	Blood		
	Bilirubin		
	Urobilinogen		
	Nitrate		
	Leukocyte esterase		
	pH		
	Specific gravity		

Table 4. contd....

Table 4. contd....

Lexicon		Deltagen	
Assay type	**Feature assayed**	**Assay type**	**Feature assayed**
Cardiology	Blood pressure		
	Heart rate		
Diagnostics	**Length**	Aging metrics	**Length**
	Weight		**Weight**
Immunology	Acute phase response		
	FACS3		
	Haematology		**Haematology**
	Ovalbumin		
	siCAM		
Neurology	Basal body temperature	**Behaviour**	
	Circadian rhythms		Rotarod test
	Functional observation battery		Tail flick test
	Hot plate		**Hot plate test**
	Inverted screen		Metrazol test
	Open field		**Open field test**
	Prepulse inhibition		**Startle response/PPI test**
	Stress-induced hyperthermia		
	Tail suspension		**Tail suspension test**
	Trace-aversive conditioning		
Oncology	Skin proliferation		
Ophthalmology	Angiogram		
	Fundus		
Pathology	**Gross**	**Pathology**	Physical exam
	Microscopic		Histopathology
Radiology	CAT scan	Densitometry	
	DEXA		Dual-energy x-ray absorptiometry
	MicroCT		
		Necropsy	
		Embryonic development	

Terms in bold represent broadly similar measurements between projects.

The US National Mouse Metabolic Phenotyping Center (MMPC) (http://www.mmpc.org) (Laughlin et al. 2012) is a network of phenotyping centres that provides standardised phenotyping of mouse lines modelling diabetes, diabetic complications, obesity and related disorders. Data from analyses are made publicly available through the web site and can be analysed and visualised there or downloaded.

A large consortium is working to decipher the mechanisms by which the immune system responds to infectious disease (http://www. systemsimmunology.org). The Scripps Research Institute, the Institute for Systems Biology and Stanford University, USA, together with The Australian National University are using multiple approaches including large-scale forward genetics (ENU mutagenesis) in the mouse to discover genes involved in innate inflammatory reactions and adaptive immune responses. Tens of thousands of mice are being subjected to a range of phenotypic screens and have thus far produced hundreds of mutants. About half of the identified genes are novel in the sense that no such phenotype had been predicted by other alleles where available. In addition to the identification of the causative mutations, all incidental mutations are also found through whole genome or exome sequencing, thereby creating a vast archive of point mutations (SNVs) in the mouse. The full list of phenotypic and incidental mutations found to date can be searched through the Mutagenetix web site (http://mutagenetix.utsouthwestern.edu), and the mutations archived in Australia can also be searched through the Australian Phenomics Facility's Missense Mutation Library (http://databases.apf.edu.au/mutations). While all mice are archived, many remain live for several months and are available for further characterisation or retrieval from the cryobank.

The Consortium for Functional Glycomics (http://www. functionalglycomics.org/) carried out phenotyping on 34 mouse lines with knockouts in genes relevant to glycobiology in the domains of histology, haematology including blood chemistry, immunology and metabolism and behaviour between 2001 and 2011. Data are provided openly for download as PDF or Excel files.

The Neuromice Consortium produced collections of mutagenised mice and screened for alterations in nervous system function and behaviour (http://165.24.81.52).

Systems Genetics Projects

Systems genetics approaches attempt to improve the mapping of phenotypic traits by simulating the conditions found in natural populations whereby individuals (represented by inbred lines in systems genetics populations) have different genetic compositions, allowing the application of mapping approaches to traits. The best-known systems genetics resource for the mouse is the Collaborative Cross (CC) (Churchill et al. 2004), started in 2004, in which strains are derived from an eight-way cross using a set of founder strains that include three wild-derived strains. Another, more recent resource is the Diversity Outbred Mouse Population (DO) (Churchill et al. 2012), which is derived from partially inbred Collaborative Cross strains and maintained by randomised outcrossing. A third resource is the Chromosome

Substitution Strains (Singer et al. 2004), in which a single chromosome of the background strain, C57BL/6J, is substituted by a chromosome from the A/J line. Ultimately, the optimal exploitation of all of these resources will depend on thorough and well-founded phenotypic characterisation. To date, some systematic phenotypic characterisation of the CC has been reported (Philip et al. 2011), along the lines of other systematic phenotyping projects although this data is available as supplementary material rather than from a database. To date, DO populations and chromosome substitution lines have been used only for mapping individual phenotypic traits.

Perspective and Future Challenges

As the first mammalian model organism for which extensive phenomics data will be collected, the laboratory mouse not only poses some unique logistical challenges but also provides some exemplary lessons for phenomics projects in other species. Below we firstly list the main areas in which significant progress has been made in mouse phenomics projects so far. We then go on to list some areas in which we believe advances will be needed to reduce the cost and improve the efficiency of high-throughput phenotyping efforts such as the IMPC.

Progress Highlights

Design of Targeted Mutations

The KOMP ES cells that form the basis of the IMPC project were carefully designed to allow them to be used for generating conditional as well as constitutive knockout mice. This means that follow-up studies can be carried out in the same genetic background as the initial phenotyping screens, maximising the transferability of information between the different phases of the characterisation of this material.

Genetic Background

Both the phenotyping and the strain production for IMPC have been carried out on the same genetic background (C57BL/6N). Consequently there should be no confounding effects of transferring ES cells into a different genetic background. All of the systematic phenotyping in the IMPC will be carried out in this background, so all data from the programme should be comparable between genes except in cases it has not been possible to produce homozygous mice (there are, however, other issues which may affect comparability—see below).

Breeding Logistics

Although the most cost-effective mammalian model organism, breeding suitably sized cohorts (in the case of EUMODIC and IMPC seven homozygous males and seven homozygous females) of experimental animals efficiently and at minimum cost is a complex undertaking. An optimised breeding schema developed by the WTSI-MGP and EUMODIC projects is described in Ayadi et al. (2012).

Design of Phenotyping Scheme

EUMODIC and IMPC aimed to carry out as many phenotyping protocols as possible in as short a time as possible. The projects aim to cover all body systems within their pipelines to maximise the utility of the data in the selection of new gene models. Minimising the timescale of the pipelines minimises the cost of phenotyping any given line, as does choosing phenotyping cohorts that are as small as possible to deliver acceptable statistical power (see below).

Phenotyping Protocols

To date, there has been an extensive process of refinement of phenotyping protocols to ensure reproducibility between phenotyping centres. Despite this, reproducibility is still limited, with many protocols producing similar (but not always identical) within-centre ranking of outcomes between strains but not similar absolute values. This may reflect a number of confounding factors including variation in housing conditions, feed, and subtle variations in experimental equipment and protocols. It is even possible that epigenetic differences reflecting the reanimation of the mice have an effect.

Informatics

Rather than introduce a data capture solution that would have to be installed at each phenotyping centre (potentially under different operating systems) and used in parallel with in-house LIMS, data capture in EUMODIC made use of existing LIMS with data being exported in a standard file format based on XML and transferred by FTP (Mallon et al. 2012). The main overhead in this process was the design of the XML format, but having developed this, the project now has the benefit of a format that is transferable to other sites and projects. EUMODIC data were then transferred to a MySQL database. Using MySQL for the database has the benefit that anyone can download and use the schema but MySQL suffers from relative instability and lack of features compared to some commercial systems, notably Oracle systems, and it is these systems which will be used for the IMPC database.

Data Quality

The EUMODIC project demonstrated the high frequency of errors in data recording. The EUMODIC XML schema was designed to suppress some kinds of error, such as incomplete data sets. However, errors such as mislabelled data sets, shifted decimal points, recordings in the wrong units, and incorrect data types were common enough that efforts to clean up data had to be increased after the data were transferred to the data centre at MRC Harwell. Reflecting this, the IMPC explicitly incorporates a DCC which will employ manual curation of incoming data to ensure errors are corrected by the originating centres.

Controls

For systematic detection of phenodeviance in experimental animals, the suitable use of controls is essential. As seasonal variation and drift in some if not all phenotypic variables occur, long-term studies need to employ suitable sets of control animals, with the same genetic background and housing conditions as the experimental animals. Ideally, experimental cohorts would be accompanied by control cohorts through all stages of the pipeline but this is prohibitively costly. Alternative approaches have been tried, including using regularly spaced control cohorts, small numbers of control animals along with experimental cohorts, and pooling control values over time to generate a reference range. No detailed analysis of these alternative approaches has been published but it is clear that an appropriate and standard approach to experimental controls needs to be adopted to minimise variability in calling phenodeviance.

Data Analysis

A critical, and difficult, issue in systematic phenotyping is to call differences in phenotypic values that represent significant phenodeviance and distinguishes those that are unimportant background noise. That this issue is unresolved is reflected in the frequency differences measured in the high-level phenotyping categories between EUMODIC and WTSI-MGP as reported in Table 4 of Ayadi et al. (2012). A simple χ^2 analysis indicates that where identical tests are reported as being carried out between projects, 6 of 9 showed significantly different frequencies of positives ($P < 0.05$ after Bonferroni correction). This frequency was slightly lower (4 of 9) for different or similar but not identical tests. The origins of these differences are no doubt complex, probably reflecting a combination of different local conditions in the phenotyping centres and differences in the statistical analysis, but it is clear that better standardisation is required before consistency is achieved.

The calling of positive differences between strains goes beyond the statistical treatment of the data. Differences may be statistically significant but small (due to small variances) and these differences may not be considered biologically important. The EuroPhenome web site employed a second criterion beyond statistical significance for the calling of positive differences in the form of effect size, Cohen's d (the ratio of the difference of means to the pooled standard deviation). This allows users to filter results on that basis but is broadly unsatisfactory because it introduces subjectivity into the selection of annotations. More broadly, the provision of data analysis tools such as those available at the Mouse Phenome Database, EuroPhenome and the MMPC allows users to evaluate results and allows them to assess how confident they are about the annotations presented to them, especially if the annotations are automatic.

Ontologies

The standard ontology for representing mouse phenotypes is the Mammalian Phenotype ontology (Smith and Eppig 2012). This ontology is good for representing the existence of phenodeviance but cannot represent non-deviant phenotypes or provide detail of the results of measurements. An alternative approach is the EQ approach which combines the PATO ontology, which describes qualities of measured entities, with ontologies describing those entities (Gkoutos et al. 2005). The application of the EQ approach to annotating mouse mutations was developed during EUMODIC (Beck et al. 2009) but it has yet to be fully implemented.

Data Dissemination

Most phenomics projects disseminate data via web sites. Often this dissemination is in the form of reports, which may be Excel files, comma or tab-delimited text files or even PDFs. The EuroPhenome site extends this by providing access via BioMart (Smedley et al. 2009), which allows simple linking of remotely located databases. Through this, EUMODIC data are made available through the BioMart central portal MartView interface (http://www.biomart.org/biomart/martview/) (Haider et al. 2009) which allows joint querying of a growing number of databases. EuroPhenome also provides summary annotations, in the form of ontology terms, to the Mouse

Genome Informatics database at the Jackson Laboratory (Blake et al. 2011). This allows access to users via the most-used mouse database.

Future Issues

For the future, while many issues have been addressed by mouse phenomics projects, some issues remain to be fully addressed:

Reproducibility

Few phenotyping procedures provide indistinguishable results when applied at different sites. This reduces the utility of the data in large-scale analyses. The causes of this lack of reproducibility are manifold and include lack of appropriate detail in protocol descriptions, variation in housing and feed (the latter being particularly important for physiological measures such as blood chemistry), and subtle aspects such as different approaches to animal handling by experimenters. It may be that some of these sources of variation cannot be eliminated, but further progress towards reproducibility between centres needs to be a focus of future efforts.

Data Quality

The IMPC approach of employing a DCC is state-of-the-art at present. However, as a DCC is necessarily remote from data generation and capture process, requiring faulty data to be corrected by the generating centre, there is potential for better error correction at the point of origin using improved software. The relative cost benefit of such an approach needs to be assessed, however.

Controls and Experimental Design

Like most biological experiments, phenomics experiments have not so far incorporated planning for data analysis from the start, i.e., they have been treated as "biological experiments" rather than "data experiments". A particular risk is a drive towards sensitivity at all costs, when increased sensitivity may just mean an increased frequency of false positives. Careful analysis of the use of control animals, the statistical approach to calling phenodeviance, and the significance and effect size cut-offs to be used need to be assessed systematically.

Data Access

Although approaches to data access have been superficially good in mouse phenomics projects, full sets of data including appropriate controls are not easy to obtain. This may seem unnecessary but, apart from being in the spirit of open data sharing (Schofield et al. 2009, Toronto International Data Release Workshop Authors et al. 2009), making full data sets available allows researchers not associated with projects to carry out novel analysis, for example on different ways to call phenodeviance or the use of control data. Genome sequencing projects have released raw data since their inception and phenomics data should follow that lead and develop means to fully share raw phenotyping data. It will be important for future databases to comply with developing standards, such as the BioDBcore criteria; see Gaudet et al. (2011).

Semantic Integration

A key issue for using mouse phenotype data for the identification of models of human disease is linking the data in a meaningful way to human disease data. There are still significant barriers to doing this, most notably the lack of a genotype-phenotype database for humans (an issue being tackled by the Human Variome Project (Howard et al. 2012)). However, some progress has been made in linking phenotypes and diseases across species (Chen et al. 2012) and the next step will be to build semantic systems linking phenotypes and disease to molecular function and underlying pathways.

Molecular Phenotyping

At the inception of phenomics projects, characterisation of gene expression patterns was too expensive to contemplate as part of a systematic project. With a new generation of vectors incorporating the *LacZ* gene, *LacZ* assessment of gene expression patterns is now a practical proposition and forms part of the IMPC project but this does not address the effects of gene mutation on expression levels of other genes. This assessment is confounded by the need to carry out analysis on different tissues and, potentially, different life stages. However, as High-Throughput Sequencing technology progresses, with decreasing cost and requirement for less material, measurement of gene expression profiles is likely to become affordable. Looking further into the future, other omics technologies, such as proteomics (Cheley et al. 2006, Howorka et al. 2004), are likely to undergo technological revolutions in accuracy and cost. In particular, measurement of protein expression levels would provide an invaluable accompaniment to RNA expression measurements. Current phenotyping efforts also lack access to age-related phenotypes, including cancer, characterisation of embryonic lethals, and whole-body imaging.

Smarter Phenotyping

Current phenotyping pipelines are designed for thoroughness but as a consequence carry out many tests that produce non-significant results. As data emerge from these projects, more information, based on network analysis of the data (e.g., Baker et al. (2009), Espinosa and Hancock (2011), Guan et al. (2008)), will emerge on the underlying structure of the data. This should allow predictions to be made about the types of test that are more likely to produce significant results. This kind of analysis may lead to smarter phenotyping pipelines, which omit some unnecessary tests and substitute them with others, and more broadly a redrawing of pipelines to make them more informative for comparable effort.

Return to ENU?

The availability of High-Throughput Sequencing opens up a further opportunity to take full advantage of the resources built up by the ENU mutagenesis projects (Bull et al. 2013). By combining full information on mutations and other changes in mutagenised genomes with detailed phenotyping, it may be possible to map higher-order genetic interactions in mice bearing relatively subtle mutations rather than null alleles. Sequencing and phenotyping of systems genetics strains could provide similar data. The availability of such an approach may lead to a second generation of mouse phenomics projects providing much deeper information on the genetics underlying phenotypes and disease.

Available Resources for Mouse Phenomics

The following table summarises online resources referred to in the text, excluding the web addresses for mouse clinics, which are listed in Table 1.

Resource	URL
Australian Phenomics Network	http://www.australianphenomics.org.au/
CASIMIR	http://www.casimir.org.uk
CASIMIR/Interphenome briefing paper "Legal Issues on Phenotype Data Accessibility"	http://www.casimir.org.uk/fullstory.php?storyid=76
Charles River	http://www.criver.com
Consortium for Functional Glycomics	http://www.functionalglycomics.org/
CreZOO database	http://bioit.fleming.gr/crezoo/; http://www.creline.org/search_cre_mice
Deltagen and Lexicon phenotype data and protocols	http://www.informatics.jax.org/external/ko/
EUMODIC	http://www.eumodic.org
EUMORPHIA	http://www.eumorphia.org
European Mouse Phenotyping Resource of Standardised Screens	http://empress.har.mrc.ac.uk/
European Mutant Mouse Archive (EMMA)	http://www.emmanet.org
EuroPhenome	http://www.europhenome.org
Federation of International Mouse Resources (FIMRe)	http://www.fimre.org
Graphical view of EUMODIC phenotyping pipelines 1 & 2	http://empress.har.mrc.ac.uk/viewempress/ index.php?map=EUMODIC+Pipeline
IEG Munich	http://www.helmholtz-muenchen.de/en/ieg/
IMPRESS database	https://www.mousephenotype.org/impress
Infrafrontier	http://www.infrafrontier.eu
Infrafrontier LIMS report	http://www.infrafrontier.eu//internal/docs/workpackages/WP7/D7.2b
International Knockout Mouse Consortium (IKMC)	http://www.knockoutmouse.org
International Mouse Phenotyping Consortium	https://www.mousephenotype.org/
International Mouse Strain Resource (IMSR)	http://www.findmice.org
Interphenome	http://www.interphenome.org
Jackson Laboratory, Maine, USA	http://www.jax.org/
Knockout Mouse Project (KOMP)	http://www.komp.org
Macquarie University, Sydney	http://mq.edu.au/
MartView	http://www.biomart.org/biomart/martview/
Missense Mutation Library (Australian SNV archive)	http://databases.apf.edu.au/mutations
Mouse Genome Informatics gene count	http://www.informatics.jax.org/genes.shtml

Mouse Mutagenesis for Developmental Defects, Baylor College of Medicine, USA	http://www.nih.gov/science/models/mouse/resources/fmmpdd.html
Mouse Phenome Database	http://phenome.jax.org
Mouse Phenome Project 1999 Workshop	http://phenome.jax.org/db/q?rtn=docs/workshop1999
Mouse Phenome Project priority strains	http://phenome.jax.org/db/q?rtn=strains/ search&reqpanel=mpd_priority
MRC Harwell	http://www.har.mrc.ac.uk/
Mutagenetix Project at the University of Texas Southwestern Medical Center, Dallas, USA	http://mutagenetix.utsouthwestern.edu/
Nathan Shock Center of Excellence in the Basic Biology of Aging	http://agingmice.jax.org/index.php/research/center-research/
National Mouse Metabolic Phenotyping Centers (MMPC)	http://www.mmpc.org
NIH Neurogenomics Project, Northwestern University, Illinois	http://genomics.northwestern.edu/neuro/
RIKEN, Japan	http://www.brc.riken.jp/lab/gsc/mouse/
Systems Approach to Immunology	http://www.systemsimmunology.org
Taconic	http://www.taconic.com
Toronto Centre for Modelling Human Disease, Canada	http://www.cmhd.ca/

References Cited

Andrews, T.D., B. Whittle, M.A. Field, B. Balakishnan, Y. Zhang, Y. Shao, V. Cho, M. Kirk, M. Singh, Y. Xia, J. Hager, S. Winslade, G. Sjollema, B. Beutler, A. Enders and C.C. Goodnow. 2012. Massively parallel sequencing of the mouse exome to accurately identify rare, induced mutations: an immediate source for thousands of new mouse models. Open Biol. 2(5): 120061.

Ashburner, M., C.A. Ball, J.A. Blake, D. Botstein, H. Butler, J.M. Cherry, A.P. Davis, K. Dolinski, S.S. Dwight, J.T. Eppig, M.A. Harris, D.P. Hill, L. Issel-Tarver, A. Kasarskis, S. Lewis, J.C. Matese, J.E. Richardson, M. Ringwald, G.M. Rubin and G. Sherlock. 2000. Gene ontology: tool for the unification of biology. The Gene Ontology Consortium. Nature genetics 25(1): 25–29.

Austin, C.P., J.F. Battey, A. Bradley, M. Bucan, M. Capecchi, F.S. Collins, W.F. Dove, G. Duyk, S. Dymecki, J.T. Eppig, F.B. Grieder, N. Heintz, G. Hicks, T.R. Insel, A. Joyner, B.H. Koller, K.C. Lloyd, T. Magnuson, M.W. Moore, A. Nagy, J.D. Pollock, A.D. Roses, A.T. Sands, B. Seed, W.C. Skarnes, J. Snoddy, P. Soriano, D.J. Stewart, F. Stewart, B. Stillman, H. Varmus, L. Varticovski, I.M. Verma, T.F. Vogt, H. von Melchner, J. Witkowski, R.P. Woychik, W. Wurst, G.D. Yancopoulos, S.G. Young and B. Zambrowicz. 2004. The knockout mouse project. Nature Genetics 36(9): 921–924.

Ayadi, A., M.C. Birling, J. Bottomley, J. Bussell, H. Fuchs, M. Fray, V. Gailus-Durner, S. Greenaway, R. Houghton, N. Karp, S. Leblanc, C. Lengger, H. Maier, A.M. Mallon, S. Marschall, D. Melvin, H. Morgan, G. Pavlovic, E. Ryder, W.C. Skarnes, M. Selloum, R. Ramirez-Solis, T. Sorg, L. Teboul, L. Vasseur, A. Walling, T. Weaver, S. Wells, J.K. White, A. Bradley, D.J. Adams, K.P. Steel, M. Hrabe de Angelis, S.D. Brown and Y. Herault. 2012. Mouse large-scale phenotyping initiatives: overview of the European Mouse Disease

Clinic (EUMODIC) and of the Wellcome Trust Sanger Institute Mouse Genetics Project. Mamm Genome 23(9-10): 580–586.

Baker, E.J., J.J. Jay, V.M. Philip, Y. Zhang, Z. Li, R. Kirova, M.A. Langston and E.J. Chesler. 2009. Ontological Discovery Environment: a system for integrating gene-phenotype associations. Genomics 94(6): 377–387.

Beck, T., H. Morgan, A. Blake, S. Wells, J.M. Hancock and A.-M. Mallon. 2009. Practical application of ontologies to annotate and analyse large scale raw mouse phenotype data. BMC Bioinformatics 10(Suppl 5): S2.

Blake, A., K. Pickford, S. Greenaway, S. Thomas, A. Pickard, C.M. Williamson, N.C. Adams, A. Walling, T. Beck, M. Fray, J. Peters, T. Weaver, S.D. Brown, J.M. Hancock and A.M. Mallon. 2010. MouseBook: an integrated portal of mouse resources. Nucleic Acids Res. 38(Database Issue): D593–D599.

Blake, J.A., C.J. Bult, J.A. Kadin, J.E. Richardson, J.T. Eppig and Mouse Genome Database Group. 2011. The Mouse Genome Database (MGD): premier model organism resource for mammalian genomics and genetics. Nucleic Acids Res. 39(Database issue): D842–D848.

Bogue, M.A., S.C. Grubb, T.P. Maddatu and C.J. Bult. 2007. Mouse Phenome Database (MPD). Nucleic Acids Res. 35(Database Issue): D643–D649.

Brown, S.D. and M.W. Moore. 2012a. The International Mouse Phenotyping Consortium: past and future perspectives on mouse phenotyping. Mamm Genome 23(9-10): 632–640.

Brown, S.D. and M.W. Moore. 2012b. Towards an encyclopaedia of mammalian gene function: the International Mouse Phenotyping Consortium. Dis. Model Mech. 5(3): 289–292.

Brown, S.D.M., P. Chambon, M. Hrabé de Angelis and EUMORPHIA_Consortium. 2005. EMPReSS: Standardised phenotype screens for functional annotation of the mouse genome. Nat. Genet. 37(11): 1155.

Bull, K.R., A.J. Rimmer, O.M. Siggs, L.A. Miosge, C.M. Roots, A. Enders, E.M. Bertram, T.L. Crockford, B. Whittle, P.K. Potter, M.M. Simon, A.M. Mallon, S.D. Brown, B. Beutler, C.C. Goodnow, G. Lunter and R.J. Cornall. 2013. Unlocking the bottleneck in forward genetics using whole-genome sequencing and identity by descent to isolate causative mutations. PLoS Genet. 9(1): e1003219.

Cheley, S., H. Xie and H. Bayley. 2006. A genetically encoded pore for the stochastic detection of a protein kinase. Chem. Bio. Chem. 7(12): 1923–1927.

Chen, C.K., C.J. Mungall, G.V. Gkoutos, S.C. Doelken, S. Köhler, B.J. Ruef, C. Smith, M. Westerfield, P.N. Robinson, S.E. Lewis, P.N. Schofield and D. Smedley. 2012. MouseFinder: Candidate disease genes from mouse phenotype data. Hum. Mutat. 33(5): 858–866.

Church, D., L. Goodstadt, L. Hillier, M. Zody, S. Goldstein, X. She, C. Bult, R. Agarwala, J. Cherry, M. DiCuccio, W. Hlavina, Y. Kapustin, P. Meric, D. Maglott, Z. Birtle, A. Marques, T. Graves, S. Zhou, B. Teague, K. Potamousis, C. Churas, M. Place, J. Herschleb, R. Runnheim, D. Forrest, J. Amos-Landgraf, D. Schwartz, Z. Cheng, K. Lindblad-Toh, E. Eichler, C. Ponting and Mouse Genome Sequencing Consortium. 2009. Lineage-specific biology revealed by a finished genome assembly of the mouse. PLoS Biology 7(5): e1000112.

Churchill, G.A., D.C. Airey, H. Allayee, J.M. Angel, A.D. Attie, J. Beatty, W.D. Beavis, J.K. Belknap, B. Bennett, W. Berrettini, A. Bleich, M. Bogue, K.W. Broman, K.J. Buck, E. Buckler, M. Burmeister, E.J. Chesler, J.M. Cheverud, S. Clapcote, M.N. Cook, R.D. Cox, J.C. Crabbe, W.E. Crusio, A. Darvasi, C.F. Deschepper, R.W. Doerge, C.R. Farber, J. Forejt, D. Gaile, S.J. Garlow, H. Geiger, H. Gershenfeld, T. Gordon, J. Gu, W. Gu, G. de Haan, N.L. Hayes, C. Heller, H. Himmelbauer, R. Hitzemann, K. Hunter, H.C. Hsu, F.A. Iraqi, B. Ivandic, H.J. Jacob, R.C. Jansen, K.J. Jepsen, D.K. Johnson, T.E. Johnson, G. Kempermann, C. Kendziorski, M. Kotb, R.F. Kooy, B. Llamas, F. Lammert, J.M. Lassalle, P.R. Lowenstein, L. Lu, A. Lusis, K.F. Manly, R. Marcucio, D. Matthews, J.F. Medrano, D.R. Miller, G. Mittleman, B.A. Mock, J.S. Mogil, X. Montagutelli, G. Morahan, D.G. Morris, R. Mott, J.H. Nadeau, H. Nagase, R.S. Nowakowski, B.F. O'Hara, A.V. Osadchuk, G.P. Page, B. Paigen, K. Paigen, A.A. Palmer, H.J. Pan, L. Peltonen-Palotie, J. Peirce, D. Pomp, M. Pravenec,

D.R. Prows, Z. Qi, R.H. Reeves, J. Roder, G.D. Rosen, E.E. Schadt, L.C. Schalkwyk, Z. Seltzer, K. Shimomura, S. Shou, M.J. Sillanpää, L.D. Siracusa, H.W. Snoeck, J.L. Spearow, K. Svenson, L.M. Tarantino, D. Threadgill, L.A. Toth, W. Valdar, F.P. de Villena, C. Warden, S. Whatley, R.W. Williams, T. Wiltshire, N. Yi, D. Zhang, M. Zhang, F. Zou and Complex Trait Consortium. 2004. The Collaborative Cross, a community resource for the genetic analysis of complex traits. Nat. Genet. 36(11): 1133–1137.

Churchill, G.A., D.M. Gatti, S.C. Munger and K.L. Svenson. 2012. The diversity outbred mouse population. Mamm Genome 23(9–10): 713–718.

Collins, F.S., J. Rossant and W. Wurst. 2007. A mouse for all reasons. Cell 128(1): 9–13.

Cong, L., F. Ran, D. Cox, S. Lin, R. Barretto, N. Habib, P. Hsu, X. Wu, W. Jiang, L. Marraffini and F. Zhang. 2013. Multiplex genome engineering using CRISPR/Cas systems. Science 339(6121): 819–823.

Donahue, L.R., M. Hrabe de Angelis, M. Hagn, C. Franklin, K.C. Lloyd, T. Magnuson, C. McKerlie, N. Nakagata, Y. Obata, S. Read, W. Wurst, A. Hörlein and M.T. Davisson. 2012. Centralized mouse repositories. Mamm Genome 23(9–10): 559–571.

Eppig, J.T., J.A. Blake, C.J. Bult, J.A. Kadin, J.E. Richardson and the Mouse Genome Database Group. 2011. The Mouse Genome Database (MGD): comprehensive resource for genetics and genomics of the laboratory mouse. Nucleic Acids Res. 40(Database Issue): D881–D886.

Eppig, J.T. and M. Strivens. 1999. Finding a mouse: the International Mouse Strain Resource (IMSR). Trends Genet. 15(2): 81–82.

Espinosa, O. and J.M. Hancock. 2011. A gene-phenotype network for the laboratory mouse and its implications for systematic phenotyping. PLoS One 6(5): e19693.

FIMRe Board of Directors. 2006. FIMRe: Federation of International Mouse Resources: Global Networking of Resource Centers. Mammalian Genome 17(5): 363–364.

Gailus-Durner, V., H. Fuchs, L. Becker, I. Bolle, M. Brielmeier, J. Calzada-Wack, R. Elvert, N. Ehrhardt, C. Dalke, T.J. Franz, E. Grundner-Culemann, S. Hammelbacher, S.M. Holter, G. Holzlwimmer, M. Horsch, A. Javaheri, S.V. Kalaydjiev, M. Klempt, E. Kling, S. Kunder, C. Lengger, T. Lisse, T. Mijalski, B. Naton, V. Pedersen, C. Prehn, G. Przemeck, I. Racz, C. Reinhard, P. Reitmeir, I. Schneider, A. Schrewe, R. Steinkamp, C. Zybill, J. Adamski, J. Beckers, H. Behrendt, J. Favor, J. Graw, G. Heldmaier, H. Hofler, B. Ivandic, H. Katus, P. Kirchhof, M. Klingenspor, T. Klopstock, A. Lengeling, W. Muller, F. Ohl, M. Ollert, L. Quintanilla-Martinez, J. Schmidt, H. Schulz, E. Wolf, W. Wurst, A. Zimmer, D.H. Busch and M.H. de Angelis. 2005. Introducing the German Mouse Clinic: open access platform for standardized phenotyping. Nat. Methods 2(6): 403–404.

Gaudet, P., A. Bairoch, D. Field, S.A. Sansone, C. Taylor, T.K. Attwood, A. Bateman, J.A. Blake, C.J. Bult, J.M. Cherry, R.L. Chisholm, G. Cochrane, C.E. Cook, J.T. Eppig, M.Y. Galperin, R. Gentleman, C.A. Goble, T. Gojobori, J.M. Hancock, D.G. Howe, T. Imanishi, J. Kelso, D. Landsman, S.E. Lewis, I.K. Mizrachi, S. Orchard, B.F. Ouellette, S. Ranganathan, L. Richardson, P. Rocca-Serra, P.N. Schofield, D. Smedley, C. Southan, T.W. Tan, T. Tatusova, P.L. Whetzel, O. White and C. Yamasaki. 2011. Towards BioDBcore: a community-defined information specification for biological databases. Nucleic Acids Res. 39(Database Issue): D7–D10.

Gkoutos, G.V., E.C.J. Green, A.M. Mallon, J.M. Hancock and D. Davidson. 2004. Building mouse phenotype ontologies. Pac. Symp. Biocomputing 9: 178–189.

Gkoutos, G.V., E.C.J. Green, A.-M. Mallon, J.M. Hancock and D. Davidson. 2005. Using ontologies to describe mouse phenotypes. Genome Biology 6(1): R8.

Green, E.C.J., G.V. Gkoutos, H. Lad, A. Blake, J. Weekes and J.M. Hancock. 2005. EMPReSS: European Mouse Phenotyping Resource for Standardised Screens. Bioinformatics 21(12): 2930–2931.

Gruenberger, M., R. Alberts, D. Smedley, M. Swertz, P. Schofield, K. Schughart and The CASIMIR Consortium. 2010. Towards the integration of mouse databases—definition and implementation of solutions to two use-cases in mouse functional genomics. BMC Res. Notes 3(1): 16.

Guan, Y., C.L. Myers, R. Lu, I.R. Lemischka, C.J. Bult and O.G. Troyanskaya. 2008. A genomewide functional network for the laboratory mouse. PLoS Comput. Biol. 4(9): e1000165.

Haider, S., B. Ballester, D. Smedley, J. Zhang, P. Rice and A. Kasprzyk. 2009. BioMart Central Portal—unified access to biological data. Nucleic Acids Res. 37(Web Server issue): W23–W27.

Hancock, J.M., P.N. Schofield, C. Chandras, M. Zouberakis, V. Aidinis, D. Smedley, N. Rosenthal, K. Schughart and The CASIMIR Consortium. 2008. CASIMIR: Coordination and Sustainability of International Mouse Informatics Resources. In 8th IEEE International Conference on Bioinformatics and Bioengineering 382–387.

Howard, H.J., A. Beaudet, V. Gil-da-Silva Lopes, M. Lyne, G. Suthers, P. Van den Akker, K. Wertheim-Tysarowska, P. Willems and F. Macrae. 2012. Disease-specific databases: Why we need them and some recommendations from the Human Variome Project Meeting, May 28, 2011. Am. J. Med. Genet. A 158A(11): 2763–2766.

Howorka, S., J. Nam, H. Bayley and D. Kahne. 2004. Stochastic detection of monovalent and bivalent protein-ligand interactions. Angewandte Chemie International Edition 43(7): 842–846.

Karp, N.A., D. Melvin, R.F. Mott and Sanger Mouse Genetics Project. 2012. Robust and sensitive analysis of mouse knockout phenotypes. PLoS One 7(12): e52410.

Keane, T.M., L. Goodstadt, P. Danecek, M.A. White, K. Wong, B. Yalcin, A. Heger, A. Agam, G. Slater, M. Goodson, N.A. Furlotte, E. Eskin, C. Nellaker, H. Whitley, J. Cleak, D. Janowitz, P. Hernandez-Pliego, A. Edwards, T.G. Belgard, P.L. Oliver, R.E. McIntyre, A. Bhomra, J. Nicod, X. Gan, W. Yuan, L. van der Weyden, C.A. Steward, S. Bala, J. Stalker, R. Mott, R. Durbin, I.J. Jackson, A. Czechanski, J.A. Guerra-Assuncao, L.R. Donahue, L.G. Reinholdt, B.A. Payseur, C.P. Ponting, E. Birney, J. Flint and D.J. Adams. 2011. Mouse genomic variation and its effect on phenotypes and gene regulation. Nature 477(7364): 289–294.

Kollmus, H., R. Post, M. Brielmeier, J. Fernandez, H. Fuchs, C. McKerlie, L. Montoliu, P.J. Otaegui, M. Rebelo, H. Riedesel, J. Ruberte, R. Sedlacek, M.H. de Angelis and K. Schughart. 2012. Structural and functional concepts in current mouse phenotyping and archiving facilities. J. Am. Assoc. Lab Anim. Sci. 51(4): 418–435.

Landrette, S.F., J.C. Cornett, T.K. Ni, M.W. Bosenberg and T. Xu. 2011. piggyBac transposon somatic mutagenesis with an activated reporter and tracker (PB-SMART) for genetic screens in mice. PLoS One 6(10): e26650.

Laughlin, M.R., K.C. Lloyd, G.W. Cline and D.H. Wasserman. 2012. NIH Mouse Metabolic Phenotyping Centers: the power of centralized phenotyping. Mamm Genome 23(9–10): 623–631.

Leduc, M.S., R.S. Hageman, Q. Meng, R.A. Verdugo, S.W. Tsaih, G.A. Churchill, B. Paigen and R. Yuan. 2010. Identification of genetic determinants of IGF-1 levels and longevity among mouse inbred strains. Aging Cell 9(5): 823–836.

Maddatu, T.P., S.C. Grubb, C.J. Bult and M.A. Bogue. 2012. Mouse Phenome Database (MPD). Nucleic Acids Res. 40(Database Issue): D887–D894.

Mallon, A.-M., A. Blake and J.M. Hancock. 2008. EuroPhenome and EMPReSS: online mouse phenotyping resource. Nucleic Acids Res. 36: D715–D718.

Mallon, A.M., V. Iyer, D. Melvin, H. Morgan, H. Parkinson, S.D. Brown, P. Flicek and W.C. Skarnes. 2012. Accessing data from the International Mouse Phenotyping Consortium: state of the art and future plans. Mamm Genome 23(9-10): 641–652.

Morgan, H., T. Beck, A. Blake, H. Gates, N. Adams, G. Debouzy, S. Leblanc, C. Lengger, H. Maier, D. Melvin, H. Meziane, D. Richardson, S. Wells, J. White, J. Wood, M.H. de Angelis, S.D. Brown, J.M. Hancock and A.M. Mallon. 2010. EuroPhenome: a repository for high-throughput mouse phenotyping data. Nucleic Acids Res. 38(Database Issue): D577–D585.

Mouse Phenotype Database Integration Consortium. 2007. Integration of mouse phenome data resources. Mamm Genome 18(3): 157–163.

Paigen, K. and J.T. Eppig. 2000. A mouse phenome project. Mamm Genome 11(9): 715–717.

Petkova, S.B., R. Yuan, S.W. Tsaih, W. Schott, D.C. Roopenian and B. Paigen. 2008. Genetic influence on immune phenotype revealed strain-specific variations in peripheral blood lineages. Physiol. Genomics 34(3): 304–314.

Philip, V.M., G. Sokoloff, C.L. Ackert-Bicknell, M. Striz, L. Branstetter, M.A. Beckmann, J.S. Spence, B.L. Jackson, L.D. Galloway, P. Barker, A.M. Wymore, P.R. Hunsicker, D.C. Durtschi, G.S. Shaw, S. Shinpock, K.F. Manly, D.R. Miller, K.D. Donohue, C.T. Culiat, G.A. Churchill, W.R. Lariviere, A.A. Palmer, B.F. O'Hara, B.H. Voy and E.J. Chesler. 2011. Genetic analysis in the Collaborative Cross breeding population. Genome Res. 21(8): 1223–1238.

Premsrirut, P.K., L.E. Dow, S.Y. Kim, M. Camiolo, C.D. Malone, C. Miething, C. Scuoppo, J. Zuber, R.A. Dickins, S.C. Kogan, K.R. Shroyer, R. Sordella, G.J. Hannon and S.W. Lowe. 2011. A rapid and scalable system for studying gene function in mice using conditional RNA interference. Cell 145(1): 145–158.

Rosenthal, N. and S. Brown. 2007. The mouse ascending: perspectives for human-disease models. Nat. Cell Biol. 9(9): 993–999.

Schofield, P.N., T. Bubela, T. Weaver, L. Portilla, S.D. Brown, J.M. Hancock, D. Einhorn, G. Tocchini-Valentini, M. Hrabe de Angelis, N. Rosenthal and CASIMIR Rome Meeting participants. 2009. Post-publication sharing of data and tools. Nature 461(7261): 171–173.

Schofield, P.N., J. Eppig, E. Huala, M. Hrabe de Angelis, M. Harvey, D. Davidson, T. Weaver, S. Brown, D. Smedley, N. Rosenthal, K. Schughart, V. Aidinis, G. Tocchini-Valentini and J.M. Hancock. 2010. Sustaining the data and bioresource commons. Science 330(6004): 592–593.

Singer, J.B., A.E. Hill, L.C. Burrage, K.R. Olszens, J. Song, M. Justice, W.E. O'Brien, D.V. Conti, J.S. Witte, E.S. Lander and J.H. Nadeau. 2004. Genetic dissection of complex traits with chromosome substitution strains of mice. Science 304(5669): 445–448.

Sinke, A.P., C. Caputo, S.W. Tsaih, R. Yuan, D. Ren, P.M. Deen and R. Korstanje. 2011. Genetic analysis of mouse strains with variable serum sodium concentrations identifies the Nalcn sodium channel as a novel player in osmoregulation. Physiol. Genomics 43(5): 265–270.

Smedley, D., S. Haider, B. Ballester, R. Holland, D. London, G. Thorisson and A. Kasprzyk. 2009. BioMart-biological queries made easy. BMC Genomics 10: 22.

Smedley, D., A. Oellrich, S. Köhler, B. Ruef, Sanger Mouse Genetics Project, M. Westerfield, P. Robinson, S. Lewis and C. Mungall. 2013. PhenoDigm: analyzing curated annotations to associate animal models with human diseases. Database (Oxford) 2013: bat025.

Smedley, D., M.A. Swertz, K. Wolstencroft, G. Proctor, M. Zouberakis, J. Bard, J.M. Hancock and P. Schofield. 2008. Solutions for data integration in functional genomics: a critical assessment and case study. Briefings in Bioinformatics 9(6): 532–544.

Smith, C.L. and J.T. Eppig. 2012. The Mammalian Phenotype Ontology as a unifying standard for experimental and high-throughput phenotyping data. Mamm Genome 23(9-10): 653–668.

Smith, C.L., C.A. Goldsmith and J.T. Eppig. 2005. The Mammalian Phenotype Ontology as a tool for annotating, analyzing and comparing phenotypic information. Genome Biol. 6(1): R7.

Sundberg, J.P., A. Berndt, B.A. Sundberg, K.A. Silva, V. Kennedy, R. Bronson, R. Yuan, B. Paigen, D. Harrison and P.N. Schofield. 2011. The mouse as a model for understanding chronic diseases of aging: the histopathologic basis of aging in inbred mice. Pathobiology of Aging & Age-related Diseases 1: 7179.

Sung, Y.H., I.-J. Baek, D.H. Kim, J. Jeon, J. Lee, K. Lee, D. Jeong, J.-S. Kim and H.-W. Lee. 2013. Knockout mice created by TALEN-mediated gene targeting. Nature Biotechnol. 31(1): 23–24.

Taylor, C.F., D. Field, S.A. Sansone, J. Aerts, R. Apweiler, M. Ashburner, C.A. Ball, P.A. Binz, M. Bogue, T. Booth, A. Brazma, R.R. Brinkman, A. Michael Clark, E.W. Deutsch, O. Fiehn, J. Fostel, P. Ghazal, F. Gibson, T. Gray, G. Grimes, J.M. Hancock, N.W. Hardy, H. Hermjakob, R.K. Julian, Jr., M. Kane, C. Kettner, C. Kinsinger, E. Kolker, M. Kuiper, N.L. Novere, J. Leebens-Mack, S.E. Lewis, P. Lord, A.M. Mallon, N. Marthandan, H. Masuya, R. McNally, A. Mehrle, N. Morrison, S. Orchard, J. Quackenbush, J.M. Reecy, D.G. Robertson, P. Rocca-Serra, H. Rodriguez, H. Rosenfelder, J. Santoyo-Lopez, R.H. Scheuermann, D. Schober, B. Smith, J. Snape, C.J. Stoeckert, Jr., K. Tipton, P. Sterk, A. Untergasser, J. Vandesompele and S. Wiemann. 2008. Promoting coherent minimum reporting guidelines for biological and biomedical investigations: the MIBBI project. Nat. Biotechnol. 26(8): 889–896.

Toronto International Data Release Workshop Authors, E. Birney, T.J. Hudson, E.D. Green, C. Gunter, S. Eddy, J. Rogers, J.R. Harris, S.D. Ehrlich, R. Apweiler, C.P. Austin, L. Berglund, M. Bobrow, C. Bountra, A.J. Brookes, A. Cambon-Thomsen, N.P. Carter, R.L. Chisholm, J.L. Contreras, R.M. Cooke, W.L. Crosby, K. Dewar, R. Durbin, S.O. Dyke, J.R. Ecker, K. El Emam, L. Feuk, S.B. Gabriel, J. Gallacher, W.M. Gelbart, A. Granell, F. Guarner, T. Hubbard, S.A. Jackson, J.L. Jennings, Y. Joly, S.M. Jones, J. Kaye, K.L. Kennedy, B.M. Knoppers, N.C. Kyrpides, W.W. Lowrance, J. Luo, J.J. MacKay, L. Martín-Rivera, W.R. McCombie, J.D. McPherson, L. Miller, W. Miller, D. Moerman, V. Mooser, C.C. Morton, J.M. Ostell, B.F. Ouellette, J. Parkhill, P.S. Raina, C. Rawlings, S.E. Scherer, S.W. Scherer, P.N. Schofield, C.W. Sensen, V.C. Stodden, M.R. Sussman, T. Tanaka, J. Thornton, T. Tsunoda, D. Valle, E.I. Vuorio, N.M. Walker, S. Wallace, G. Weinstock, W.B. Whitman, K.C. Worley, C. Wu, J. Wu and J. Yu. 2009. Prepublication data sharing. Nature 461(7261): 168–170.

Tsaih, S.W., M.G. Pezzolesi, R. Yuan, J.H. Warram, A.S. Krolewski and R. Korstanje. 2010. Genetic analysis of albuminuria in aging mice and concordance with loci for human diabetic nephropathy found in a genome-wide association scan. Kidney Int. 77(3): 201–210.

Waterston, R.H., K. Lindblad-Toh, E. Birney, J. Rogers, J.F. Abril, P. Agarwal, R. Agarwala, R. Ainscough, M. Alexandersson, P. An, S.E. Antonarakis, J. Attwood, R. Baertsch, J. Bailey, K. Barlow, S. Beck, E. Berry, B. Birren, T. Bloom, P. Bork, M. Botcherby, N. Bray, M.R. Brent, D.G. Brown, S.D. Brown, C. Bult, J. Burton, J. Butler, R.D. Campbell, P. Carninci, S. Cawley, F. Chiaromonte, A.T. Chinwalla, D.M. Church, M. Clamp, C. Clee, F.S. Collins, L.L. Cook, R.R. Copley, A. Coulson, O. Couronne, J. Cuff, V. Curwen, T. Cutts, M. Daly, R. David, J. Davies, K.D. Delehaunty, J. Deri, E.T. Dermitzakis, C. Dewey, N.J. Dickens, M. Diekhans, S. Dodge, I. Dubchak, D.M. Dunn, S.R. Eddy, L. Elnitski, R.D. Emes, P. Eswara, E. Eyras, A. Felsenfeld, G.A. Fewell, P. Flicek, K. Foley, W.N. Frankel, L.A. Fulton, R.S. Fulton, T.S. Furey, D. Gage, R.A. Gibbs, G. Glusman, S. Gnerre, N. Goldman, L. Goodstadt, D. Grafham, T.A. Graves, E.D. Green, S. Gregory, R. Guigo, M. Guyer, R.C. Hardison, D. Haussler, Y. Hayashizaki, L.W. Hillier, A. Hinrichs, W. Hlavina, T. Holzer, F. Hsu, A. Hua, T. Hubbard, A. Hunt, I. Jackson, D.B. Jaffe, L.S. Johnson, M. Jones, T.A. Jones, A. Joy, M. Kamal, E.K. Karlsson, D. Karolchik, A. Kasprzyk, J. Kawai, E. Keibler, C. Kells, W.J. Kent, A. Kirby, D.L. Kolbe, I. Korf, R.S. Kucherlapati, E.J. Kulbokas, D. Kulp, T. Landers, J.P. Leger, S. Leonard, I. Letunic, R. Levine, J. Li, M. Li, C. Lloyd, S. Lucas, B. Ma, D.R. Maglott, E.R. Mardis, L. Matthews, E. Mauceli, J.H. Mayer, M. McCarthy, W.R. McCombie, S. McLaren, K. McLay, J.D. McPherson, J. Meldrim, B. Meredith, J.P. Mesirov, W. Miller, T.L. Miner, E. Mongin, K.T. Montgomery, M. Morgan, R. Mott, J.C. Mullikin, D.M. Muzny, W.E. Nash, J.O. Nelson, M.N. Nhan, R. Nicol, Z. Ning, C. Nusbaum, M.J. O'Connor, Y. Okazaki, K. Oliver, E. Overton-Larty, L. Pachter, G. Parra, K.H. Pepin, J. Peterson, P. Pevzner, R. Plumb, C.S. Pohl, A. Poliakov, T.C. Ponce, C.P. Ponting, S. Potter, M. Quail, A. Reymond, B.A. Roe, K.M. Roskin, E.M. Rubin, A.G. Rust, R. Santos, V. Sapojnikov, B. Schultz, J. Schultz, M.S. Schwartz, S. Schwartz, C. Scott, S. Seaman, S. Searle, T. Sharpe, A. Sheridan, R. Shownkeen, S. Sims, J.B. Singer, G. Slater, A. Smit, D.R. Smith, B. Spencer, A. Stabenau, N. Stange-Thomann, C. Sugnet, M. Suyama, G. Tesler, J. Thompson, D. Torrents, E. Trevaskis, J. Tromp, C. Ucla, A. Ureta-Vidal, J.P. Vinson,

A.C. Von Niederhausern, C.M. Wade, M. Wall, R.J. Weber, R.B. Weiss, M.C. Wendl, A.P. West, K. Wetterstrand, R. Wheeler, S. Whelan, J. Wierzbowski, D. Willey, S. Williams, R.K. Wilson, E. Winter, K.C. Worley, D. Wyman, S. Yang, S.P. Yang, E.M. Zdobnov, M.C. Zody and E.S. Lander. 2002. Initial sequencing and comparative analysis of the mouse genome. Nature 420(6915): 520–562.

Wilkinson, P., J. Sengerova, R. Matteoni, C.K. Chen, G. Soulat, A. Ureta-Vidal, S. Fessele, M. Hagn, M. Massimi, K. Pickford, R.H. Butler, S. Marschall, A.M. Mallon, A. Pickard, M. Raspa, F. Scavizzi, M. Fray, V. Larrigaldie, J. Leyritz, E. Birney, G.P. Tocchini-Valentini, S. Brown, Y. Herault, L. Montoliu, M. Hrabé de Angelis and D. Smedley. 2010. EMMA-mouse mutant resources for the international scientific community. Nucleic Acids Res. 38(Database issue): D570–D576.

Xing, S., S.W. Tsaih, R. Yuan, K.L. Svenson, L.M. Jorgenson, M. So, B.J. Paigen and R. Korstanje. 2009. Genetic influence on electrocardiogram time intervals and heart rate in aging mice. Am. J. Physiol. Heart Circ. Physiol. 296(6): H1907–H1913.

Yuan, R., S.W. Tsaih, S.B. Petkova, C.M. de Evsikova, S. Xing, M.A. Marion, M.A. Bogue, K.D. Mills, L.L. Peters, C.J. Bult, C.J. Rosen, J.P. Sundberg, D.E. Harrison, G.A. Churchill and B. Paigen. 2009. Aging in inbred strains of mice: study design and interim report on median lifespans and circulating IGF1 levels. Aging Cell 8(3): 277–287.

4

Phenotyping Zebrafish

Elisabeth M. Busch-Nentwich

Introduction

Zebrafish occupy a variety of fresh water environments (Engeszer et al. 2007) in their natural habitat in South and Southeast Asia. They thrive in a broad range of temperatures, conductivities and flow rates, including 18°C cold, rapid flowing rivers and 36°C warm and stagnant pools on rice fields, thus demonstrating great tolerance to different environmental conditions. They gather in shallow calm areas to mate and the females lay hundreds of eggs that are fertilised externally.

At 28.5°C, the standard laboratory temperature for raising zebrafish embryos, the fertilised egg starts to divide after 40 minutes (Kimmel et al. 1995). Within 24 hours post-fertilisation (hpf), the body plan, which is typical of a vertebrate embryo, is established including eyes, a compartmentalised brain, otic placodes, and a beating heart. By 26 hpf, the embryo reacts to a tactile stimulus by flicking its tail, thereby employing slow muscle fibres. A day or so later the embryos begin to hatch and by 120 hpf they have developed into free swimming and feeding fry. They exhibit all the hallmarks of a vertebrate. For example, they have a fully functional digestive system comprising intestines, pancreas and liver, a circulatory system with arteries, veins and lymphatic vessels, and ears with semicircular canals. At this stage, the fry also display hunting and flight behaviour and react to sound and light. During this period of development, and sometime

Wellcome Trust Sanger Institute, Hinxton, CB10 1SA, UK.

beyond, the young zebrafish remain translucent, which, together with the extrauterine development, makes the zebrafish highly accessible to imaging and manipulation. Given typical laboratory growth conditions, zebrafish reach sexual maturity in about three months and, if maintained under optimal conditions, the fish can produce hundreds of offspring on a weekly basis well into their second year of life.

Taken together, the combination of numerous offspring and a short generation time, the robustness of the adults, comparatively inexpensive and easy husbandry and, importantly the ease of imaging, make the zebrafish a very attractive model to use for vertebrate genetics and phenotype analysis. Crucially, access to early embryonic processes in zebrafish complements similar studies in mice where early stages of embryogenesis are less accessible.

George Streisinger at the University of Oregon was the first to use zebrafish to study vertebrate development in the early 1980s (Chakrabarti et al. 1983, Felsenfeld et al. 1990, Grunwald and Streisinger 1992). He generated the first laboratory zebrafish strain by crossing a pet shop variety, strain "A" with wild-caught zebrafish imported from India, strain "B". The resulting AB strain is still used over 30 years later. Moreover, the first mutants that he described, which affect a variety of developmental pathways, continue to be an important resource used in many more recent studies (Dooley et al. 2012, Lamason et al. 2005, Parichy et al. 1999).

The first systematic genome-wide phenotype-driven screens of N-ethyl-N-nitrosourea (ENU) mutagenised zebrafish families were conducted during the early 1990s in the laboratories of Christiane Nüsslein-Volhard in Tübingen, Germany and Wolfgang Driever in Boston, MA (Driever et al. 1996, Haffter et al. 1996). These screens established and defined groups of mutants and their phenotypes and were published as a special issue of *Development* in 1996. Both screens focused on early development and identified a plethora of mostly recessive mutants that are still studied and for which many of the affected genes are being identified. Equally significant to the mutants are the methods of phenotyping that were developed during the screens. Since then many laboratories have conducted classical genetic screens isolating mutants not only in processes of embryonic development, but also metabolism (Ho et al. 2004), behaviours such as drug seeking (Darland and Dowling 2001), sleep (Yokogawa et al. 2007) and anxiety (Egan et al. 2009), and juvenile and adult morphological phenotypes such as pigmentation (Parichy and Turner 2003) and bone formation (Harris et al. 2008).

In 2001, the Wellcome Trust Sanger Institute started the zebrafish genome-sequencing project. With the current assembly, Zv9, the zebrafish reference genome sequence is, like the human and mouse genomes, represented by a series of overlapping large-insert clones, each finished to a high quality. Crucially, this includes the annotation of protein-coding genes (Collins et al. 2012) which has enabled the development of efficient mutation detection strategies (Kettleborough et al. 2013). Together with the recent development of targeted knockout approaches, the annotated zebrafish reference genome allows the systematic disruption of every protein-coding gene.

Importantly, the full potential and value of a resource of disruptive mutations in every gene are only realised when the mutations are interrogated for their phenotypic consequences in a systematic and unbiased manner. The two initial genetic screens and subsequent years of zebrafish research have established a wealth of phenotyping methods, which are very amenable to a genome-wide scale. In this review, we describe the recent developments on the way to a complete description of phenotypic consequences associated with specific gene disruptions in zebrafish.

Overview

Mutagenesis Methods

A large range of methods and technologies has been employed for zebrafish research to produce disruptive lesions in genes. Methods that are useful for a genome-wide approach are discussed below.

Gene-targeted Mutagenesis

Traditionally, targeted transgenic approaches to gene alteration have been fairly small scale in zebrafish (Trinh le et al. 2011) when compared to other model organisms such as mouse and *Drosophila melanogaster*. In *Drosophila*, the Gal4/UAS system allows for spatial and temporal control of gene function (Brand and Perrimon 1993, Duffy et al. 1998). Since the system was pioneered, thousands of Gal4 and UAS lines have been established and are available through the *Drosophila* stock centres. While conditional overexpression of dominant-negative, dominant-positive and wild type forms of a given gene is a powerful way of examining gene function, it has been of limited use for zebrafish as a method for the disruption of gene function. The Gal4/UAS system is mainly applied to genetic fate mapping and conditional transgene expression (Distel et al. 2009).

The availability of embryonic stem cells in mouse has facilitated the development of sophisticated transgenic gene-targeting strategies (Thomas and Capecchi 1987). A recently published resource has so far generated 9,000 tagged conditional alleles in ES cells (Skarnes et al. 2011). However, a more complete study of such alleles *in vivo* requires the generation of mutant mice, which is very expensive. Similar approaches such as the flip-trap system have been taken in zebrafish, but these do not readily reach genome-wide scale.

Targeted Genome Editing. Several methods for the targeted editing of genomes have been developed in the last few years in quick succession. Zinc finger endonucleases (Bibikova et al. 2002, Doyon et al. 2008, Lloyd et al. 2005) were the first system to gain traction in zebrafish. Since then TAL effector nucleases (TALENs) have proven to be even more amenable to systematic and automated design strategies. TALENs use the DNA recognition domain of transcription activator-like (TAL) effectors in combination with the endonuclease domain of the restriction enzyme Fok1. Like zinc finger (ZFNs), TALENs create double strand breaks with subsequent repair that can induce mutations. The DNA recognition motif possesses a core of two amino acids that recognises single bases in the DNA. By a linear combination of the motifs, a sequence-specific DNA recognition domain can be easily generated. Recently, several groups have published protocols for "TALEN toolboxes" that make it comparatively straightforward to generate TALENs targeting the genes of interest (Cermak et al. 2011, Christian et al. 2010, Huang et al. 2011, Reyon et al. 2012, Sander et al. 2011). Additionally, off-target effects seem to be negligible.

A third approach, the CRISPR-Cas9 system, which uses a bacterial mechanism for the silencing of foreign nucleic acids, has been shown to be effective in zebrafish as well, but has not been tested on a large scale yet (Hwang et al. 2013).

In summary, although significant progress has been made recently in gene targeting in zebrafish, these techniques are too expensive to implement for each of the more than 26,000 zebrafish protein-coding genes.

Random Mutagenesis

Retroviruses or transposons are used to generate randomly inserted DNA constructs in the genome, which can disrupt gene function. Several groups have used this approach to generate mutants. An often-cited advantage of insertional mutagenesis for forward genetics screens is the ability to identify the affected gene quickly by using the insertion itself as a tag for genotypic identification. Also, by including green fluorescent protein (GFP) in the insertion it is possible simultaneously to modify gene function and examine expression of the affected gene. The overall hit rate is, however, much lower than with ENU, which can mutate 1 in every 200,000 base pairs (bp). It has been shown that one can expect on average 10 insertions per heterozygous F1 fish (Varshney et al. 2013, Wang et al. 2007). Of 15,223 integration sites that were mapped to the genome, 7,896 (52%) integrated into genes, with the majority (88%) targeting introns. Exons were affected by 12% (963) of integrations. Extrapolating from pilot data (Wang et al. 2007), about 20% of the insertions are expected to affect mRNA levels.

Targeting-induced local lesions in genomes (TILLING) was pioneered in *Arabidopsis* using a mismatch nicking enzyme, Cel1, to identify rare and induced SNPs (McCallum et al. 2000). This approach was adapted for the use in zebrafish using sequencing for mutation detection (Wienholds et al. 2002, Wienholds et al. 2003). Later several groups and consortia (Moens, Solnica-Krezel, Postlethwait, Zebrafish Mutation Project) established the technique and have produced several thousand mutants (Kettleborough et al. 2011).

As for a classical genetics screen the basis of this approach is to generate ENU-mutagenised libraries. Spermatogonial mutagenesis can be achieved relatively easily in zebrafish by repeatedly immersing sedated adult males in an ENU solution. Mutagenised males are then crossed to wild-type females to generate the library consisting of F1 families. These can be kept as live families to use directly for outcrossing or sperm can be cryopreserved to create an archive. For TILLING, these families were screened for mutations by PCR amplifying 5' exons of genes of interest across several thousand individual F1 DNA samples. Founders carrying mutations of interest were outcrossed for phenotypic analysis.

Crucially, each F1 fish carries about 10–15 nonsense and essential splice mutations and about 150 missense mutations (Kettleborough et al. 2013). With TILLING these mutations were never identified and instead seen as a necessary evil, or "background mutations", which had to be overcome by multiple rounds of outcrossing to clean up the background for phenotypic analysis. Falling sequencing costs and the advent of next-generation sequencing have made it feasible, both experimentally and financially, to identify the vast majority of induced mutations in the genes of each individual fish. The method employed by the Zebrafish Mutation Project (ZMP) is to capture exons of an individual fish using RNA baits designed against all annotated zebrafish genes, and then tag and analyse the enriched exomes using Illumina sequencing. Following this approach the ZMP has over the course of two years identified over 17,000 non-sense and essential splice site mutations in over 10,000 genes, covering 38% of zebrafish genes (Kettleborough et al. 2013).

Finally, a fast genotyping method, KASP genotyping, in combination with robotics makes it possible to genotype individual fish or embryos for all detected mutations in a rapid and reliable manner (Cuppen 2007). This is paramount to the establishment of correlations between detected phenotypes and genotypes.

A combination of rapid mutation detection in ENU-mutagenised fish and a targeted approach such as TALENs make it now feasible to generate a knockout allele for every protein-coding gene.

Phenotyping Methods

Phenomics is, by definition, a field with its boundaries set by the researcher's imagination (Butte and Kohane 2006). The depth to which a particular phenotype is analysed is unlimited and in a genome-wide approach it is instrumental to strike a balance between the detail and accuracy of phenotypic data and the necessary throughput. It is desirable to acquire as much information as possible about the consequences of a loss of function allele; so one approach would be to subject each mutant to a series of standardised assays. The nature of a phenotype, however, might make certain tests obsolete; for example, to test for a flight reflex in a mutant that is immotile due to a motoneuronal defect, is in the best case a wasted effort and in the worst case a wrong result. A tiered approach that channels mutants into phenotyping paths dependent on previous results is therefore more efficient. What are the features of a phenotyping approach that are suitable for a high-throughput project?

Embryonic Development

Early embryonic development was the focus of the first systematic genome-wide phenotypic screens in zebrafish. For each F2 family, about six clutches, covering 95% of induced mutations, were obtained by random F2 in-crosses. During the first five days of development, each clutch was then examined under a dissecting microscope for morphological defects and behavioural abnormalities. The developmental phenotypes scored included gastrulation defects, dorsal-ventral patterning, neural patterning, and eye and heart development. Functional assays included motility, hearing and balance.

Even though these observational assays represent very simple means of examination, they allow an experienced researcher to rapidly process large numbers of embryos and clutches and thereby identify gross as well as subtle phenotypes. Following a standard protocol of developmental milestone checks, tissues and behaviours can produce a comprehensive description of embryonic development.

Many groups have employed *in situ* staining methods using antibodies or antisense mRNA hybridisation to enhance their ability to detect phenotypes. This is very suited to automation and high-throughput as staining can be performed and analysed automatically in a 96-well format (Walker et al. 2012). The staining itself, however, limits the number of processes interrogated and thus makes the method more suited for specialised assays limited to a subset of mutations.

Automatic image analysis has recently gained a lot of interest for the analysis of not only labelled substructures of the zebrafish embryo, but also three-dimensional imaging of live embryos. There are, for example, microscopy platforms with software that can automatically detect and image regions of embryos in 96-well plates, a format that is highly amenable for DNA extraction and subsequent genotyping (Peravali et al. 2011).

One drawback of such a system is that embryos in a 96-well plates are difficult to orient to capture standardised images. Thus, others have designed a system that loads live embryos in a capillary, taken either from a multi-well plate or from a reservoir such as a petri dish (Pardo-Martin et al. 2010). The capillary is held between a brightfield and a confocal microscope and can be automatically rotated with the help of a high-speed camera to orient the embryo. After imaging, the live embryo can be dispensed back into a multi-well dish. This screening platform makes it possible to analyse large numbers of individual embryos at cellular resolution and different time points in a multi-well format. Zebrafish embryos survive and develop normally in 96-well plates. The ability to hold zebrafish embryos in 96-well plates and the fact that the embryos are sturdy enough to be handled by automatic fluidics systems make them suitable for automatic large-scale screens like no other vertebrate model.

Fry Behaviour

Zebrafish fry exhibit complex behaviours. They react to touch, sound, smell, food and optical stimuli with a large range of responses that can be detected and quantified (Colwill and Creton 2011, Wolman and Granato 2012).

Motility. Touch response in zebrafish embryos is established very early in development. Before one day-post-fertilisation (dpf), embryos will react to a tactile stimulus to the head or tail with a lateral tail flip, which can be used as a proxy for neuromuscular function (Saint-Amant and Drapeau 1998). From about 48 hpf onwards, the slow-twitch muscle response in a 24-hpf embryo is replaced by a fast-twitch muscle-mediated burst of movement that propels the fish.

A change of response to a tactile stimulus can have several causes and the nature of the altered response can indicate what process has been affected. In previous screens, hundreds of motility mutants in several classes were isolated (Granato et al. 1996). Apart from mutants with obvious skeletal or muscular defects, mutants with very specific motoneuronal defects were isolated. For example, *space cadet* mutants fail to execute a proper flight response upon tactile or acoustic stimulation, but initiate unilateral tail bends instead (Lorent et al. 2001). These behaviours have been shown to originate from faulty projections of Mauthner neurons. In contrast, *accordion* mutants contract bilaterally, such that their body shortens instead of bending from side to side, which is due to a lack of contralateral neuronal inhibition. A third group of mutants, such as *techno trousers*, hyper-react to stimuli with movements that are more intense in amplitude as well as duration, which also suggests a defect in neuronal inhibition.

From three dpf onwards, fry react to a vibrational stimulus, such as a tap on the dish, with a startle response consisting of a short forward burst. Embryos with impaired hearing as opposed to motility defects are easily identified, when the startle reflex is absent, but the fry respond to touch. If balance is affected as well, fry will swim in a circular motion and are unable to keep themselves upright.

Interestingly, with repeated tapping the startle response diminishes in wild-type embryos, meaning that embryos are capable of non-associative learning.

All these motility behaviours can be very easily elicited with a simple tap to the dish and high-speed cameras can help to capture and differentiate the responses.

Vision. Zebrafish adapt their pigmentation according to the brightness of their environment by expanding or contracting the melanosomes in their black pigment cells. Blind fry at five dpf thus appear much darker than their wild-type siblings. Some blind mutants, however, have pigmentation defects and therefore cannot be identified this way. For more subtle vision defects, tests have been developed that examine the fish's ability to detect moving objects or reaction to weak and strong visual stimuli (Burgess et al. 2010). From four dpf onwards, fry will follow moving objects with their eyes and this optokinetic response can be examined under a dissecting microscope (Fleisch and Neuhauss 2006). For this purpose, larvae are immobilised in a viscous methylcellulose solution, placed inside a rotating drum with black and white stripes and their eye movements are observed. This test can be refined further to test for colour vision, contrast sensitivity and acuity; however, the need for mounting individual embryos makes this assay less amenable to a genome-scale project (Renninger et al. 2011).

From seven dpf onwards, fry tend to follow moving objects. This behaviour, called optomotor response, is used to test for blindness by placing the dish on a screen that displays moving stripes. Fry with normal vision will gather at one end of the dish trying to follow (or escape from) the moving stripes. This is again a very simple test well suited for a high-throughput approach (Pelkowski et al. 2011).

Sleep. Circadian activity is another behaviour that can be measured in large numbers of individual fry in 96-well-based automated tracking systems. Provided the developing embryo is exposed to an appropriate light-dark cycle, by five dpf, zebrafish have an established wake-rest cycle (Appelbaum et al. 2009, Prober et al. 2006). At night, larvae show strongly reduced movement, rest in periods of several minutes and show a heightened arousal threshold similar to mammals and *Drosophila*. A mutant in the receptor for hypocretin, a neuropeptide regulating arousal, is homozygous viable, but displays disrupted sleep patterns at night, which can already be picked up at the larval stage (Yokogawa et al. 2007).

Chemical Screens

The option to raise zebrafish embryos in 96-well plates makes it possible to test pharmacologically active compounds *in vivo* on a scale that was previously only achievable in cell culture or invertebrate models (Creton 2009, Peterson et al. 2004). Screens have been done to test the effect of psychoactive drugs on the rest-wake cycle and locomotion in zebrafish fry (Rihel et al. 2010). It was possible to cluster behaviours and show that related compounds elicit the same response. Indeed, poorly characterised drugs were found to cluster with known compounds and could thus be assigned a potential physiological activity.

Chemical screening can be used either to interrogate the consequences of gene loss on the activity of already characterised compounds or to screen larger chemical libraries for new compounds with a modifying effect on either developmental or behavioural changes due to mutations (Laggner et al. 2012). In either case, the experimental effort is probably too large to justify incorporation in a primary phenotype analysis pipeline.

Metabolism

The endocrine system and the glands involved in hormone production are largely conserved between mammals and zebrafish (Lohr and Hammerschmidt 2011, McGonnell and Fowkes 2006). Equally, energy metabolism is similar enough to make zebrafish a valid model for these processes (Imrie and Sadler 2010). In the light of the growing urgency to tackle common diseases such as cardiovascular disease and type 2 diabetes, it makes sense to integrate assays in a zebrafish phenotyping project that elucidate the role that genes play in the disease aetiology.

Lipid uptake in the intestine and gall bladder can be visualised in zebrafish larvae using PED6, a live, fluorescent dye that becomes unquenched after cleavage by phospholipase A_2 (Anderson et al. 2011, Hama et al. 2009). Reduced levels of fluorescence can be easily scored, and indicate a morphological defect or a fault of the uptake or metabolism of acyl chains. In a similar fashion, fluorescently labelled free fatty acids and cholesterols visualise the uptake and distribution of these molecules in the larval body (Minchin and Rawls 2011).

The development and distribution of adipose tissue can be examined in fixed tissue either through *in situ* mRNA hybridisation for markers of adipocytes, staining of fat using Oil Red O in postembryonic, juvenile and adult stages, or in live animals using the lipophilic dye Nile Red. The live dye has the advantage that it enables the monitoring of the consequences of different feeding conditions in juvenile stages. These staining methods are straightforwardly incorporated into a high-throughput pipeline. The readout, however, may be ambiguous limiting the utility of such assays to selected alleles.

In contrast to carnivorous teleosts, omnivorous zebrafish use carbohydrates in a similar manner to mammals (Eames et al. 2010). It has been shown that a zebrafish's response to fasting also correlates well with the mammalian carbohydrate metabolism (Craig and Moon 2011). Upon fasting, glycogen stores in the liver get depleted and transcript levels of gluconeogenic enzymes go up such that zebrafish can maintain blood glucose levels for up to three weeks of fasting. While these methods are at the moment not suitable for application genome-wide, because blood glucose measurements are only possible in adult fish and quantitative transcript analysis is too labour intensive, there is still a strong case for the inclusion of these tests in the second tier of phenotyping.

Adult Phenotypes

Identifying and studying adult phenotypes is more challenging due to the increased space and maintenance requirement. Furthermore, examining adults for anything other than gross morphology is much more labour intensive. Moreover, because of the limitations on the number of individuals tested, the tested traits need to be robust and ideally highly penetrant so that they can be distinguished from phenotypes caused by non-genetic effects. Consequently, there have been only a few genetic screens to identify mutations affecting juvenile and adult stages for processes other than gross morphology or fertility.

Similarly, in a genome-wide reverse genetics approach it is not practical to submit large numbers of families to a large variety of phenotypic tests. Identified alleles, however, can be cherry-picked according to gene expression patterns, knowledge gleaned from other organisms or to generate models for human disease genes. Especially with respect to human disease, the study of juvenile and adult phenotypes is very desirable. A distinct advantage of reverse genetics in adults is that the mutation that is to be interrogated is known. Thus, genotyping can be used to identify and sort the adults and hence more robustly associate an expected phenotype to the loss of function allele.

A first and very simple, yet informative, assay is to test whether homozygous mutants, which are indistinguishable from their siblings at five dpf, survive to adulthood. If in-crosses are raised, individual fish can be genotyped at different stages to monitor survival. This is a very straightforward read-out for whether a gene plays a role during postembryonic development, which enables further studies on gene function outside the context of early lethality.

Fertility/Parental Effect. Several groups have conducted classic genetic screens on the effect of maternal- and paternal-recessive mutations (Pelegri and Mullins 2011, Wagner et al. 2004). Whereas the subsequent identification of the affected mutation is not easy, the initial discovery of parental effect mutations is relatively straightforward. Through several rounds of inbreeding, mutations are driven to homozygosity and in-crosses are grown to adulthood. When a fish homozygous for a highly penetrant mutation affecting fertilisation or embryonic development is crossed to another fish (be it mutant or wild-type), the vast majority of offspring will be affected. In general the affected processes for maternal effect mutations are oogenesis, egg activation, fertilisation and cytokinesis, while paternal effect mutations affect germ cell euploidy and centriole inheritance of the egg (Lindeman and Pelegri 2010, Yabe et al. 2007).

Morphology. A fair number of screens have looked at body morphology in adults, such as pigmentation, scale and fin formation and regeneration, skeletal and eye development and general body shape. Observations can be made either on morphology and behaviour of the live fish, or adults can be subjected to whole-mount staining for bone or cartilage. Alternatively, juvenile and adult animals can be sectioned and photographed for automatic image analysis. Traditional histological approaches are one option, but without automation they are too slow for a large-scale project. Another possibility is to use synchrotron micron-scale computer tomography (MicroCT) to image whole zebrafish; however, technologies such as robotic sample preparation and high resolution imaging of specimens as large as adult zebrafish still need further development (Cheng et al. 2011).

Behaviour. Increasingly, adult zebrafish are used to study intricate behaviours such as aggression, anxiety, learning and memory (Norton and Bally-Cuif 2010). As social animals that seek the proximity of their peers, zebrafish are very useful in probing the conservation of genetic determinants of social behaviour. These behaviours are more complex to evaluate, but robust assays have been established and it will be interesting to examine alleles in genes linked to human psychiatric conditions (Brennan 2011). For example, it has been demonstrated that loss of function of *fgfr1a* leads to an increase of aggression, boldness and exploration in adult zebrafish (Norton et al. 2011).

Learning and memory can be tested in a fairly high-throughput manner in a T-maze or shuttle box set-up where fish learn to negotiate a maze or choose compartments in a tank to collect a reward or to avoid an unpleasant stimulus (Norton and Bally-Cuif 2010). Positive stimuli such as food, certain drugs or conspecifics, or adverse stimuli such as certain odours, sudden light or mild electric shocks delivered together with a neutral stimulus such as colour can be used in associative learning experiments. Several groups have shown that zebrafish perceive drugs such as cocaine (Darland and Dowling 2001), morphine (Cachat et al. 2010), nicotine (Petzold et al. 2009), ethanol (Peng et al. 2009) and amphetamines (Webb et al. 2009) as a reward and that zebrafish can develop drug dependency. Some have isolated mutants with an altered sensitivity to amphetamines, cocaine and morphine. For example, while wild-type zebrafish recognise morphine as a reward in a place preference test, carriers of the *too few* mutation are insensitive to morphine but still show place preference when rewarded with food (Lau et al. 2006). This demonstrates a separation of the two reward systems even though both are mediated through opioid signalling.

Multi-tiered Genome-wide Phenotyping

The zebrafish research community recognises that the high-quality genome reference sequence together with the advances in sequencing at continually falling prices make it now possible to generate a knockout allele for every of the about 26,000 zebrafish genes. The ZMP at the Wellcome Trust Sanger Institute has already generated alleles in close to 50% of zebrafish genes and expects to have generated at least one allele for each protein-coding gene within the next four years. This presages the first phenotypic annotation of a vertebrate genome. Several meetings have been held with the aim to establish a collaborative strategy for a comprehensive array of embryonic and adult phenotypic assays. The most efficient approach will be to conduct the first tier of phenotyping at the institutes where mutations are generated and then distribute alleles to other places where more specialised second-tier analysis can be carried out.

The first layer of phenotypic analysis is a morphological and basic behavioural description during the first five days of development. Examination even only at dissecting microscope level gives a very good indication of a gene's function during embryonic development. There are important differences between classical genetic screens and an unbiased reverse genetics project that influence the way phenotyping has to be approached, even though the basic observations are the same. Classical screens are focused on a certain set of phenotypes and once clutches are identified as harbouring the desired phenotypes in a Mendelian fashion these phenotypes are then followed through further crosses for phenotypic analysis and eventual identification of the affected gene. Thus, the researcher initially ignores and then actively eradicates by outcrossing any other phenotypes.

With a genome-wide reverse genetic approach, the researcher does not necessarily know what to expect and is thus relatively unbiased in their approach. In the case of random ENU mutagenesis with subsequent exome resequencing, the majority of induced mutations is known. While this affords some *a priori* knowledge for some mutations, the researcher is still faced with many partially overlapping phenotypes. Each of these phenotypes needs to be assigned to a mutant genotype. With a good mutagenesis rate, each F1 carrier contains on average 10–15 potentially deleterious nonsense and essential splice site mutations and about 150 missense mutations, making it very difficult to correlate phenotypes and genotypes. One obvious solution would be to isolate each mutation by multiple rounds of outcrossing and genotyping. It is clear, however, that this is neither practical nor financially viable. The ZMP has therefore developed a phenotyping approach that allows simultaneous analysis of all alleles present in an F2 family without the need for outcrossing (Kettleborough et al. 2013, Dooley et al. 2013).

High-throughput Multi-allelic Phenotype Analysis

Previous experience with phenotyping of alleles identified through exon resequencing of mutagenised fish and data from mouse suggest that only about 20–30% of gene knockouts cause an embryonic lethal phenotype. This insight led to the development of a two-step process, in which the first step identifies non-phenotypic alleles and the second step establishes correlations between observed phenotypes and genotypes. In this process, only the nonsense and essential splice site mutations are examined for two reasons. Firstly, it is prohibitively expensive to follow over 150 mutations in each family. Secondly, it can be expected that any genes with phenotypic missense alleles will be captured again either as nonsense or essential splice site mutations later, which will then be subjected to the phenotyping pipeline.

Nonsense, splice and missense mutations are identified in F1 DNA by exome enrichment followed by Illumina sequencing. F2 families are generated either through outcrossing of live F1 fish or through *in vitro* fertilisation using cryopreserved F1 sperm. Six to 12 pairs are randomly mated to produce clutches of at least 150 embryos. The parents are fin-clipped for DNA preparation and genotyped for all nonsense and essential splice site mutations identified in the F1. Embryos are grown to five dpf, any phenotypic embryos are discarded and 48 phenotypically normal embryos are collected from each clutch for DNA preparation. Embryos are then genotyped only for the alleles that are heterozygous in both parents. Thus, only alleles that are 25% homozygous mutant in the offspring and thus can potentially give rise to a recessive phenotype are subsequently analysed. If the genotyping results show for a certain allele that 25% of the phenotypically normal embryos are homozygous mutant, the conclusion is that that allele does not cause a phenotype that is detectable with any of the assays used. If there are reduced numbers of homozygous mutant embryos among the phenotypically normal embryos and the ratio between homozygous wild-type and heterozygous is 1:2, this suggests that the allele causes a phenotype. These alleles are then taken forward to the next step in which genotype-phenotype correlations are established.

In the second step, carriers for the potentially phenotypic alleles are incrossed to obtain multiple clutches. These clutches are carefully examined at multiple time points up to five dpf following a standardised protocol. Phenotypes are documented in images and free text description. If the allele causes a highly penetrant phenotype and there were no genotyping errors in the parents, it can be expected that all clutches display one common phenotype among the on average one or two Mendelian phenotypes observed in each clutch. In any case, for each clutch, non-phenotypic and embryos with any observed phenotype are collected for DNA preparation. Finally, the DNA of these embryos is genotyped for each specific mutant allele. Phenotypes that are correlated with a genotype are then described using entity and quality ontologies to generate a searchable phenotype dataset.

It is important to note that this method relies on the ability to genotype a large number of samples quickly, for a variety of genotypes. Robotic automation is pivotal to secure the necessary throughput of DNA and reaction preparation, as well as bioinformatic analysis of the results. These methods are well established and transferrable.

Using this approach, about 20–30 alleles can be analysed comfortably by a single person per week. The ZMP has shown that this way a phenotype can be detected for about 7% of alleles. The phenotypes scored include morphology as well as behavioural responses (motility observations to check for motoneuronal or muscular defects and tests for reflexive responses such as the startle reflex upon an acoustic stimulus and escape response following touch). It is clear that this first pass analysis only captures highly penetrant and early acting mutations. By employing a few other relatively simple assays it may be possible to increase the rate of phenotype detection within the first tier.

Temperature Sensitivity. Several groups have shown that some mutant alleles affecting a variety of developmental and cellular processes are sensitive to temperature (Berghmans et al. 2005, Dick et al. 2000, Johnson et al. 2011, Nechiporuk et al. 2003, Parichy and Turner 2003, Tian et al. 2003). Specifically, some phenotypes increase in severity and penetrance or only become apparent when embryos are raised or adult fish are kept at a different temperature. It therefore may be beneficial to split clutches and raise them at two different temperatures in the first round of analysis to increase the chances of detecting phenotypes.

Juvenile Survival. Survival of fry to adulthood is a simple but very clear indicator of gene function during post-embryonic development. Incrosses containing homozygous mutants with no discernible phenotype at five dpf can be grown to adulthood and genotyped at certain time points to monitor survival. Alleles that have reduced viability can then be selectively subjected to a more detailed analysis in second tier assays.

Adult Screening. The same two-step approach described for embryonic phenotyping can be applied to F3 adults to correlate genotypes with phenotypes. Easily assessed phenotypes are morphology, general growth, fin regeneration, and basic behaviours such as startle response. Additionally, failure to fertilise or early developmental arrest of whole clutches can identify parental effect mutations.

Second-tier Analysis

More complex and detailed analysis is carried out on only a subset of the identified alleles. Alleles that warrant a more in-depth analysis can be selected based on *a priori* knowledge about the affected gene or results of the first-tier screen. If alleles are chosen for the genes they affect, one should consider that this biases the analysis towards known genes, potentially preventing the discovery of unexpected gene functions, thus weakening the genome-wide approach. Due to the nature of the mutation discovery and the high mutagenic load, however, in each experiment multiple known alleles are examined simultaneously. One can therefore choose certain families for analysis based on interesting alleles, but would at the same time analyse all other alleles present in the chosen family.

Transcriptome Analysis. The ZMP already submits all alleles that display a phenotype during the first five days of development to a further round of analysis. First of all, F2 carriers are outcrossed and another generation is raised to confirm the phenotype-genotype correlation on larger numbers of homozygous embryos. Phenotypic and non-phenotypic embryos are then collected to measure transcript expression differences associated with the loss of function of the affected gene. The method used is transcript counting by next-generation sequencing of 3' ends of reverse transcribed mRNA. With falling sequencing prices and refined methodology it is likely that this analysis will be incorporated eventually in the first-tier analysis. Embryos could be subjected to transcriptome analysis independent from a visual phenotype to try and capture transcriptional change either prior to the phenotype or to record changes that do not lead to a visually scorable change.

Conclusion

The zebrafish has one very clear advantage over any other vertebrate model. It gives us the opportunity to rapidly phenotype large numbers of disruptive alleles in a lot of individuals. This is afforded by the general robustness and short generation time of zebrafish, coupled with large numbers of offspring, *ex vivo* fertilisation and development and easy access for imaging throughout embryonic development. Furthermore, husbandry is very cost- and space-effective. These characteristics, together with a high-quality reference genome, make it feasible to functionally annotate a vertebrate genome with whole organism analysis of gene disruption.

During the last two decades, a diverse set of phenotypic analysis methods has been developed in zebrafish. Thanks to the translucency and *ex vivo* development of the larvae it is instrumentally simple to produce detailed phenotypic data of embryonic development. This is an area where zebrafish research perfectly complements mouse research, because it is difficult to obtain comparable early embryonic data.

The International Knockout Mouse Consortium (IKMC) is working to mutate every protein-coding gene in mouse using gene trapping and gene targeting in embryonic stem cells. According to the website, the consortium has so far disrupted about 18,000 genes in ES cells and generated 2,098 mice from them. An international phenotyping effort, piloted by the European Mouse Disease Clinic (EUMODIC) and carried on by the International Mouse Phenotyping Consortium (IMPC), is planning to comprehensively phenotype 20,000 mouse knockout lines over the next 10 years. Thus far, about 560 knockout mice have been phenotyped and the data made available (http://www.mousephenotype.org/). There is no doubt about the importance of mouse research for human disease. A zebrafish phenome project, however, will produce rich data at a fraction of the cost and can serve as an entry point to more in-depth phenotypic analysis using mammalian model systems for genes relevant for human development and disease.

The Zebrafish Mutation Project at the Wellcome Trust Sanger Institute has started a phenomics approach that will produce disruptive alleles for all 26,000 protein-coding genes in zebrafish with annotated embryonic phenotype data for about 9,000 genes within the next five years. In addition, the zebrafish community is in the process of establishing an international effort that will broaden the spectrum of analysed phenotypes to include a range of adult phenotypes and more comprehensive imaging. This dataset will provide a rich resource for not only the zebrafish research community, but more importantly, for any researcher or clinician wanting to learn about the consequences of loss of gene function. Moreover, the image and transcript data will be available for future analysis using tools that might not have been developed yet at the point of acquisition.

References Cited

Anderson, J.L., J.D. Carten and S.A. Farber. 2011. Zebrafish lipid metabolism: from mediating early patterning to the metabolism of dietary fat and cholesterol. Methods Cell Biol. 101: 111–141.

Appelbaum, L., G.X. Wang, G.C. Maro, R. Mori, A. Tovin, W. Marin, T Yokogawa, K Kawakami, S.J. Smith, Y. Gothilf, E. Mignot and P. Mourrain. 2009. Sleep-wake regulation and hypocretin-melatonin interaction in zebrafish. Proc. Natl. Acad. Sci. USA 106: 21942–21947.

Berghmans, S., R.D. Murphey, E. Wienholds, D. Neuberg, J.L. Kutok, C.D. Fletcher, J.P. Morris, T.X. Liu, S. Schulte-Merker, J.P. Kanki, R. Plasterk, L.I. Zon and A.T. Look. 2005. tp53 mutant zebrafish develop malignant peripheral nerve sheath tumors. Proc. Natl. Acad. Sci. USA 102: 407–412.

Bibikova, M., M. Golic, K.G. Golic and D. Carroll. 2002. Targeted chromosomal cleavage and mutagenesis in Drosophila using zinc-finger nucleases. Genetics 161: 1169–1175.

Brand, A.H. and N. Perrimon. 1993. Targeted gene expression as a means of altering cell fates and generating dominant phenotypes. Development 118: 401–415.

Brennan, C.H. 2011. Zebrafish behavioural assays of translational relevance for the study of psychiatric disease. Rev. Neurosci. 22: 37–48.

Burgess, H.A., H. Schoch and M. Granato. 2010. Distinct retinal pathways drive spatial orientation behaviors in zebrafish navigation. Curr. Biol. 20: 381–386.

Butte, A.J. and I.S. Kohane. 2006. Creation and implications of a phenome-genome network. Nat. Biotechnol. 24: 55–62.

Cachat, J., P. Canavello, M. Elegante, B. Bartels, P. Hart, C. Bergner, R. Egan, A. Duncan, D. Tien, A. Chung, K. Wong, J. Goodspeed, J. Tan, C. Grimes, S. Elkhayat, C. Suciu, M. Rosenberg, K.M. Chung, F. Kadri, S. Roy, S. Gaikwad, A. Stewart, I. Zapolsky, T. Gilder, S. Mohnot, E. Beeson, H. Amri, Z. Zukowska, R.D. Soignier and A.V. Kalueff. 2010. Modeling withdrawal syndrome in zebrafish. Behav. Brain Res. 208: 371–376.

Cermak, T., E.L. Doyle, M. Christian, L. Wang, Y. Zhang, C. Schmidt, J.A. Baller, N.V. Somia, A.J. Bogdanove and D.F. Voytas. 2011. Efficient design and assembly of custom TALEN and other TAL effector-based constructs for DNA targeting. Nucleic Acids Res. 39: e82.

Chakrabarti, S., G. Streisinger, F. Singer and C. Walker. 1983. Frequency of gamma-ray induced specific locus and recessive lethal mutations in mature germ cells of the zebrafish, BRACHYDANIO RERIO. Genetics 103: 109–123.

Cheng, K.C., X. Xin, D.P. Clark and P. La Riviere. 2011. Whole-animal imaging, gene function, and the Zebrafish Phenome Project. Curr. Opin. Genet. Dev. 21: 620–629.

Christian, M., T. Cermak, E.L. Doyle, C. Schmidt, F. Zhang, A. Hummel, A.J. Bogdanove and D.F. Voytas. 2010. Targeting DNA double-strand breaks with TAL effector nucleases. Genetics 186: 757–761.

Collins, J.E., S. White, S.M. Searle and D.L. Stemple. 2012. Incorporating RNA-seq data into the zebrafish Ensembl genebuild. Genome Research 22: 2067–2078.

Colwill, R.M. and R. Creton. 2011. Imaging escape and avoidance behavior in zebrafish larvae. Rev. Neurosci. 22: 63–73.

Craig, P.M. and T.W. Moon. 2011. Fasted zebrafish mimic genetic and physiological responses in mammals: a model for obesity and diabetes? Zebrafish 8: 109–117.

Creton, R. 2009. Automated analysis of behavior in zebrafish larvae. Behav. Brain Res. 203: 127–136.

Cuppen, E. 2007. Genotyping by Allele-Specific Amplification (KASPar). CSH Protoc. 2007: pdb prot4841.

Darland, T. and J.E. Dowling. 2001. Behavioral screening for cocaine sensitivity in mutagenized zebrafish. Proc. Natl. Acad. Sci. USA 98: 11691–11696.

Dick, A., M. Hild, H. Bauer, Y. Imai, H. Maifeld, A.F. Schier, W.S. Talbot, T. Bouwmeester and M. Hammerschmidt. 2000. Essential role of Bmp7 (snailhouse) and its prodomain in dorsoventral patterning of the zebrafish embryo. Development 127: 343–354.

Distel, M., M.F. Wullimann and R.W. Koster. 2009. Optimized Gal4 genetics for permanent gene expression mapping in zebrafish. Proc. Natl. Acad. Sci. USA 106: 13365–13370.

Dooley, C.M., H. Schwarz, K.P. Mueller, A. Mongera, M. Konantz, S.C. Neuhauss, C. Nusslein-Volhard and R. Geisler. 2012. Slc45a2 and V-ATPase are regulators of melanosomal pH homeostasis in zebrafish, providing a mechanism for human pigment evolution and disease. Pigment Cell & Melanoma Research.

Dooley, C.M., C. Scahill, F. Fényes, R.N. Kettleborough, D.L. Stemple, E.M. Busch-Nentwich. 2013. Multi-allelic phenotyping—A systematic approach for the simultaneous analysis of multiple induced mutations. Methods. PMID: 23624102.

Doyon, Y., J.M. McCammon, J.C. Miller, F. Faraji, C. Ngo, G.E. Katibah, R. Amora, T.D. Hocking, L. Zhang, E.J. Rebar, P.D. Gregory, F.D. Urnov and S.L. Amacher. 2008. Heritable targeted gene disruption in zebrafish using designed zinc-finger nucleases. Nat. Biotechnol. 26: 702–708.

Driever, W., L. Solnica-Krezel, A.F. Schier, S.C. Neuhauss, J. Malicki, D.L. Stemple, D.Y. Stainier, F. Zwartkruis, S. Abdelilah, Z. Rangini, J. Belak and C. Boggs. 1996. A genetic screen for mutations affecting embryogenesis in zebrafish. Development 123: 37–46.

Duffy, J.B., D.A. Harrison and N. Perrimon. 1998. Identifying loci required for follicular patterning using directed mosaics. Development 125: 2263–2271.

Eames, S.C., L.H. Philipson, V.E. Prince and M.D. Kinkel. 2010. Blood sugar measurement in zebrafish reveals dynamics of glucose homeostasis. Zebrafish 7: 205–213.

Egan, R.J., C.L. Bergner, P.C. Hart, J.M. Cachat, P.R. Canavello, M.F. Elegante, S.I. Elkhayat, B.K. Bartels, A.K. Tien, D.H. Tien, S. Mohnot, E. Beeson, E. Glasgow, H. Amri, Z. Zukowska and A.V. Kalueff. 2009. Understanding behavioral and physiological phenotypes of stress and anxiety in zebrafish. Behav. Brain Res. 205: 38–44.

Engeszer, R.E., L.B. Patterson, A.A. Rao and D.M. Parichy. 2007. Zebrafish in the wild: a review of natural history and new notes from the field. Zebrafish 4: 21–40.

Felsenfeld, A.L., C. Walker, M. Westerfield, C. Kimmel and G. Streisinger. 1990. Mutations affecting skeletal muscle myofibril structure in the zebrafish. Development 108: 443–459.

Fleisch, V.C. and S.C. Neuhauss. 2006. Visual behavior in zebrafish. Zebrafish 3: 191–201.

Granato, M., F.J. van Eeden, U. Schach, T. Trowe, M. Brand, M. Furutani-Seiki, P. Haffter, M. Hammerschmidt, C.P. Heisenberg, Y.J. Jiang, D.A. Kane, R.N. Kelsh, M.C. Mullins, J. Odenthal and C. Nusslein-Volhard. 1996. Genes controlling and mediating locomotion behavior of the zebrafish embryo and larva. Development 123: 399–413.

Grunwald, D.J. and G. Streisinger. 1992. Induction of recessive lethal and specific locus mutations in the zebrafish with ethyl nitrosourea. Genet. Res. 59: 103–116.

Haffter, P., M. Granato, M. Brand, M.C. Mullins, M. Hammerschmidt, D.A. Kane, J. Odenthal, F.J. van Eeden, Y.J. Jiang, C.P. Heisenberg, R.N. Kelsh, M. Furutani-Seiki, E. Vogelsang, D. Beuchle, U. Schach, C. Fabian and C. Nusslein-Volhard. 1996. The identification of genes with unique and essential functions in the development of the zebrafish, Danio rerio. Development 123: 1–36.

Hama, K., E. Provost, T.C. Baranowski, A.L. Rubinstein, J.L. Anderson, S.D. Leach and S.A. Farber. 2009. *In vivo* imaging of zebrafish digestive organ function using multiple quenched fluorescent reporters. Am. J. Physiol. Gastrointest. Liver Physiol. 296: G445–G453.

Harris, M.P., N. Rohner, H. Schwarz, S. Perathoner, P. Konstantinidis and C. Nusslein-Volhard. 2008. Zebrafish eda and edar mutants reveal conserved and ancestral roles of ectodysplasin signaling in vertebrates. PLoS Genet. 4: e1000206.

Ho, S.Y., J.L. Thorpe, Y. Deng, E. Santana, R.A. DeRose and S.A. Farber. 2004. Lipid metabolism in zebrafish. Methods Cell Biol. 76: 87–108.

Huang, P., A. Xiao, M. Zhou, Z. Zhu, S. Lin and B. Zhang. 2011. Heritable gene targeting in zebrafish using customized TALENs. Nat. Biotechnol. 29: 699–700.

Hwang, W.Y., Y. Fu, D. Reyon, M.L. Maeder, S.Q. Tsai, J.D. Sander, R.T. Peterson, J.R. Yeh and J.K. Joung. 2013. Efficient genome editing in zebrafish using a CRISPR-Cas system. Nat. Biotechnol.

Imrie, D. and K.C. Sadler. 2010. White adipose tissue development in zebrafish is regulated by both developmental time and fish size. Dev. Dyn. 239: 3013–3023.

Johnson, S.L., A.N. Nguyen and J.A. Lister. 2011. mitfa is required at multiple stages of melanocyte differentiation but not to establish the melanocyte stem cell. Dev. Biol. 350: 405–413.

Kettleborough, R.N., E. Bruijn, F. Eeden, E. Cuppen and D.L. Stemple. 2011. High-throughput target-selected gene inactivation in zebrafish. Methods Cell Biol. 104: 121–127.

Kettleborough, R.N.W., E.M. Busch-Nentwich, S.A. Harvey, C.M. Dooley, E. de Bruijn, F. van Eeden, I. Sealy, R.J. White, C. Herd, I.J. Nijman, F. Fényes, S. Mehroke, C. Scahill, R. Gibbons, N. Wali, S. Carruthers, A. Hall, J. Yen, E. Cuppen and D.L. Stemple. 2013. A systematic genome-wide analysis of zebrafish protein-coding gene function. Nature (In press).

Kimmel, C.B., W.W. Ballard, S.R. Kimmel, B. Ullmann and T.F. Schilling. 1995. Stages of embryonic development of the zebrafish. Dev. Dyn. 203: 253–310.

Laggner, C., D. Kokel, V. Setola, A. Tolia, H. Lin, J.J. Irwin, M.J. Keiser, C.Y. Cheung, D.L. Minor, Jr., B.L. Roth, R.T. Peterson and B.K. Shoichet. 2012. Chemical informatics and target identification in a zebrafish phenotypic screen. Nat. Chem. Biol. 8: 144–146.

Lamason, R.L., M.A. Mohideen, J.R. Mest, A.C. Wong, H.L. Norton, M.C. Aros, M.J. Jurynec, X. Mao, V.R. Humphreville, J.E. Humbert, S. Sinha, J.L. Moore, P. Jagadeeswaran, W. Zhao, G. Ning, I. Makalowska, P.M. McKeigue, D. O'Donnell, R. Kittles, E.J. Parra, N.J. Mangini, D.J. Grunwald, M.D. Shriver, V.A. Canfield and K.C. Cheng. 2005. SLC24A5, a putative cation exchanger, affects pigmentation in zebrafish and humans. Science 310: 1782–1786.

Lau, B., S. Bretaud, Y. Huang, E. Lin and S. Guo. 2006. Dissociation of food and opiate preference by a genetic mutation in zebrafish. Genes, Brain, and Behavior 5: 497–505.

Lindeman, R.E. and F. Pelegri. 2010. Vertebrate maternal-effect genes: Insights into fertilization, early cleavage divisions, and germ cell determinant localization from studies in the zebrafish. Mol. Reprod. Dev. 77: 299–313.

Lloyd, A., C.L. Plaisier, D. Carroll and G.N. Drews. 2005. Targeted mutagenesis using zinc-finger nucleases in Arabidopsis. Proc. Natl. Acad. Sci. USA 102: 2232–2237.

Lohr, H. and M. Hammerschmidt. 2011. Zebrafish in endocrine systems: recent advances and implications for human disease. Annu. Rev. Physiol. 73: 183–211.

Lorent, K., K.S. Liu, J.R. Fetcho and M. Granato. 2001. The zebrafish space cadet gene controls axonal pathfinding of neurons that modulate fast turning movements. Development 128: 2131–2142.

McCallum, C.M., L. Comai, E.A. Greene and S. Henikoff. 2000. Targeted screening for induced mutations. Nat. Biotechnol. 18: 455–457.

McGonnell, I.M. and R.C. Fowkes. 2006. Fishing for gene function—endocrine modelling in the zebrafish. J. Endocrinol. 189: 425–439.

Minchin, J.E. and J.F. Rawls. 2011. *In vivo* analysis of white adipose tissue in zebrafish. Methods Cell Biol. 105: 63–86.

Nechiporuk, A., K.D. Poss, S.L. Johnson and M.T. Keating. 2003. Positional cloning of a temperature-sensitive mutant emmental reveals a role for sly1 during cell proliferation in zebrafish fin regeneration. Dev. Biol. 258: 291–306.

Norton, W. and L. Bally-Cuif. 2010. Adult zebrafish as a model organism for behavioural genetics. BMC Neurosci. 11: 90.

Norton, W.H., K. Stumpenhorst, T. Faus-Kessler, A. Folchert, N. Rohner, M.P. Harris, J. Callebert and L. Bally-Cuif. 2011. Modulation of Fgfr1a signaling in zebrafish reveals a genetic basis for the aggression-boldness syndrome. J. Neurosci. 31: 13796–13807.

Pardo-Martin, C., T.Y. Chang, B.K. Koo, C.L. Gilleland, S.C. Wasserman and M.F. Yanik. 2010. High-throughput *in vivo* vertebrate screening. Nat. Methods 7: 634–636.

Parichy, D.M., J.F. Rawls, S.J. Pratt, T.T. Whitfield and S.L. Johnson. 1999. Zebrafish sparse corresponds to an orthologue of c-kit and is required for the morphogenesis of a subpopulation of melanocytes, but is not essential for hematopoiesis or primordial germ cell development. Development 126: 3425–3436.

Parichy, D.M. and J.M. Turner. 2003. Zebrafish puma mutant decouples pigment pattern and somatic metamorphosis. Dev. Biol. 256: 242–257.

Pelegri, F. and M.C. Mullins. 2011. Genetic screens for mutations affecting adult traits and parental-effect genes. Methods Cell Biol. 104: 83–120.

Pelkowski, S.D., M. Kapoor, H.A. Richendrfer, X. Wang, R.M. Colwill and R. Creton. 2011. A novel high-throughput imaging system for automated analyses of avoidance behavior in zebrafish larvae. Behav. Brain Res. 223: 135–144.

Peng, J., M. Wagle, T. Mueller, P. Mathur, B.L. Lockwood, S. Bretaud and S. Guo. 2009. Ethanol-modulated camouflage response screen in zebrafish uncovers a novel role for cAMP and extracellular signal-regulated kinase signaling in behavioral sensitivity to ethanol. J. Neurosci. 29: 8408–8418.

Peravali, R., J. Gehrig, S. Giselbrecht, D.S. Lutjohann, Y. Hadzhiev, F. Muller and U. Liebel. 2011. Automated feature detection and imaging for high-resolution screening of zebrafish embryos. Biotechniques 50: 319–324.

Peterson, R.T., S.Y. Shaw, T.A. Peterson, D.J. Milan, T.P. Zhong, S.L. Schreiber, C.A. MacRae and M.C. Fishman. 2004. Chemical suppression of a genetic mutation in a zebrafish model of aortic coarctation. Nat. Biotechnol. 22: 595–599.

Petzold, A.M., D. Balciunas, S. Sivasubbu, K.J. Clark, V.M. Bedell, S.E. Westcot, S.R. Myers, G.L. Moulder, M.J. Thomas and S.C. Ekker. 2009. Nicotine response genetics in the zebrafish. Proc. Natl. Acad. Sci. USA 106: 18662–18667.

Prober, D.A., J. Rihel, A.A. Onah, R.J. Sung and A.F. Schier. 2006. Hypocretin/orexin overexpression induces an insomnia-like phenotype in zebrafish. J. Neurosci. 26: 13400–13410.

Renninger, S.L., H.B. Schonthaler, S.C. Neuhauss and R. Dahm. 2011. Investigating the genetics of visual processing, function and behaviour in zebrafish. Neurogenetics 12: 97–116.

Reyon, D., S.Q. Tsai, C. Khayter, J.A. Foden, J.D. Sander and J.K. Joung. 2012. FLASH assembly of TALENs for high-throughput genome editing. Nat. Biotechnol.

Rihel, J., D.A. Prober, A. Arvanites, K. Lam, S. Zimmerman, S. Jang, S.J. Haggarty, D. Kokel, L.L. Rubin, R.T. Peterson and A.F. Schier. 2010. Zebrafish behavioral profiling links drugs to biological targets and rest/wake regulation. Science 327: 348–351.

Saint-Amant, L. and P. Drapeau. 1998. Time course of the development of motor behaviors in the zebrafish embryo. J. Neurobiol. 37: 622–632.

Sander, J.D., L. Cade, C. Khayter, D. Reyon, R.T. Peterson, J.K. Joung and J.R. Yeh. 2011. Targeted gene disruption in somatic zebrafish cells using engineered TALENs. Nat. Biotechnol. 29: 697–698.

Skarnes, W.C., B. Rosen, A.P. West, M. Koutsourakis, W. Bushell, V. Iyer, A.O. Mujica, M. Thomas, J. Harrow, T. Cox, D. Jackson, J. Severin, P. Biggs, J. Fu, M. Nefedov, P.J. de Jong, A.F. Stewart and A. Bradley. 2011. A conditional knockout resource for the genome-wide study of mouse gene function. Nature 474: 337–342.

Thomas, K.R. and M.R. Capecchi. 1987. Site-directed mutagenesis by gene targeting in mouse embryo-derived stem cells. Cell 51: 503–512.

Tian, J., C. Yam, G. Balasundaram, H. Wang, A. Gore and K. Sampath. 2003. A temperature-sensitive mutation in the nodal-related gene cyclops reveals that the floor plate is induced during gastrulation in zebrafish. Development 130: 3331–3342.

Trinh le, A., T. Hochgreb, M. Graham, D. Wu, F. Ruf-Zamojski, C.S. Jayasena, A. Saxena, R. Hawk, A. Gonzalez-Serricchio, A. Dixson, E. Chow, C. Gonzales, H.Y. Leung, I. Solomon, M. Bronner-Fraser, S.G. Megason and S.E. Fraser. 2011. A versatile gene trap to visualize and interrogate the function of the vertebrate proteome. Genes Dev. 25: 2306–2320.

Varshney, G.K., J. Lu, D. Gildea, H. Huang, W. Pei, Z. Yang, S.C. Huang, D.S. Schoenfeld, N. Pho, D. Casero, T. Hirase, D.M. Mosbrook-Davis, S. Zhang, L.E. Jao, B. Zhang, I.G. Woods, S. Zimmerman, A.F. Schier, T. Wolfsberg, M. Pellegrini, S.M. Burgess and S. Lin. 2013. A large-scale zebrafish gene knockout resource for the genome-wide study of gene function. Genome Res. 23(4): 727–35.

Wagner, D.S., R. Dosch, K.A. Mintzer, A.P. Wiemelt and M.C. Mullins. 2004. Maternal control of development at the midblastula transition and beyond: mutants from the zebrafish II. Dev. Cell 6: 781–790.

Walker, S.L., J. Ariga, J.R. Mathias, V. Coothankandaswamy, X. Xie, M. Distel, R.W. Koster, M.J. Parsons, K.N. Bhalla, M.T. Saxena and J.S. Mumm. 2012. Automated reporter quantification *in vivo*: high-throughput screening method for reporter-based assays in zebrafish. PLoS One 7: e29916.

Wang, D., L.E. Jao, N. Zheng, K. Dolan, J. Ivey, S. Zonies, X. Wu, K. Wu, H. Yang, Q. Meng, Z. Zhu, B. Zhang, S. Lin and S.M. Burgess. 2007. Efficient genome-wide mutagenesis of zebrafish genes by retroviral insertions. Proc. Natl. Acad. Sci. USA 104: 12428–12433.

Webb, K.J., W.H. Norton, D. Trumbach, A.H. Meijer, J. Ninkovic, S. Topp, D. Heck, C. Marr, W. Wurst, F.J. Theis, H.P. Spaink and L. Bally-Cuif. 2009. Zebrafish reward mutants reveal novel transcripts mediating the behavioral effects of amphetamine. Genome Biology 10: R81.

Wienholds, E., S. Schulte-Merker, B. Walderich and R.H. Plasterk. 2002. Target-selected inactivation of the zebrafish rag1 gene. Science 297: 99–102.

Wienholds, E., F. van Eeden, M. Kosters, J. Mudde, R.H. Plasterk and E. Cuppen. 2003. Efficient target-selected mutagenesis in zebrafish. Genome Research 13: 2700–2707.

Wolman, M. and M. Granato. 2012. Behavioral genetics in larval zebrafish: Learning from the young. Dev. Neurobiol. 72: 366–372.

Yabe, T., X. Ge and F. Pelegri. 2007. The zebrafish maternal-effect gene cellular atoll encodes the centriolar component sas-6 and defects in its paternal function promote whole genome duplication. Dev. Biol. 312: 44–60.

Yokogawa, T., W. Marin, J. Faraco, G. Pezeron, L. Appelbaum, J. Zhang, F. Rosa, P. Mourrain and E. Mignot. 2007. Characterization of sleep in zebrafish and insomnia in hypocretin receptor mutants. PLoS Biol. 5: e277.

5

Systematic Cell Phenotyping

Jean-Karim Hériché

Introduction

High-throughput sequencing and computerized genome annotation combined with proteomics approaches have contributed to the identification of most of the basic components of the eukaryotic cell. However, despite this, we are still far from understanding how a cell functions. To reach this goal, we need to understand how cellular processes are generated from interactions between the cell's components. To cell biologists, phenotypes are thus of more interest than genomes because they are the observations that need explaining. Characterization of cellular phenotypes in gene perturbation experiments is the standard way of establishing causal relationships between genes and phenotypes and provides the most reliable way of assigning functions to genes. Phenotyping cells has thus long provided insights into cellular functions. However, the cost of phenotyping being higher than that of genotyping, gene identification has largely outpaced gene function characterization. As a result, 13 years after the completion of the human genome project, less than a third of the human protein-coding genes have been experimentally assigned a function. This points to phenotyping becoming the bottleneck in the

Cell Biology and Biophysics Unit, European Molecular Biology Laboratory, Meyerhofstrasse 1, 69117 Heidelberg, Germany.
Email: heriche@embl.de

production of knowledge in cell biology and calls for more systematic approaches. Therefore, one way of doing systematic cell phenotyping is to identify all genes contributing to one biological process. Another way to be systematic is to simultaneously study a range of phenotypes. Although cellular phenotypes can be defined at many different levels such as global gene expression changes or proteome maps, since the invention of the first compound microscopes, eukaryotic cells have traditionally been described using light microscopy. The development of molecular genetics techniques to label molecules inside the cell has produced a wealth of insights into cellular processes, in particular with the use of genetically encoded fluorescent proteins (e.g., GFP (Green Fluorescent Protein) tags) allowing live cell imaging. Developments in fluorescence microscopy combined with automation now allow researchers to do high-throughput cell-based experiments. As traditional manual annotation of phenotypes can't keep up with data generation, computerized analyses have been implemented. This chapter will present an overview of the tools, methods and challenges of doing cellular phenotyping experiments on a large scale with a particular focus on large-scale image-based screens by RNA interference (RNAi) as they currently represent the main providers of systematic phenotypic data at the cellular level. As this chapter will show, running a pipeline from assay design to automated microscopy to data analysis is far from being a mindless robotic approach to cell biology but requires both rigorous and creative thinking in a multidisciplinary context.

Interfering RNAs as Perturbation Reagents

Linking genes to phenotypes is typically done by interfering with normal gene function in the cells. Before the advent of RNAi, cDNA-driven protein overexpression was the perturbation of choice in phenotyping experiments and there is still interest in the approach. For example, Shindoh et al. (2012) screened multiple cDNA libraries to identify genetic variants with oncogenic or drug-resistance properties. However, the consequences of overexpression of a protein are often ill-defined as phenotypes obtained by overexpression can result from either a gain of function that may or may not be related to the normal function of the protein or a loss of function if the overexpressed protein acts as a dominant-negative mutant.

In light of this, RNAi has become the preferred way of producing phenotypes. RNAi is a posttranscriptional gene-silencing mechanism by which double-stranded RNAs (dsRNAs) introduced into cells cause degradation or translational repression of homologous mRNAs (Meister and Tuschl 2004). Intracellular processing of long dsRNAs produces pools of small (21–23 nucleotides) double-stranded interfering RNAs (siRNAs). Based on the thermodynamic properties of the siRNA, one strand (called the guide strand) is incorporated into RNA-induced silencing complexes (RISCs) which then bind complementary mRNAs. While *Drosophila* cells can be treated directly with long dsRNAs, dsRNAs longer than 30 bp can trigger a global non-specific response in mammalian cells (Gantier and Williams 2007). To circumvent this issue, application of RNAi in mammalian cells can be done using either chemically synthesized siRNAs or vector-based short hairpin RNAs (shRNAs). shRNAs are designed to be analogous to miRNAs and, once synthesized by the cells, are presumably processed by the miRNA pathway.

RNAi has revolutionized cell-based loss of function experiments by providing an easy way of doing reverse genetics in cell culture. However, unlike gene knockouts, RNAi generally produces hypomorphic phenotypes. This is not necessarily a drawback as allelic series of hypomorphic mutants are valuable tools in classical genetics of model organisms. In particular, hypomorphic series can be used to genetically separate different molecular or biological functions of a gene but because different interfering RNAs targeting the same gene can have different knock-down efficiencies or even target different subsets of transcripts, the corresponding phenotypes can be quantitatively and qualitatively different.

Given its mechanism of action, RNAi is prone to so-called off-target effects by which phenotypes are generated independently of the knock-down of the intended target gene. The traditional genetics way of proving the link between gene and phenotype is to revert the phenotype by expressing a wild-type gene product in the mutant background. This phenotypic rescue strategy is also applicable to RNAi using an RNAi-resistant version of the target gene (Kittler et al. 2005, Neumann et al. 2010) but is not widespread. An alternative strategy consists in using RNAi-independent approaches such as analyzing orthologous mutants in model organisms. Because these approaches tend to require lengthy follow-up experiments with a lower throughput, practitioners rely instead on phenotype reproducibility by multiple interfering RNAs targeting different regions of the gene on the assumption that different dsRNAs have different off-target effects. While this can be done in high-throughput mode, there are two reasons which make this a less-than-ideal approach. First comes the variable hypomorphic nature of the reagents as mentioned above. Unless a significant number of different dsRNAs per gene are tested, one may fail to confirm the connection between gene and phenotype with no way of deciding if the original observation was due to off-target effects. Second comes the possibility of observing the same phenotype via independent off-target effects. In particular, it is expected that non-specific phenotypes like cell viability would be more sensitive to off-target effects because these phenotypes can be generated by interfering with many independent processes or pathways. However, more specific phenotypes are also not immune to this problem. For example, MAD2 has been reported as a common off-target cause of mitotic spindle assembly checkpoint phenotypes (Hübner et al. 2010).

There are several ways to mitigate the off-target issue. One strategy relies on the assumption that off-target effects would be diluted by combining multiple interfering RNAs against the same target. In systems that tolerate them, like *Drosophila* cells, long dsRNAs (> 400 bp) that are converted intracellularly into pools of siRNAs are typically used. The equivalent strategy in mammalian cells is to use pools of siRNAs either chemically produced or generated by *in vitro* digestion of long dsRNAs (endoribonuclease-prepared siRNAs or esiRNAs) (Myers et al. 2006, Kittler et al. 2007). Another strategy relies on chemical modifications of siRNAs in order, for example, to favour the guide strand to reduce potential gene silencing induced by the passenger strand. Finally, better dsRNA design exploiting a better understanding of the mechanisms producing off-target effects may lead to a further reduction of the problem (Fisher et al. 2012).

While genome-wide libraries are available, many laboratories restrict their work to subsets of genes. For human cells, the most used libraries cover about 700 protein kinases. Libraries covering so-called druggable genes are also available although the coverage of a few thousand genes varies somewhat between suppliers. Other smaller libraries cover specific molecular functions such as phosphatases, proteases or transcription factors. In a few cases, when biologists are interested in a more systematic understanding of a biological process but can't afford a genome-wide library, small custom-made libraries are built from candidate genes after extensive database searches and literature curation. This candidate gene selection can be automated and thus made more efficient. In particular, gene function prediction algorithms can be used to build RNAi libraries targeting a particular biological process. The fact that this is not commonly done may be due to either cell biologists not being aware of such algorithms or cell biologists not trusting the algorithms because of lack of experimental demonstration that such an approach does indeed produce libraries enriched in genes with the expected phenotypes. Although custom libraries tend to be more expensive per siRNA than off-the-shelf ones, a smaller targeted library could turn out to be less expensive than a larger generic library especially when the cost per hit or the total cost of screening is taken into account.

With dsRNAs being sequence-based reagents, target gene inference is highly dependent on the genome annotations used. With constantly evolving genome annotations and genomes of cell lines used in experiments soon to become available, availability of the sequences of all the reagents is key to interpretation and reuse of data. For example, sequences are needed to evaluate off-target effects of previous studies in light of new algorithms or knowledge and in a systematic context, all sequences of reagents including those producing negative results are of interest. Data repositories will also need reagents' sequences to integrate different data sets. It is therefore appalling that several vendors of siRNA libraries oppose any public disclosure of the reagents they sell and even more appalling that some scientists accept such unscientific behaviour.

Experimental Systems and Reporters

High-throughput cellular phenotyping was initially implemented as a strategy for identifying active chemicals from very large compound libraries. In this context, because of the need for speed, assays have generally been limited to simple, single parameter cellular phenotypes (Inglese et al. 2007). This is typically done through the use of a reporter gene. For example, the first genome-scale RNAi screens in *Drosophila* cells used respectively a quantitative luciferase-based assay to measure cellular ATP levels as a proxy for cell number (Boutros et al. 2004) and β-galactosidase under the promoter of the antimicrobial peptide Dipt to monitor response to bacterial cell wall components (Foley and O'Farrell 2004). While Boutros et al. (2004) used a plate reader to produce quantitative results, Foley and O'Farrell (2004) performed a visual qualitative screen using a microscope.

However, fluorescence is the standard reporter used in microscopy-based cellular assays. Its use in a systematic phenotyping strategy applied to mammalian cells could be traced to attempts at systematically identifying subcellular localization patterns of proteins using a collection of GFP-tagged human cDNAs (Simpson et al. 2000) with roughly a thousand proteins currently annotated from visual inspection of the microscopy slides and available in the LIFEdb database (Mehrle et al. 2006). With similar goals, the Human Protein Atlas project (Uhlén et al. 2010) uses immunofluorescence to systematically localize human proteins. It currently provides subcellular localizations for roughly one half of human protein-coding genes. Annotations were initially derived from visual inspection of confocal images of immunofluorescent staining of various cell lines and have recently been complemented by automated image analysis (Li et al. 2012). A limitation of current automated methods for subcellular localization annotation is their inability to reliably identify patterns resulting from a mixture of subcellular localizations. And while it is possible to make inferences about protein function from systematic subcellular localizations (e.g., Hutchins et al. 2010), available information on subcellular localization generally lacks spatial and temporal resolution to make any reliable prediction.

Because microscopes uniquely allow to simultaneously monitor many parameters, from intensity measurements to morphological changes, systematic phenotyping benefits from maximizing the information content of the images. To this end, fluorescence is also typically used to monitor expression of various cellular markers. For example, the first large-scale microscopy-based RNAi screen in cells surveyed about a thousand genes for their contribution to the shape of *Drosophila* S2 cells by looking at fixed cells fluorescently labelled for tubulin, actin and DNA (Kiger et al. 2003). Many subsequent assays have similarly relied on detecting pattern variations of fluorescent markers (e.g., Mukherji et al. 2006, Goshima et al. 2007, Winograd-Katz et al. 2009, Neumann et al. 2010).

While most assays use cells from established cell lines, it is possible to do RNAi screens in mammalian primary cells like mouse embryonic stem cells (Westerman et al. 2011, Yang et al. 2012) or human macrophages (Ley et al. 2013). Primary cells from *Drosophila* embryos have also been proposed as an experimental system for large scale RNAi screens (Bai et al. 2009).

Perhaps surprisingly given the wide range of potential markers available, few biological processes have been interrogated in a systematic or large-scale approach in cells. Table 1 gives an overview of some of the RNAi screens addressing these processes. A majority of studies concern three cellular processes of direct medical relevance: host-pathogen interactions, oncogenic signaling pathways and DNA-damage response. The typical reporters used in host-pathogen RNAi screens are modified pathogens expressing GFP although pathogen-sensitive reporter genes can also be used. Oncogenic signaling pathways are usually studied using pathway-responsive reporter genes although monitoring of proliferation or cell survival in a sensitized context has also been reported. Monitoring DNA-damage response is typically done using antibodies against pathway effectors with at least one study using a reporter gene engineered to respond to homologous recombination. Fewer studies focus on basic cellular processes such as cell viability and proliferation, cell cycle, protein secretion, cell adhesion and ciliogenesis.

It is, however, possible to monitor several distinct biological processes by interrogating multiple markers (Fuchs et al. 2010, Singh et al. 2010) or characterizing genetic interactions with respect to a single parameter (Horn et al. 2011) in phenotypic profiling experiments.

Although many biological processes are dynamic by nature, almost all high-throughput experiments use fixed time point assays. However, these can lead to false negatives or even erroneous conclusions. One study in particular has demonstrated the feasibility and usefulness of high-throughput time-lapse imaging to characterize mitotic phenotypes in a genome-wide screen (Neumann et al. 2010). By analyzing arbitrary time points from this time-lapse screen, one can see that fixed time point assays can easily leave many genes undetected. In addition, depending on the assay, some phenotypes can only be interpreted in a temporal context. For example, time-resolved phenotypes allow the distinction between primary effects and secondary consequences such as distinguishing cytokinesis

Table 1. Overview of some large-scale cell-based RNAi screens.

Biological process addressed	Reporters	Cell type	Size (genes)	Image analysis	Number of descriptors	Statistical analysis	Screen reference
Host-pathogen interactions	LacZ under control of the Dipt promoter	*Drosophila* S2	~7,000	Visual inspection	1	None	Foley and O'Farrell 2004
Host-pathogen interactions	*Listeria monocytogenes* expressing GFP	*Drosophila* S2	~14,000	Visual inspection	1	None	Agaisse et al. 2005
Host-pathogen interactions	*Candida albicans* expressing GFP, Hoechst 33258	*Drosophila* S2	~7,000	Visual inspection	1	None	Stroschein-Stevenson et al. 2006
Host-pathogen interactions	Flu virus expressing VSVG-luciferase	*Drosophila* S2	~13,000	Plate reader	1	Z-score	Hao et al. 2008
Host-pathogen interactions	Assay1: Hoechst 33342 and antibody against p24Gag; assay2: Tat-dependent β-galactosidase	Human HeLa TZM-bl	~21,000	Assay1: Automated; assay2: plate reader	1	Mean and standard deviation	Brass et al. 2008
Host-pathogen interactions	HIV expressing β-galactosidase	Human HeLa P4/R5	~20,000	Plate reader	1	Strictly standardized mean differences	Zhou et al. 2008
Host-pathogen interactions	Assay1: DAPI and virus-specific antibody, assay2: influenza virus-specific luciferase construct	Human A549 and 293T	~23,000	Assay1: Automated; assay2: plate reader	1	B-score	Karlas et al. 2010
Host-pathogen interactions	*Salmonella* Typhimurium expressing GFP	Human HeLa	~7,000	Automated	1	Z-score	Misselwitz et al. 2011
Host-pathogen interactions	VAVC expressing GFP, DAPI	Human HeLa	~7,000	Automated	1	Z-score	Mercer et al. 2012
Host-pathogen interactions	*Salmonella typhimurium* expressing GFP, Hoechst 33342	Human MCF-7	~20,000	Automated	1	Z-score	Thornbrough et al. 2012

Host-pathogen interactions	Francisella tularensis expressing GFP	Human THP-1	~38,500	Flow cytometry	1	None	Zhou et al. 2012
Signaling pathways	Hh-responsive luciferase	Drosophila cl-8	~6,000	Plate reader	1	Mean and standard deviation	Lum et al. 2003
Signaling pathways	Luciferase under control of the Draf promoter	Drosophila Kc167	~13,000	Plate reader	1	Z-score	Müller et al. 2005
Signaling pathways	Wn-responsive luciferase	Drosophila cl-8	~13,000	Plate reader	1	Z-score	DasGupta et al. 2005
Signaling pathways	antibody against phosphorylated ERK(Rolled)	Drosophila S2R+	~20,000	Plate reader	1	Z-score	Friedman and Perrimon 2006
Signaling pathways	TCF/LEF driven luciferase	Human HeLa	~21,000	Plate reader	1	Z-score	Tang et al. 2008
Signaling pathways	Cell proliferation as measured by microarray hybridization	Human DLD-1	unavailable (~32,000 transcripts)	Microarray	1	t-test	Luo et al. 2009
Signaling pathways	Cell survival	Mouse GL261	~28,000	None	1	None	Sheng et al. 2010
Signaling pathways	GFP under control of the ID2 promoter	Human HaCaT	220	Automated	1	Median	Wu et al. 2011
Signaling pathways	NFκB-responsive GFP	Human HEK293	~21,000	Flow cytometry	1	Z-score	Gewurz et al. 2012
Signaling pathways	Alamar Blue staining	Human HFF	~3,300	Plate reader	1	Z-score	Toyoshima et al. 2012
DNA damage response	antibody against 53BP1, Hoechst 33258	Human U2OS	~21,000	Automated and visual inspection	1	Z-score	Doil et al. 2009

Table 1. contd....

Table 1. contd.

Biological process addressed	Reporters	Cell type	Size (genes)	Image analysis	Number of descriptors	Statistical analysis	Screen reference
DNA damage response	antibody against phosphorylated KAP1	Human HeLa	~2,300	Visual inspection	1	None	Lovejoy et al. 2010
DNA-damage response	Relative abundance of shRNAs	Human U2OS	~32,000 transcripts				
DNA damage response	antibody against phosphorylated histone H3, DAPI	*Drosophila* S2R+	~13,500	Automated and visual inspection	1	None	Kondo and Perrimon 2011
DNA damage response	DR-GFP, Hoechst	Human U2OS	~21,000	Automated	1	Mean and standard deviation	Adamson et al. 2012
DNA damage response	antibody against 53BP1, Hoechst	Human HeLa and U2OS	~1,350	Automated	1	Z-score	Moudry et al. 2012
DNA damage response	GFP-RNF168	Human U2OS	~1,350	Automated	1	Z-score	Gudjonsson et al. 2012
Proliferation/viability	Luciferase-based assay for ATP levels	*Drosophila* Kc167 and S2R+	~19,500	Plate reader	1	Z-score	Boutros et al. 2004
Proliferation/viability	mitochondrial function by WST-1 assay	Human HeLa	~5,300	Plate reader	1	None	Kittler et al. 2004
Proliferation/viability	DAPI	Human U2OS	~24,400	Automated	8	Z-score	Mukherji et al. 2006
Cell cycle	dsRed and Hoechst 33342	Human HeLa	780	Visual inspection	1	None	Draviam et al. 2007

Process	Markers	Cell type	Number	Method	N	Statistic	Reference
Cell cycle	DAPI and antibodies against tubulin, γ-tubulin, phosphorylated histone H3	*Drosophila* S2	~14,500	Automated and visual inspection	9	Bootstrap p-values	Goshima et al. 2007
Cell cycle	Histone H2B fused to GFP	Human HeLa	~21,000	Automated and visual inspection	16	Median	Neumann et al. 2010
Protein secretion	Luciferase fused to a signal peptide	*Drosophila* S2	~14,000	Plate reader	1	Z-score	Wendler et al. 2010
Protein secretion	tsO45G fused to CFP, DAPI	Human HeLa	~21,000	Automated	1	Deviation from control	Simpson et al. 2012
Cell adhesion	Paxillin fused to YFP	Human HeLa	~1,100	Automated and visual inspection	8	Z-score	Winograd-Katz et al. 2009
Ciliogenesis	Smoothened fused to GFP, DAPI	Human RPE	~8,000	Automated	1	Mean and standard deviation	Kim et al. 2010

defects resulting from chromosome segregation problems from cytokinesis defects directly caused by gene knock-downs. Many fluorescent assays that have been developed to monitor biochemical reactions in live cells (VanEngelenburg and Palmer 2008, Johnsson 2009) could, in principle, be applied to large-scale phenotyping experiments. Although robustness of live cell assays is an issue when scaling up, the main obstacle to the production of time-resolved phenotypes lies with the more complex data processing and analysis that this type of experiments requires.

Cell Arrays

Given the size of RNAi libraries, any systematic phenotyping study requires processing hundreds to tens of thousands of samples. While plate readers are adequate for simple assays, more detailed phenotyping uses a light microscope. However, standard light microscopes have stages that are typically designed to accommodate one small glass slide but the traditional one-sample-per-slide approach clearly doesn't have the required throughput. To increase throughput, stages have been devised that can hold multi-well plates. However, plates still have a limited density of wells with most studies using plates with either 96 or 384 wells. Alternatively, multiple samples can be placed on a standard microscopy slide. In these so-called cell arrays, cells are grown on small circular spots arranged at regular interval on a glass slide (Oode et al. 2000, Ziauddin and Sabatini 2001). The spots are defined by coating the glass surface with a substance promoting cell adhesion, typically gelatine, and eventually containing perturbation reagents such as transfection media with cDNAs (Ziauddin and Sabatini 2001) or interfering RNAs (Erfle et al. 2004). Although a high density of spots can be achieved using contact spotters, available technology commonly produces arrays of 384 spots per slide mirroring the 384-well plates used by liquid-handling robots. Throughput can be further increased by using adapters that can hold multiple slides on the microscope stage. Cell arrays require less material than multi-well plates which is particularly advantageous when using costly chemicals such as transfection reagents. In cell arrays, cells are grown in the same environment (a single culture chamber) with the only differences resulting from the perturbation (e.g., transfection) associated with each spot. This results in a more homogeneous treatment of the cells than in multi-well plates where well to well variations and inhomogeneities across the plate have often been problems that could complicate statistical data analysis, although these issues with plates can often be controlled by using an automatic cell dispenser and by careful optimization of the staining procedures.

Concerns about cross-spot contamination by migrating cells have led to cell arrays with widely spaced spots (500 µm to 1.1 mm), therefore limiting spot density. Recent developments of the cell array technology have tried to address concerns about cross-spot contamination by creating arrays with microwells (Reymann et al. 2009) or reducing cell adhesion in interspot regions (Rantala et al. 2011) leading to larger arrays and a further increase in spot density. However, a recent study found that cross-spot contamination in cell arrays used for solid-phase reverse transfection doesn't seem to occur through cell migration but most likely at the seeding stage when cells can get transfected before settling down and adhering at their final position (Fengler et al. 2012). Nonetheless, the level of interspot contamination observed appeared low (up to 7% in the case of HeLa cells) and cell line-specific, presumably in proportion to the time it takes for cells to become fully adherent. In this study, levels of contamination between spots could be controlled experimentally by a carefully timed post-seeding washing step. Depending on the assay, appropriate statistical analysis could also be used such as detecting plate-specific clusters of phenotypes. The type of assay is also a marginal consideration affecting cell array density because the desired useful number of cells per spot is governed by the combination of spot size, the sizes of the cells and their proliferation properties under the experimental conditions tested. For example, observing a low-frequency event such as mitosis in an unsynchronized population would require looking at more cells which in turn may require a larger spot size.

Automated Light Microscopy

Cell arrays reduce the time taken by sample manipulation by allowing multiple samples on the microscope stage but the objective still has to be positioned and focused on each sample in turn which is also quite time-consuming when done manually. In addition, since high-magnification objectives have a reduced field of view, imaging multiple positions inside each spot or well may be needed to increase the number of recorded cells. Since most microscopes in common use already had motorized stages by the time automation was becoming necessary, a natural development was to implement software to control the stage and automatically direct the objective to each position of the plate or slide that requires imaging and to coordinate stage movement with image acquisition. This type of software is now commonly available on wide-field microscopes used in screening applications. Selection of fluorescence filters can also be automated to allow image acquisition in multiple channels at each position. Automatic focus control is also now part of most systems. Most automated microscopes use image-based autofocusing in which the focal plane is determined by analyzing z-stacks of images (Geusebroek et al. 2000, Shen et al. 2006). This approach has the two disadvantages of exposing the samples to potentially damaging amounts of light and of being slow. This is not much of concern with fixed samples but, in time-lapse imaging of live cells, software autofocusing is usually used only once before starting image acquisition with the hope that stage drift will be minimal during the course of the experiment. A less common but faster and less-damaging alternative is hardware-based autofocusing which locates the bottom of the slide or plate by the light it reflects. The cell's focal plane is then found by addition of a predefined z-offset.

In more recent developments, image analysis has been coupled to microscope control to enable complex image-based assays (Conrad et al. 2011). In this configuration, low-resolution images of the samples are analyzed online and their content automatically classified. If an object of interest is detected, the software reconfigures the microscope to execute a more complex imaging routine; otherwise it continues scanning the samples. When coupled with automatic liquid handling on the microscope stage, the system can be adapted to automatically change the cell culture medium as well as simultaneously reconfiguring the microscope upon detection of a cellular event of interest.

While the automations described above are currently used for epifluorescence imaging, other imaging modalities such as fluorescence correlation spectroscopy (Kim et al. 2007) will also be automated in the future.

Automated Image Analysis

Although automated computational processing of fluorescence microscopy images had already been introduced to the cell biology community in 1998 as a way of automating identification of subcellular structures in mammalian cells (Boland et al. 1998), the standard approach to image-based phenotyping still relied for a long time on trained biologists to browse image sets in search of objects or events of interest and annotating them for phenotypes. To this day, manual annotation is still considered a gold standard despite being prone to errors and biases. In particular, many biologists are unaware of various phenomena that can influence the outcome of any visually driven annotation process such as sequential effects by which the current annotation is strongly biased by previously seen images (Miller 1956, Stewart et al. 2005). In addition, visual inspection doesn't produce consistent data because two biologists can often produce different annotations of the same images. Finally, manual annotation is not quantitative because the human eye is poor at detecting quantitative changes of low magnitude. It is clear for example that the assay described in Simpson et al. (2012), which consists in evaluating the fraction of a fluorescent marker that reaches the cell surface by quantitatively comparing marker intensity at the cell surface to the total intensity, cannot be reliably scored visually.

However, the main reason for not processing data manually is simply that it is a costly approach that doesn't scale because automated microscopes produce images faster than can be annotated by a person. For example, the study of Neumann et al. (2010) produced close to 19 million images. Spending just one minute on each image, a biologist would need over 30 years to look at all of them. This observation leads to the development of software for automatic analysis of microscopy images (reviewed by Eliceiri et al. 2012) and the development of the field of bioimage informatics. A typical automated image-processing pipeline starts by some preprocessing steps involving noise reduction, sharpness enhancement and background subtraction followed by segmentation of the objects of interest such as nuclei or cells. Because it defines the objects that will be analyzed, segmentation is an important part of image analysis pipelines. In analyzing cellular images, segmentation is usually done by thresholding methods with various refinements to distinguish touching objects. Application of this simple approach is facilitated by the use of adequate markers, such as DNA dyes, that provide well-defined object boundaries. For more complex situations, supervized machine learning methods can be applied to classify individual pixels or voxels into different objects (Sommer et al. 2011). In the next steps, each object is described by a vector of quantitative features. Although the field started with using a small number of relatively simple features such as area and average intensity (e.g., Perlman et al. 2004, Conrad et al. 2004), later works introduced more complex expressions to describe shapes, textures and granulometries (Carpenter et al. 2006, Walter et al. 2010). Most image analysis software can compute several hundred features per object. While many of them are clearly redundant, the assumption is that using a rich combination of features is useful because it allows detection of hidden or complex phenomena. However, this has the potential to create a "large p, small n" problem where algorithms used in subsequent analysis can become unstable. Avoiding this situation requires using a sufficient number of objects (i.e., on the order of the number of features used) or applying a dimensionality reduction algorithm.

Phenotypic Profiling

When not reduced to counting objects or measuring global fluorescence intensity, automated image annotation has to deal with producing quantitative descriptions of the images. Such descriptions take the form of profiles which are multidimensional vectors providing a summary of the objects derived from each image or sample. Profiling is particularly useful when the phenotypes are not known *a priori* as, for example, in drug characterization. Another advantage of phenotypic profiling with a large number of descriptors is that it potentially reduces the impact of off-target effects in RNAi screens because it seems unlikely that different genes would affect all measured parameters in the same way. There are multiple ways of generating phenotypic profiles, the simplest being the generation of a statistical summary such as the mean for each feature. As a more robust alternative to the mean, Perlman et al. (2004) proposed to use the nonparametric Kolmogorov–Smirnov statistic against a control population of cells to summarize each measurement. However, such summaries discard potentially useful information present in the heterogeneity of the cell population. To remedy this situation, Slack et al. (2008) proposed to characterize each cell population as a mixture of Gaussian distributions. Other unsupervised machine learning algorithms can also be applied to cluster segmented objects into categories. The main problem with these approaches resides in the potential difficulty in assigning biological meaning to the categories generated by the algorithms. Despite this, unsupervized learning methods have already been successfully applied to the detection of mitotic phases (Zhong et al. 2012).

Image annotation is also often a classification problem where biologists decide to which category an object belongs. For example, annotations might be concerned with whether a particular marker is nuclear or whether the cell is in a particular phase of the cell cycle. So profiles could be generated from user-defined classes instead of automatically determined categories. This approach is also amenable to automation through the use of pattern recognition algorithms. The most common methods used for this task apply supervized machine learning algorithms to produce classifiers and the most popular algorithm used for this is the support vector machine (Ben-Hur et al. 2008). Classifiers rely on user-supplied examples of the different classes to learn feature values that discriminate between the different classes and apply these to the classification of unlabeled objects. A good representation of phenotypic variability in the training set is critical for classifier performance with well-trained classifiers having accuracies in the range of 80–90%. The need for a representative training set is a drawback of the supervized machine learning methods because it often requires extensive manual annotation of segmented objects by trained biologists. This can be a lengthy iterative process in large projects where originally unknown or subtle phenotypes have to be accounted for (Jones et al. 2009).

The main use of phenotypic profiles relies on the assumption that similar phenotypes have similar causes. For example, in drug screening, uncharacterized compounds are predicted to have a similar mechanism of action as a characterized compound with a similar profile. In RNAi-induced loss-of-function experiments, similar phenotypes are taken as indicative of functional relationships between the targeted genes. This has been successfully used to predict gene function from similar phenotypic profiles in genome-wide RNAi screens (Fuchs et al. 2010, Neumann et al. 2010).

Using phenotypic profiles to group together similar samples is typically done through the application of clustering algorithms. Efficient clustering relies on the use of a metric to evaluate similarity or dissimilarity between profiles. There are many metrics to choose from to compare phenotypic profiles, the most commonly used being the Euclidean distance. However, Fuchs et al. (2010) reasoned that profiles of interacting proteins should be closer than those of non-interacting proteins. They use this as side information to design a metric in an approach called distance metric learning (Xing et al. 2003).

Clustering high-dimensional profiles can be problematic because of the phenomenon of distance concentration. This phenomenon is the counter-intuitive observation that as the number of dimensions increases the distance between data points tends towards a constant or, put another way, the distance between the farthest and the closest points tends to 0 which would obviously render any comparison meaningless (Kabàn 2012). In practice, this problem is often mitigated by the existence of some structure in the data and the choice of a metric, such as the cosine similarity, known to concentrate at a slow rate. However, when the profiles contain a large fraction of dimensions generated by independent and identically distributed variables, a dimensionality reduction or feature selection step becomes necessary before comparing profiles.

Comparing time-resolved phenotypic profiles is relatively less straightforward as it involves computing a measure of similarity between multidimensional time series. A classical method for comparing time series is dynamic time warping, an algorithm originally used for word recognition (Rabiner and Juang 1993, Müller 2007). Dynamic time warping finds the best alignment between two series by locally stretching or compressing them to make them look as similar as possible. With dynamic time warping, the distance between the aligned series is computed as the sum of the distances between the aligned points. In its original form, dynamic time warping is relatively expensive to compute. Alternatively, Walter et al. (2010) view time series of phenotypic profiles as trajectories that can be approximated by a low number of vectors in a multidimensional space. To compute pairwise distances, the trajectories are first approximated by two vectors using a least squares error approach then a Hausdorff-type distance between the pairs of vectors is computed. Although faster to compute, this method requires the user to make assumptions about the data in order to select the appropriate number of vectors.

Despite the large number of available clustering algorithms, most studies use agglomerative hierarchical clustering. This simple method starts by assigning each sample to its own cluster then iteratively merges the closest clusters until all samples are in one cluster. As no decision is made by the algorithm to isolate defined clusters, the output is visualized as a dendrogram from which users can select clusters. As an alternative to dendrograms, clusters of samples have also been visually selected from 2D graphs using self-organizing maps (Bakal et al. 2007) or multidimensional scaling (Fuchs et al. 2010).

Statistical Data Analysis

The combination of automated microscopy and automated image analysis can produce large data sets. Deriving meaningful conclusions from such data may not be obvious. Even when the output is a single number per image, one would still want to objectively characterize the underlying phenotype. In a screening context, the goal of this analysis step is to identify treatments resulting in interesting phenotypes, usually called 'hits', a term inherited from the pharmaceutical industry drug screening activity. In many cases, phenotyping is reduced to an outlier detection problem: hits are identified because they produce images with a content sufficiently different from a reference. This reference is usually a negative control or treatment which is known not to affect the process under observation. In RNAi screens, negative control treatments typically use double stranded RNAs that do not target any gene.

Because outliers can be caused by artefacts unrelated to the process being studied, it is important to identify and correct problems in the data before proceeding with statistical analysis. Although it could be automated, quality control relies heavily on visual inspection of the controls on each array or plate and on global visualization methods such as heatmap overviews of plates to detect strong biases like edge effects.

Assuming that quality control has been performed, the next step in data analysis is to make all data comparable using a normalization procedure that removes systematic errors (Malo et al. 2006, Wiles et al. 2008, Birmingham et al. 2009, Dragiev et al. 2012). It is important to note that statistical analysis is greatly facilitated by proper experimental design. For example, random distribution of treatments across arrays and plates is recommended to control biases introduced when similar reagents are grouped together. At this stage, statistics such as the Z-factor (Zhang et al. 1999) can be used to assess assay performance if suitable negative and positive controls are available. Because the Z-factor measures the spread of values between negative and positive controls, it is only meaningful when the strength of the positive controls is close to what is expected from the, as yet unidentified, hits. The definition of hits is extremely variable because it generally consists in deciding which values of the measurements are different enough from negative controls to be considered biologically meaningful. Although some biologists limit further analysis to a small number at the top of a ranked list, systematic approaches require exhaustive and objective identification of meaningful phenotypes. To this end, a diverse array of statistical methods is available to partition the data into hits and non-hits (Malo et al. 2006, Birmingham et al. 2009) with the Z-score being the most widely used. When phenotyping is limited to a single variable, most hits are expected to be false positives (Stone et al. 2007) due to experimental errors and to treatments affecting the assay by mechanisms not related to the process being studied. Experimental errors can be controlled by using independent replicates of each treatment although many researchers find it impractical for cost or logistic reasons. The second cause of false positives can be best controlled by using multidimensional phenotypic descriptors (Stone et al. 2007) because treatments affecting unrelated processes will be less likely to generate the same phenotypic output over multiple readouts.

One of the difficulties of statistical analysis in a phenotyping context is that statistically significant doesn't always mean biologically relevant. Because assessing the biological relevance of the original findings usually requires costly and lengthy follow-up experiments, practitioners have been concerned about false positives but no study has yet provided a systematic evaluation of its false positive rate. Instead, researchers devise so-called validated hit lists which they believe to be enriched in true positives. Although one could think that the validation process would directly address biological relevance, in practice, validation refers to methods used to increase confidence in the causal relationship between treatments and phenotypes. In RNAi screens, the most popular validation method is to demonstrate reproducibility of the phenotype(s) with more than one interfering RNA targeting the same gene.

In contrast to false positive rates, false negative rates have attracted considerably less attention but a few attempts have been made to quantify them. For RNAi screens in *Drosophila* cells, false negative rates are estimated at between 8 and 34% (Liu et al. 2009, Booker et al. 2011) while one RNAi screen in HeLa cells has an estimated false negative rate of 25% (Neumann et al. 2010). The immediate economic cost of false negatives is often neglected because it is that of the experiments done without producing any result but a more pernicious cost to the community is that high false negative rates contribute to the low overlap between studies purporting to investigate the same biological process which some view as an argument against systematic approaches. Therefore, there could be value in reanalysing and integrating data from published studies. Nonetheless, at the moment, no genome-wide study can be considered exhaustive and may have to be replicated at some point in one form or another. Demonstrably low false negative rates would reduce the cost associated with these additional rounds of screening. It is worth noting that being too stringent with threshold values to reduce false positive rates can dramatically increase false negative rates. In some cases, false negative rates could also be lowered by using independent experimental replicates as these would allow detection of comparatively weaker but reproducible effects. Contrary to a widely held assumption, the cost of replicates in large-scale screens is only a fraction of the first pass because most reagents are bought in bulk quantities largely sufficient for several independent replicates.

Conclusion

Pipelines to systematically phenotype cells using genome-wide RNAi libraries combined with automated microscopy and image analysis are now in place. However, most large-scale projects published so far can still be considered proof-of-concept studies. With vendors now offering integrated solutions covering a number of relatively simple use cases and dedicated facilities available at some institutions to support building solutions for more complex projects, one can expect high-throughput microscopy to become more common and applied to the systematic characterization of more cellular processes. To promote dissemination and reuse of the resulting data, these studies will have to be integrated into well-structured dedicated repositories offering access to everything from images to their annotation. The current experimental approaches do, however, mostly provide lists of parts by identifying required genes. While this is a necessary step, most cellular processes are dynamic and their characterization would benefit from phenotypes with better temporal information and higher spatial resolution. One could also envision bypassing the perturbation experiment altogether and instead characterize biological processes by correlating genetic variations with cellular phenotypes in experiments combining genomic sequencing, population genetics and phenotyping of primary cells. Looking further ahead, a systems approach to cell biology would also benefit from quantitative information on cellular components. In essence, this requires answering the following questions: how much of what interacts with what, where and when? These questions could be answered using quantitative light microscopy modalities like fluorescence correlation spectroscopy and fluorescence cross-correlation spectroscopy. Automating these imaging modalities would open the door to systematic quantitative cell phenotyping.

Acknowledgements

The author thanks Dr. Beate Neumann for many discussions and critically reading the manuscript.

References Cited

Adamson, B., A. Smogorzewska, F.D. Sigoillot, R.W. King and S.J. Elledge. 2012. A genome-wide homologous recombination screen identifies the RNA-binding protein RBMX as a component of the DNA-damage response. Nat. Cell Biol. 14(3): 318–28.

Agaisse, H., L.S. Burrack, J.A. Philips, E.J. Rubin, N. Perrimon and D.E. Higgins. 2005. Genome-wide RNAi screen for host factors required for intracellular bacterial infection. Science 309(5738): 1248–51.

Bai, J., K.J. Sepp and N. Perrimon. 2009. Culture of *Drosophila* primary cells dissociated from gastrula embryos and their use in RNAi screening. Nat. Protoc. 4(10): 1502–12.

Bakal, C., J. Aach, G.M. Church and N. Perrimon. 2007. Quantitative morphological signatures define local signaling networks regulating cell morphology. Science 316(5832): 1753–6.

Ben-Hur, A., C.S. Ong, S. Sonnenburg, B. Schölkopf and G. Rätsch. 2008. Support vector machines and kernels for computational biology. PLoS Comput. Biol. 4(10): e1000173.

Birmingham, A., L.M. Selfors, T. Forster, D. Wrobel, C.J. Kennedy, E. Shanks, J. Santoyo-Lopez, D.J. Dunican, A. Long, D. Kelleher, Q. Smith, R.L. Beijersbergen, P. Ghazal and C.E. Shamu. 2009. Statistical methods for analysis of high-throughput RNA interference screens. Nat. Methods 6(8): 569–75.

Boland, M.V., M.K. Markey and R.F. Murphy. 1998. Automated recognition of patterns characteristic of subcellular structures in fluorescence microscopy images. Cytometry 33(3): 366–75.

Booker, M., A.A. Samsonova, Y. Kwon, I. Flockhart, S.E. Mohr and N. Perrimon. 2011. False negative rates in *Drosophila* cell-based RNAi screens: a case study. BMC Genomics 12: 50.

Boutros, M., A.A. Kiger, S. Armknecht, K. Kerr, M. Hild, B. Koch, S.A. Haas, R. Paro and N. Perrimon. 2004. Genome-wide RNAi analysis of growth and viability in *Drosophila* cells. Science 303(5659): 832–5.

Brass, A.L., D.M. Dykxhoorn, Y. Benita, N. Yan, A. Engelman, R.J. Xavier, J. Lieberman and S.J. Elledge. 2008. Identification of host proteins required for HIV infection through a functional genomic screen. Science 319(5865): 921–6.

Carpenter, A.E., T.R. Jones, M.R. Lamprecht, C. Clarke, I.H. Kang, O. Friman, D.A. Guertin, J.H. Chang, R.A. Lindquist, J. Moffat, P. Golland and D.M. Sabatini. 2006. CellProfiler: image analysis software for identifying and quantifying cell phenotypes. Genome Biol. 7(10): R100.

Conrad, C., H. Erfle, P. Warnat, N. Daigle, T. Lörch, J. Ellenberg, R. Pepperkok and R. Eils. 2004. Automatic identification of subcellular phenotypes on human cell arrays. Genome Res. 14(6): 1130–6.

Conrad, C., A. Wünsche, T.H. Tan, J. Bulkescher, F. Sieckmann, F. Verissimo, A. Edelstein, T. Walter, U. Liebel, R. Pepperkok and J. Ellenberg. 2011. Micropilot: automation of fluorescence microscopy-based imaging for systems biology. Nat. Methods 8(3): 246–9.

DasGupta, R., A. Kaykas, R.T. Moon and N. Perrimon. 2005. Functional genomic analysis of the Wnt-wingless signaling pathway. Science 308(5723): 826–33.

Doil, C., N. Mailand, S. Bekker-Jensen, P. Menard, D.H. Larsen, R. Pepperkok, J. Ellenberg, S. Panier, D. Durocher, J. Bartek, J. Lukas and C. Lukas. 2009. RNF168 binds and amplifies ubiquitin conjugates on damaged chromosomes to allow accumulation of repair proteins. Cell 136(3): 435–46.

Dragiev, P., R. Nadon and V. Makarenkov. 2012. Two effective methods for correcting experimental high-throughput screening data. Bioinformatics 28(13): 1775–82.

Draviam, V.M., F. Stegmeier, G. Nalepa, M.E. Sowa, J. Chen, A. Liang, G.J. Hannon, P.K. Sorger, J.W. Harper and S.J. Elledge. 2007. A functional genomic screen identifies a role for TAO1 kinase in spindle-checkpoint signalling. Nat. Cell Biol. 9(5): 556–64.

Eliceiri, K.W., M.R. Berthold, I.G. Goldberg, L. Ibáñez, B.S. Manjunath, M.E. Martone, R.F. Murphy, H. Peng, A.L. Plant, B. Roysam, N. Stuurman, J.R. Swedlow, P. Tomancak and A.E. Carpenter. 2012. Biological imaging software tools. Nat. Methods 9(7): 697–710.

Erfle, H., J.C. Simpson, P.I. Bastiaens and R. Pepperkok. 2004. siRNA cell arrays for high-content screening microscopy. Biotechniques 37(3): 454–8, 460, 462.

Fengler, S., P.I. Bastiaens, H.E. Grecco and P. Roda-Navarro. 2012. Optimizing cell arrays for accurate functional genomics. BMC Res. Notes 5(1): 358.

Fisher, K.H., V.M. Wright, A. Taylor, M.P. Zeidler and S. Brown. 2012. Advances in genome-wide RNAi cellular screens: a case study using the *Drosophila* JAK/STAT pathway. BMC Genomics 13: 506.

Foley, E. and P.H. O'Farrell. 2004. Functional dissection of an innate immune response by a genome-wide RNAi screen. PLoS Biol. 2(8): E203.

Friedman, A. and N. Perrimon. 2006. A functional RNAi screen for regulators of receptor tyrosine kinase and ERK signalling. Nature 444(7116): 230–4.

Fuchs, F., G. Pau, D. Kranz, O. Sklyar, C. Budjan, S. Steinbrink, T. Horn, A. Pedal, W. Huber and M. Boutros. 2010. Clustering phenotype populations by genome-wide RNAi and multiparametric imaging. Mol. Syst. Biol. 6: 370.

Gantier, M.P. and B.R. Williams. 2007. The response of mammalian cells to double-stranded RNA. Cytokine Growth Factor Rev. 18(5–6): 363–71.

Geusebroek, J.M., F. Cornelissen, A.W. Smeulders and H. Geerts. 2000. Robust autofocusing in microscopy. Cytometry 39(1): 1–9.

Gewurz, B.E., F. Towfic, J.C. Mar, N.P. Shinners, K. Takasaki, B. Zhao, E.D. Cahir-McFarland, J. Quackenbush, R.J. Xavier and E. Kieff. 2012. Genome-wide siRNA screen for mediators of NF-κB activation. Proc. Natl. Acad. Sci. USA 109(7): 2467–72.

Goshima, G., R. Wollman, S.S. Goodwin, N. Zhang, J.M. Scholey, R.D. Vale and N. Stuurman. 2007. Genes required for mitotic spindle assembly in *Drosophila* S2 cells. Science 316(5823): 417–21.

Gudjonsson, T., M. Altmeyer, V. Savic, L. Toledo, C. Dinant, M. Grøfte, J. Bartkova, M. Poulsen, Y. Oka, S. Bekker-Jensen, N. Mailand, B. Neumann, J.-K. Hériché, R. Shearer, D. Saunders, J. Bartek, J. Lukas and C. Lukas. 2012. TRIP12 and UBR5 suppress spreading of chromatin ubiquitylation at damaged chromosomes. Cell 150(4): 697–709.

Hao, L., A. Sakurai, T. Watanabe, E. Sorensen, C.A. Nidom, M.A. Newton, P. Ahlquist and Y. Kawaoka. 2008. *Drosophila* RNAi screen identifies host genes important for influenza virus replication. Nature 454(7206): 890–3.

Horn, T., T. Sandmann, B. Fischer, E. Axelsson, W. Huber and M. Boutros. 2011. Mapping of signaling networks through synthetic genetic interaction analysis by RNAi. Nat. Methods 8(4): 341–6.

Hübner, N.C., L.H. Wang, M. Kaulich, P. Descombes, I. Poser and E.A. Nigg. 2010. Re-examination of siRNA specificity questions role of PICH and Tao1 in the spindle checkpoint and identifies Mad2 as a sensitive target for small RNAs. Chromosoma 119(2): 149–65.

Hurov, K.E., C. Cotta-Ramusino and S.J. Elledge. 2010. A genetic screen identifies the Triple T complex required for DNA damage signaling and ATM and ATR stability. Genes Dev. 24(17): 193950.

Hutchins, J.R., Y. Toyoda, B. Hegemann, I. Poser, J.-K. Hériché, M.M. Sykora, M. Augsburg, O. Hudecz, B.A. Buschhorn, J. Bulkescher, C. Conrad, D. Comartin, A. Schleiffer, M. Sarov, A. Pozniakovsky, M.M. Slabicki, S. Schloissnig, I. Steinmacher, M. Leuschner, A. Ssykor, S. Lawo, L. Pelletier, H. Stark, K. Nasmyth, J. Ellenberg, R. Durbin, F. Buchholz, K. Mechtler, A.A. Hyman and J.M. Peters. 2010. Systematic analysis of human protein complexes identifies chromosome segregation proteins. Science 328(5978): 593–9.

Inglese, J., R.L. Johnson, A. Simeonov, M. Xia, W. Zheng, C.P. Austin and D.S. Auld. 2007. High-throughput screening assays for the identification of chemical probes. Nat. Chem. Biol. 3(8): 466–79.

Johnsson, K. 2009. Visualizing biochemical activities in living cells. Nat. Chem. Biol. 5(2): 63–5.

Jones, T.R., A.E. Carpenter, M.R. Lamprecht, J. Moffat, S.J. Silver, J.K. Grenier, A.B. Castoreno, U.S. Eggert, D.E. Root, P. Golland and D.M. Sabatini. 2009. Scoring diverse cellular morphologies in image-based screens with iterative feedback and machine learning. Proc. Natl. Acad. Sci. USA 106(6): 1826–31.

Kabàn, A. 2012. Non-parametric detection of meaningless distances in high dimensional data. Stat. Comput. 22(2): 375–85.

Karlas, A., N. Machuy, Y. Shin, K.P. Pleissner, A. Artarini, D. Heuer, D. Becker, H. Khalil, L.A. Ogilvie, S. Hess, A.P. Mäurer, E. Müller, T. Wolff, T. Rudel and T.F. Meyer. 2010. Genome-wide RNAi screen identifies human host factors crucial for influenza virus replication. Nature 463(7282): 818–22.

Kiger, A.A., D. Baum, S. Jones, M.R. Jones, A. Coulson, C. Echeverri and N. Perrimon. 2003. A functional genomic analysis of cell morphology using RNA interference. J. Biol. 2(4): 27.

Kim, S.A., K.G. Heinze and P. Schwille. 2007. Fluorescence correlation spectroscopy in living cells. Nat. Methods 4(11): 963–73.

Kim, J., J.E. Lee, S. Heynen-Genel, E. Suyama, K. Ono, K. Lee, T. Ideker, P. Aza-Blanc and J.G. Gleeson. 2010. Functional genomic screen for modulators of ciliogenesis and cilium length. Nature 464(7291): 1048–51.

Kittler, R., G. Putz, L. Pelletier, I. Poser, A.K. Heninger, D. Drechsel, S. Fischer, I. Konstantinova, B. Habermann, H. Grabner, M.L. Yaspo, H. Himmelbauer, B. Korn, K. Neugebauer, M.T. Pisabarro and F. Buchholz. 2004. An endoribonuclease-prepared siRNA screen in human cells identifies genes essential for cell division. Nature 432(7020): 1036–40.

Kittler, R., L. Pelletier, C. Ma, I. Poser, S. Fischer, A.A. Hyman and F. Buchholz. 2005. RNA interference rescue by bacterial artificial chromosome transgenesis in mammalian tissue culture cells. Proc. Natl. Acad. Sci. USA 102(7): 2396–401.

Kittler, R., V. Surendranath, A.K. Heninger, M. Slabicki, M. Theis, G. Putz, K. Franke, A. Caldarelli, H. Grabner, K. Kozak, J. Wagner, E. Rees, B. Korn, C. Frenzel, C. Sachse, B. Sönnichsen, J. Guo, J. Schelter, J. Burchard, P.S. Linsley, A.L. Jackson, B. Habermann and F. Buchholz. 2007. Genome-wide resources of endoribonuclease-prepared short interfering RNAs for specific loss-of-function studies. Nat. Methods 4(4): 337–44.

Kondo, S. and N. Perrimon. 2011. A genome-wide RNAi screen identifies core components of the G_2-M DNA damage checkpoint. Sci. Signal 4(154): rs1.

Ley, S., A. Weigert, J.-K. Hériché, B. Mille-Baker, R.A. Janssen and B. Brüne. 2013. RNAi screen in apoptotic cancer cell-stimulated human macrophages reveals co-regulation of IL-6/IL-10 expression. Immunobiology 218(1): 40–51.

Li, J., J.Y. Newberg, M. Uhlén, E. Lundberg and R.F. Murphy. 2012. Automated analysis and reannotation of subcellular locations in confocal images from the human protein atlas. PLoS One 7(11): e50514.

Liu, T., D. Sims and B. Baum. 2009. Parallel RNAi screens across different cell lines identify generic and cell type-specific regulators of actin organization and cell morphology. Genome Biol. 10: R26.

Lovejoy, C.A., X. Xu, C.E. Bansbach, G.G. Glick, R. Zhao, F. Ye, B.M. Sirbu, L.C. Titus, Y. Shyr and D. Cortez. 2009. Functional genomic screens identify CINP as a genome maintenance protein. Proc. Natl. Acad. Sci. USA 106(46): 19304–9.

Lum, L., S. Yao, B. Mozer, A. Rovescalli, D. Von Kessler, M. Nirenberg and P.A. Beachy. 2003. Identification of Hedgehog pathway components by RNAi in *Drosophila* cultured cells. Science 299(5615): 2039–45.

Luo, J., M.J. Emanuele, D. Li, C.J. Creighton, M.R. Schlabach, T.F. Westbrook, K.K. Wong and S.J. Elledge. 2009. A genome-wide RNAi screen identifies multiple synthetic lethal interactions with the Ras oncogene. Cell 137(5): 835–48.

Malo, N., J.A. Hanley, S. Cerquozzi, J. Pelletier and R. Nadon. 2006. Statistical practice in high-throughput screening data analysis. Nat. Biotechnol. 24(2): 167–75.

Mehrle, A., H. Rosenfelder, I. Schupp, C. del Val, D. Arlt, F. Hahne, S. Bechtel, J. Simpson, O. Hofmann, W. Hide, K.H. Glatting, W. Huber, R. Pepperkok, A. Poustka and S. Wiemann. 2006. The LIFEdb database in 2006. Nucleic Acids Res. 34(Database issue): D415–8.

Meister, G. and T. Tuschl. 2004. Mechanisms of gene silencing by double-stranded RNA. Nature 431(7006): 343–9.

Mercer, J., B. Snijder, R. Sacher, C. Burkard, C.K. Bleck, H. Stahlberg, L. Pelkmans and A. Helenius. 2012. RNAi screening reveals proteasome- and Cullin3-dependent stages in vaccinia virus infection. Cell Rep. 2(4): 1036–47.

Miller, G.A. 1956. The magical number seven, plus or minus two: some limits on our capacity for information processing. Psychological Review 63(2): 81–97.

Misselwitz, B., S. Dilling, P. Vonaesch, R. Sacher, B. Snijder, M. Schlumberger, S. Rout, M. Stark, C. von Mering, L. Pelkmans and W.D. Hardt. 2011. RNAi screen of Salmonella invasion shows role of COPI in membrane targeting of cholesterol and Cdc42. Mol. Syst. Biol. 7: 474.

Moudry, P., C. Lukas, L. Macurek, B. Neumann, J.-K. Hériché, R. Pepperkok, J. Ellenberg, Z. Hodny, J. Lukas and J. Bartek. 2012. Nucleoporin NUP153 guards genome integrity by promoting nuclear import of 53BP1. Cell Death Differ. 19(5): 798–807.

Mukherji, M., R. Bell, L. Supekova, Y. Wang, A.P. Orth, S. Batalov, L. Miraglia, D. Huesken, J. Lange, C. Martin, S. Sahasrabudhe, M. Reinhardt, F. Natt, J. Hall, C. Mickanin, M. Labow, S.K. Chanda, C.Y. Cho and P.G. Schultz. 2006. Genome-wide functional analysis of human cell-cycle regulators. Proc. Natl. Acad. Sci. USA 103(40): 14819–24.

Müller, P., D. Kuttenkeuler, V. Gesellchen, M.P. Zeidler and M. Boutros. 2005. Identification of JAK/STAT signalling components by genome-wide RNA interference. Nature 436(7052): 871–5.

Müller, M. 2007. Information Retrieval for Music and Motion. Springer-Verlag New York, Inc. Secaucus, NJ, USA.

Myers, J.W., J.T. Chi, D. Gong, M.E. Schaner, P.O. Brown and J.E. Ferrell. 2006. Minimizing off-target effects by using diced siRNAs for RNA interference. J. RNAi Gene Silencing 2(2): 181–94.

Neumann, B., T. Walter, J.-K. Hériché, J. Bulkescher, H. Erfle, C. Conrad, P. Rogers, I. Poser, M. Held, U. Liebel, C. Cetin, F. Sieckmann, G. Pau, R. Kabbe, A. Wünsche, V. Satagopam, M.H. Schmitz, C. Chapuis, G.W. Gerlich, R. Schneider, R. Eils, W. Huber, J.M. Peters, A.A. Hyman, R. Durbin, R. Pepperkok and J. Ellenberg. 2010. Phenotypic profiling of the human genome by time-lapse microscopy reveals cell division genes. Nature 464(7289): 721–7.

Oode, K., T. Furuya, K. Harada, S. Kawauchi, K. Yamamoto, T. Hirano and K. Sasaki. 2000. The development of a cell array and its combination with laser-scanning cytometry allows a high-throughput analysis of nuclear DNA content. Am. J. Pathol. 157(3): 723–8.

Perlman, Z.E., M.D. Slack, Y. Feng, T.J. Mitchison, L.F. Wu and S.J. Altschuler. 2004. Multidimensional drug profiling by automated microscopy. Science 306(5699): 1194–8.

Rabiner, L. and B.H. Juang. 1993. Fundamentals of Speech Recognition. Prentice-Hall, Upper Saddle River, NJ, USA.

Rantala, J.K., R. Mäkelä, A.R. Aaltola, P. Laasola, J.P. Mpindi, M. Nees, P. Saviranta and O. Kallioniemi. 2011. A cell spot microarray method for production of high density siRNA transfection microarrays. BMC Genomics 12: 162.

Reymann, J., N. Beil, J. Beneke, P.P. Kaletta, K. Burkert and H. Erfle. 2009. Next-generation 9216-microwell cell arrays for high-content screening microscopy. Biotechniques 47(4): 877–8.

Shen, F., L. Hodgson and K. Hahn. 2006. Digital autofocus methods for automated microscopy. Methods Enzymol. 414: 620–32.

Sheng, Z., L. Li, L.J. Zhu, T.W. Smith, A. Demers, A.H. Ross, R.P. Moser and M.R. Green. 2010. A genome-wide RNA interference screen reveals an essential CREB3L2-ATF5-MCL1 survival pathway in malignant glioma with therapeutic implications. Nat. Med. 16(6): 671–7.

Shindoh, N., A. Yoda, Y. Yoda, T.J. Sullivan, O. Weigert, A.A. Lane, N. Kopp, L. Bird, S.J. Rodig, E.A. Fox and D.M. Weinstock. 2012. Next-generation cDNA screening for oncogene and resistance phenotypes. PLoS One 7(11): e49201.

Simpson, J.C., B. Joggerst, V. Laketa, F. Verissimo, C. Cetin, H. Erfle, M.G. Bexiga, V.R. Singan, J.-K. Hériché, B. Neumann, A. Mateos, J. Blake, S. Bechtel, V. Benes, S. Wiemann, J. Ellenberg and R. Pepperkok. 2012. Genome-wide RNAi screening identifies human proteins with a regulatory function in the early secretory pathway. Nat. Cell Biol. 14(7): 764–74.

Singh, D.K., C.J. Ku, C. Wichaidit, R.J. Steininger 3rd, L.F. Wu and S.J. Altschuler. 2010. Patterns of basal signaling heterogeneity can distinguish cellular populations with different drug sensitivities. Mol. Syst. Biol. 6: 369.

Slack, M.D., E.D. Martinez, L.F. Wu and S.J. Altschuler. 2008. Characterizing Heterogeneous Cellular Responses to Perturbations. Proc. Natl. Acad. Sci. USA 105(19): 19306 11.

Sommer, C., C. Straehle, U. Koethe and F.A. Hamprecht. 2011. ilastik: Interactive Learning and Segmentation Toolkit. IEEE International Symposium on Biomedical Imaging 230–33.

Stewart, N., G.D.A. Brown and N. Chater. 2005. Absolute identification by relative judgement. Psychological Review 112: 881–911.

Stone, D.J., S. Marine, J. Majercak, W.J. Ray, A. Espeseth, A. Simon and M. Ferrer. 2007. High-throughput screening by RNA interference: control of two distinct types of variance. Cell Cycle 6(8): 898–901.

Stroschein-Stevenson, S.L., E. Foley, P.H. O'Farrell and A.D. Johnson. 2006. Identification of *Drosophila* gene products required for phagocytosis of *Candida albicans*. PLoS Biol. 4(1): e4.

Tang, W., M. Dodge, D. Gundapaneni, C. Michnoff, M. Roth and A. Lum. 2008. A genome-wide RNAi screen for Wnt/beta-catenin pathway components identifies unexpected roles for TCF transcription factors in cancer. Proc. Natl. Acad. Sci. USA 105(28): 9697–702.

Thornbrough, J.M., T. Hundley, R. Valdivia and M.J. Worley. 2012. Human genome-wide RNAi screen for host factors that modulate intracellular Salmonella growth. PLoS One 7(6): e38097.

Toyoshima, M., H.L. Howie, M. Imakura, R.M. Walsh, J.E. Annis, A.N. Chang, J. Frazier, B.N. Chau, A. Loboda, P.S. Linsley, M.A. Cleary, J.R. Park and C. Grandori. 2012. Functional genomics identifies therapeutic targets for MYC-driven cancer. Proc. Natl. Acad. Sci. USA 109(24): 9545–50.

Uhlén, M., P. Oksvold, L. Fagerberg, E. Lundberg, K. Jonasson, M. Forsberg, M. Zwahlen, C. Kampf, K. Wester, S. Hober, H. Wernerus, L. Björling and F. Ponten. 2010. Towards a knowledge-based Human Protein Atlas. Nat. Biotechnol. 28(12): 1248–50.

VanEngelenburg, S.B. and A.E. Palmer. 2008. Fluorescent biosensors of protein function. Curr. Opin. Chem. Biol. 12(1): 60–5.

Walter, T., M. Held, B. Neumann, J.-K. Hériché, C. Conrad, R. Pepperkok and J. Ellenberg. 2010. Automatic identification and clustering of chromosome phenotypes in a genome wide RNAi screen by time-lapse imaging. J. Struct. Biol. 170(1): 1–9.

Wendler, F., A.K. Gillingham, R. Sinka, C. Rosa-Ferreira, D.E. Gordon, X. Franch-Marro, A.A. Peden, J.P. Vincent and S. Munro. 2010. A genome-wide RNA interference screen identifies two novel components of the metazoan secretory pathway. EMBO J. 29(2): 304–14.

Westerman, B.A., A.K. Braat, N. Taub, M. Potman, J.H. Vissers, M. Blom, E. Verhoeven, H. Stoop, A. Gillis, A. Velds, W. Nijkamp, R. Beijersbergen, L.A. Huber, L.H. Looijenga and M. van Lohuizen. 2011. A genome-wide RNAi screen in mouse embryonic stem cells identifies Mp1 as a key mediator of differentiation. J. Exp. Med. 208(13): 2675–89.

Wiles, A.M., D. Ravi, S. Bhavani and A.J. Bishop. 2008. An analysis of normalization methods for *Drosophila* RNAi genomic screens and development of a robust validation scheme. J. Biomol. Screen 13(8): 777–84.

Winograd-Katz, S.E., S. Itzkovitz, Z. Kam and B. Geiger. 2009. Multiparametric analysis of focal adhesion formation by RNAi-mediated gene knockdown. J. Cell Biol. 186(3): 423–36.

Wu, N., D. Castel, M.A. Debily, M.A. Vigano, O. Alibert, R. Mantovani, K. Iljin, P.H. Romeo and X. Gidrol. 2011. Large scale RNAi screen reveals that the inhibitor of DNA binding 2 (ID2) protein is repressed by p53 family member p63 and functions in human keratinocyte differentiation. J. Biol. Chem. 286(23): 20870–9.

Xing, E.P., A.Y. Ng, I. Jordan and S. Russel. 2003. Distance metric learning, with application to clustering with side-information. Adv. Neural Inf. Process Syst. 15: 505–12.

Yang, S.H., T. Kalkan, C. Morrisroe, A. Smith and A.D. Sharrocks. 2012. A genome-wide RNAi screen reveals map kinase phosphatases as key ERK pathway regulators during embryonic stem cell differentiation. PLoS Genet. 8(12): e1003112.

Zhang, J.H., T.D. Chung and K.R. Oldenburg. 1999. A simple statistical parameter for use in evaluation and validation of high throughput screening assays. J. Biomol. Screen 4(2): 67–73.

Zhong, Q., A.G. Busetto, J.P. Fededa, J.M. Buhmann and D.W. Gerlich. 2012. Unsupervised modeling of cell morphology dynamics for time-lapse microscopy. Nat. Methods 9(7): 711–3.

Zhou, H., M. Xu, Q. Huang, A.T. Gates, X.D. Zhang, J.C. Castle, E. Stec, M. Ferrer, B. Strulovici, D.J. Hazuda and A.S. Espeseth. 2008. Genome-scale RNAi screen for host factors required for HIV replication. Cell Host Microbe 4(5): 495–504.

Zhou, H., G. DeLoid, E. Browning, D.J. Gregory, F. Tan, A.S. Bedugnis, A. Imrich, H. Koziel, I. Kramnik, Q. Lu and L. Kobzik. 2012. Genome-wide RNAi screen in IFN-γ-treated human macrophages identifies genes mediating resistance to the intracellular pathogen Francisella tularensis. PLoS One 7(2): e31752.

Ziauddin, J. and D.M. Sabatini. 2001. Microarrays of cells expressing defined cDNAs. Nature 411(6833): 107–10.

6

Systematic Phenotyping of Plant Development in *Arabidopsis thaliana*

Christine Granier,[1,]* Vincent Nègre[1] and Fabio Fiorani[2]

Introduction

Arabidopsis thaliana is a small flowering plant, a member of the *Brassicaceae* family with very small ellipsoid seeds (between 0.04 and 0.17 mm^2 in Herridge et al. (2011), for example). After seed germination, characterized by radicle and hypocotyl emergence, two opposite cotyledons appear and the shoot apical meristem continues the production of leaves on its flank. Successive leaves expand with a well-coordinated program combining spatial and temporal patterns of cell division and cell expansion (Gonzalez et al. 2012). Leaf primordia emerge from the shoot apical meristem and expand forming a rosette covering the soil and reaching diameters generally ranging between 1 and 10 cm depending on genotypes and cultivation

[1] Laboratoire d'Ecophysiologie des Plantes sous Stress Environnementaux UMR759, INRA-SUPAGRO, Place Viala, F-34060 Montpellier, France.
Emails: granier@supagro.inra.fr; negrev@supagro.inra.fr
[2] Institute of Bio- and Geosciences, IBG-2: Plant Sciences, Forschungszentrum Jülich GmbH, Leo-Brandt-Str., 52425 Jülich, Germany.
Email: f.fiorani@fz-juelich.de
* Corresponding author

conditions (Boyes et al. 2001). The number of leaves forming the rosette is fixed at the time when the first reproductive organs appear on the shoot apical meristem (Pouteau and Albertini 2009, Irish 2010). As a result of the elongation of the first internodes, bolting—i.e., rapid elongation of the flowering stem—occurs with the floral buds appearing in the center of the rosette. The main stem elongates to give rise to an inflorescence harboring all reproductive organs and a few cauline leaves (Gómez-Mena et al. 2001, Boyes et al. 2001). The last phase of plant development until senescence is complete includes a succession of reproductive stages from flower production to silique emergence, seed formation and silique ripening (Boyes et al. 2001). This plant cycle can be as short as one month when plants are grown in long days with 16 h to 20 h of illumination (Karlsson et al. 1993, Martinez-Zapater et al. 1994). In this respect, Arabidopsis is often described as a plant with a short life cycle. However, flowering is delayed by short-day conditions and the whole life cycle can last more than four months when plants are grown under 8 h of illumination per day (Karlsson et al. 1993, Martinez-Zapater et al. 1994). In addition, different ecotypes and mutants disrupted in photoperiod perception have a long life cycle even in long days (Lempe et al. 2005). Its small size, its simple aerial architecture, and the possibility to grow plant populations from seed to seed in less than one month in permissive conditions are big advantages to study whole plant vegetative and reproductive development and perform systematic studies with relatively large numbers of genotypes and replications. The usefulness of Arabidopsis in studies aimed at discovering the underlying genetic determinants of developmental transitions has been highlighted in many papers (Meyerowitz 1989, Meinke et al. 1998, Koornneef and Meinke 2010).

Arabidopsis root development has also been studied in detail. Based on histology and cell lineage investigation, it has been shown that the primary root axis is embryonically derived (Dolan et al. 1993). Post-embryonic development results in the formation of varying numbers of lateral roots arising from the pericycle cell layer (Dolan et al. 1993). *Arabidopsis thaliana* plants can be grown in different growth media including different types of soils, but also agar-based media and liquid media such as hydroponics systems. The latter methodologies allow both *in vivo* and *in vitro* analyses that are convenient for visualizing and measuring individual root meristems and whole root system development combined with appropriate sampling strategies (Krapp et al. 2011, De Smet et al. 2012).

Arabidopsis also offers important advantages for researches in genetics and molecular biology by its relatively small genome size compared to crops (Meinke et al. 1998, AGI 2000). Maize, for example, has a genome of approximately 2,400 Megabase pairs (Mbp), i.e., around 19 times the size of the Arabidopsis genome while the wheat genome is 16,000 Mbp, i.e., 128 times larger than the Arabidopsis one. The large and often highly repetitive crop genomes pose challenges to researchers, combining difficulty in sequencing, assembly after sequencing, as well as in the isolation and cloning of mutant loci. Most of the differences in gene number between Arabidopsis and crop species appear to result from polyploidy characteristics of several crop species' genomes, rather than from large classes of genes present in crop species that are not present in Arabidopsis. Therefore, the genes present in Arabidopsis can represent a reasonable model for the plant kingdom. Since the completion of the Arabidopsis genome sequence, gene expression has been monitored at different throughputs from specific targeted genes to whole genome-scale analyses using microarray profiling in different organs and tissue types, at different stages of plant development and in different environmental conditions (e.g., Birnbaum et al. 2003, Zeller et al. 2009, Harb et al. 2010, Skirycz et al. 2010). Extensive genetic maps of all five chromosomes, together with efficient methods for mutagenesis and plant transformation have delivered a large range of genetic and genomic resources to the plant biology research community (Alonso et al. 2003, O'Malley and Ecker 2010).

The large set of genetic resources in Arabidopsis has led different labs to develop automated high-throughput phenotyping platforms to track plant development and its response to various environmental conditions in a relatively large number of genotypes (Granier et al. 2006, Walter et al. 2007, Arvidsson et al. 2011, Zhang et al. 2012). Across these different platforms, a discrete number of phenotypic parameters can be measured routinely using non-invasive optical methods and standardized screening protocols (for recent reviews, see Furbank and Tester (2011), Fiorani et al. (2012)). These include the extraction of growth trajectories and rosette geometries via time courses of projected shoot area (Granier et al. 2006), measurements of chlorophyll fluorescence and PSII performance (Jansen et al. 2009), and measurement of leaf surface temperature with thermal cameras (Jones et al. 2009). With the knowledge gained from this model plant, it is expected to increase rapidly the identification of genes, their function and the associated processes controlling plant development and its plasticity in response to environmental challenges. The ultimate goal remains what is often referred to as 'bridging the phenotyping gap' (e.g., Houle et al. 2010, Furbank and Tester 2011) by enabling a deeper understanding of genome to phenome functional mapping for the identification of key determinants of plant productivity traits.

Frameworks of analyses as well as protocols to assess plant development are shared in the plant science community (Boyes et al. 2001, Cookson et al. 2010, Rymen et al. 2010). However, description of plant growth phenotypes is not always sufficient for reconciling datasets among groups and taking full benefits of ongoing systematic phenotyping initiatives. A first reason for this is that many phenotypic descriptors in literature are qualitative in that they address a characteristic of the plant that is not associated to a numeric value—e.g., *angustifolia* has narrow leaves whereas *rotundifolia3* has rounded leaves (Tsuge et al. 1996) and can be subjective. But, even when the phenotypes are described with quantitative data, in that they address a characteristic of the plant that is measured and supported by numeric values, the plasticity that can be encountered when the considered genotype is grown in various environmental conditions can be an obstacle in joint analyses of different datasets. An example to illustrate this point comes from a multi-site study in which three reference genotypes of *Arabidopsis thaliana* were grown in 10 laboratories using the same highly detailed protocol. The 10 labs obtained phenotypic variation within genotypes for phenotypic traits measured at the molecular, cellular and whole leaf levels, suggesting that small differences in environmental conditions and/or plant handling persisted and affected growth at different scales (Massonnet et al. 2010). This study illustrates the need for precise recording of environmental conditions, but also highlights the importance of designing protocols for measurements of phenotypic traits, to ensure replicable phenotypic plant characterization, data sharing and data comparison across different laboratories (Fabre et al. 2011).

In this chapter, we review the framework of analyses that have been developed and are publicly available, and that can be adopted towards systematic shoot and root development phenotyping in *Arabidopsis thaliana*. Here we focus primarily on the accurate description of macroscopic phenotypic traits of both shoot and root development. However, we link to studies addressing the underlying organization of cellular dynamics contributing to organ expansion. Also, we will present how some of these studies have led to a better understanding of how plants develop and respond to a varying environment often characterized by limited resources for growth. Finally, we will stress the necessity to capture and store phenotypic observations within appropriate data management systems to enable sharing common tools, and data reuse for meta-analyses. In this respect we review in particular efforts towards the adoption of controlled vocabularies and ontologies for an efficient and systematic phenotyping within a data sharing perspective aiming at quantitative descriptions of Arabidopsis phenotypes.

Overview of the Field

Systematic Shoot Growth Phenotyping in Arabidopsis thaliana

Shoot development is affected by hormonal, nutritional and environmental cues. Combined with this level of regulation are the inherent genetic differences in number, size and form of shoot organs such as leaves, flowers, inflorescence branches, siliques and seeds. These factors give rise to the extraordinary diversity in plant architecture found in nature and that is also observed within the *Arabidopsis thaliana* species both in accessions that present a large natural diversity and in mutant collections. Identifying the genetic control of shoot growth and the developmental regulators involved requires a precise characterization of growth phenotypes using robust descriptors of shoot organ's size and overall geometry, rates of production and growth rates. Because shoot development is the sum of discrete morphogenetic structures that appear at regular positions at the shoot apical meristem flanks over time, spatial and temporal descriptors are commonly used to phenotype shoot growth. In most studies, morphometric traits are recorded in adult plants but in a few cases, these datasets are completed with temporal indicators, i.e., specific time points at which visible events contributing to final shoot morphometry take place. As in other flowering plants, Arabidopsis shoot development is characterized by three distinct phases—a juvenile vegetative phase, an adult vegetative phase that is reproductively competent and a reproductive phase (Poethig 2003).

Phenotyping Leaf Production: From Initiation at the Shoot Apical Meristem to Emerging Leaves Forming a Whole Rosette

During shoot development, leaves are initiated at regular intervals as primordia at specific sites on the shoot apical meristem. The leaf primordium is defined as a group of founder cells that develop further into a leaf after a series of coordinated division and expansion (Horiguchi et al. 2006a,b, Vanhaeren et al. 2010, Gonzalez et al. 2012). This is the first manifestation of the leaf that is not visible by naked eye. The rhythm at which leaf primordia are produced can affect the final number of leaves as shown in different mutants or different environmental conditions (Cookson et al. 2005). It is often but not strictly related to the size of the shoot apical meristem (Skirycz et al. 2010). Precisely measuring the number of initiated leaves and shoot apical meristem size is not easy. It requires the careful dissection of the rosettes using a binocular or a microscope at high magnification and removal of the cotyledons and subsequent leaves one by one with a scalpel until all primordia and the whole meristem are visible (Cookson et al. 2010). Measurements of leaf initiation rate, i.e., the slope of the relationship between the numbers of initiated leaves and time, or plastochron—the reciprocal of leaf initiation rate—requires sequential destructive measurements over time increasing the number of replicates to be grown together in a single experiment (Cookson et al. 2010). These tasks are even more tedious in *Arabidopsis thaliana* than in other plants due to its small size and, as a consequence, quantitative datasets with phenotypic characterization of these variables are often restricted to a limited number of genotypes and/or environmental conditions (Cookson et al. 2005, Vanhaeren et al. 2010, Skirycz et al. 2010).

Because of the technical constraints mentioned above, leaf production is often scored by counting the number of visible (or emerged) leaves either at a given date after sowing, or a given stage such as at the end of rosette development just before flowering (Fig. 1, Pérez-Pérez et al. 2002, Tisné et al. 2008, Méndez-Vigo et al. 2010, Massonnet et al. 2011). When the number

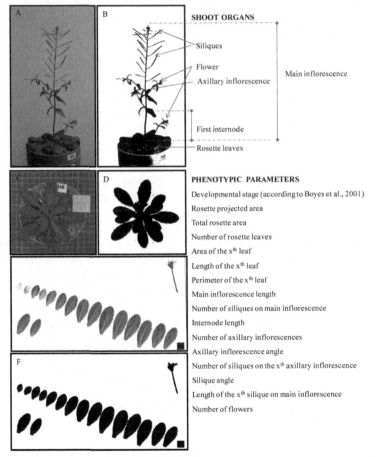

Figure 1. Non-exhaustive list of two-dimensional shoot phenotypic traits extracted from lateral view images (A, B), top view images (C, D) and scans of dissected shoots (E, F). The different shoot organs that are visible on the images are indicated on the left of Figure B and the list of quantitative traits is given below. Depending on the traits, they can be extracted automatically with macros developed on image analysis software, by counting organs manually, or tracing organ length, width or contours.

Color image of this figure appears in the color plate section at the end of the book.

of emerged leaves is scored over time during plant development, the rate of leaf emergence (i.e., the number of leaves emerged per unit of time) and its reciprocal, the phyllochron (i.e., the time elapsed between the emergence of two successive leaves) can be calculated (Cookson et al. 2010). Datasets encompassing the total number of rosette leaves, leaf emergence rate and/or phyllochron are currently available in the Arabidopsis literature for high number of genotypes grown in controlled conditions (Pérez-Pérez et al. 2002, Méndez-Vigo et al. 2010). Analyses of these datasets identified a diversity of natural alleles controlling the rate of leaf production in Arabidopsis, most of them also affecting the total number of rosette leaves and/or flowering time (Méndez-Vigo et al. 2010).

Phenotyping the Spatial Arrangement of Shoot Organs

An additional phenotypic trait characterizing leaf production is related to the spatial arrangement of the successive leaves on a plant and is called phyllotaxy. Phyllotaxy puzzles plant biologists and modelers because of the strikingly regular position of leaves. A group of undifferentiated cells, forming the shoot apical meristem, plays a crucial role in the formation of this pattern by initiating organ primordia on its flanks in a temporally and spatially controlled manner. In *Arabidopsis thaliana*, many mutants with abnormal phyllotactic patterns have been described (Byrne et al. 2003, Smith and Hake 2003, Bhatt et al. 2004, Peaucelle et al. 2007). Their patterns were generally measured on fully grown plants, measuring angles between successive petiole and internode lengths, but in a few cases they were also determined at the primordial stage, measuring angles between successive primordia by microscopic observations (Peaucelle et al. 2007). Because floral primordia and flowers are comparatively easier to score visually and more accessible than young leaf primordia or even mature leaves organized in rosettes with largely overlapping leaves, much more research on the spatial arrangements of shoot organs have been performed on reproductive organs. This has led to a deviation of the definition of the word phyllotaxy that is now used both for leaves and floral organs at least in *Arabidopsis thaliana* (Peaucelle et al. 2007). By phenotyping different genotypes with altered phyllotaxy, it was shown that variability in phyllotaxy is sometimes associated with altered meristem size, meristem organization, or plastochron length. However, variation in phyllotaxy also results, in some genotypes, from processes occurring after organ emergence (Peaucelle et al. 2007). By phenotyping spatial arrangement of successive leaves and flowers, the mechanisms controlling organ positioning in the shoot apical meristem are beginning to be unraveled highlighting a crucial role for auxin and miRNA (Reinhardt et al. 2003, Peaucelle et al. 2007).

Phenotyping Leaf Expansion: From Leaf Initiation to Final Size and Shape

Development of a leaf starts as soon as it is initiated as a primordium and, as a consequence, a first phase of leaf expansion from its initiation to its emergence is hidden in the apical bud. Because growth is exponential during this period, changes in leaf expansion rate during this time, or changes in the initial size of the primordium, can have strong impacts on final leaf size (Granier and Tardieu 1999a,b). However, as mentioned before for processes associated with leaf initiation, studies comparing leaf growth phenotypes during those early phases are rare because of the technical constraints associated with destructive measurements, and the number of plants that have to be grown together to have sufficient replicates at each sampling time point (Cookson et al. 2005). Recent advances in microscopy and image analyses after 3D reconstruction now allow assessing these variables. They revealed that subtle differences in shoot apical meristem volumes contribute to differences in rosette expansion rate observed in a few genotypes (Vanhaeren et al. 2010). The throughput of these measurements is typically low and consequently we cannot currently envisage large experimental campaigns to compare high numbers of genotypes grown together (e.g., hundreds) by measuring these variables.

In most high-throughput phenotypic analyses, leaf expansion is analyzed by comparing the sizes and/or shape of visible leaves at a given time after sowing or at pre-set developmental stages (Fig. 1, Juenger et al. 2005, Pérez-Pérez et al. 2011, Massonnet et al. 2011). In large collections of natural variants or mutant genotypes, i.e., hundreds of variants, genotypes have been classified according to their leaf shape and/or size (Berná et al. 1999, Serrano-Cartagena et al. 1999, Juenger et al. 2005, Micol 2009). Genes or quantitative trait loci that control leaf development have been identified in those high-throughput screens (Juenger et al. 2005, Tisné et al. 2008, Pérez-Pérez et al. 2011). Size-related lamina variables and the length of the vein network were positively correlated in large collections of genotypes, suggesting that leaf growth and vein patterning share at least some genetic controls (Pérez-Pérez et al. 2011). A user-friendly web interface, LIMANI, has been developed for the automatic analysis of venation patterns in Arabidopsis leaves with the aim to improve the interaction between vascular patterning and leaf growth (Dhondt et al. 2012).

Nonetheless, only dynamic analyses of leaf development over time could reveal variability in leaf growth phenotypes in a few situations (Lièvre et al. 2013). An identical final leaf size can result from different leaf growth trajectories. If plants are imaged at regular intervals, their leaf area can be plotted against time and this produces a sigmoid curve from which a few phenotypic traits can be extracted, such as relative expansion rate, absolute expansion rate, initial expansion rate and duration of expansion (Cookson et al. 2010). In certain circumstances, an increase in the duration of expansion can compensate for a reduction in leaf expansion rate. This is the case for Arabidopsis genotypes that do not reduce their final leaf area when grown in growth-limiting conditions (Cookson and Granier 2006, Aguirrezabal et al. 2006). Additionally, only kinematic analysis could reveal the phenotypes of transgenic plants over-expressing a D-type cyclin which do not show visible leaf phenotypic differences at full rosette size but have higher rates of expansion during a shorter period of growth (Cockcroft et al. 2000). Consequently, the most pertinent level of leaf development investigation to identify growth phenotypes is the characterization of the patterns of growth. Image acquisition in high-throughput automated phenotyping platforms will certainly increase the characterization of leaf growth dynamics in high number of genotypes. Also, high-resolution 3D imaging of excised leaf discs is amenable to high-throughput analysis of growth dynamics (Biskup et al. 2009). In parallel, different computer-based tools are designed for rapid and large-scale analyses of leaf shape and size variation on images of plants that are raised in standard laboratory conditions (Walter et al. 2007, Weight and Parnham 2008, Bylesjö et al. 2008).

Phenotyping Growth-associated Cellular Processes in Leaves

Routine methods are used in many groups for leaf phenotype assessment at a cellular level using bright field or differential interference contrast microscopy. Cell density, cell area and cell number are determined on epidermal peels or paradermal views of cleared leaves. These measurements are mainly limited to the upper epidermis because of technical constraints but also because the epidermis is often considered as the tissue physically limiting whole leaf expansion (Pérez-Pérez et al. 2011). In many mutants, strong changes in mesophyll cell numbers do not significantly affect leaf size, whereas the size of epidermal cells is positively correlated with leaf area (González-Bayón et al. 2006, Pérez-Pérez et al. 2011). With these methods, cellular behaviors underlying leaf shape and size variations have been assessed in large collections of mutants, populations of recombinant inbred lines and naturally occurring accessions, giving insights into the genetic control of these variables but also to their coordination and role during leaf development (Micol 2006, Horiguchi et al. 2006a,b, Tisné et al. 2008, Pérez-Pérez et al. 2011, Massonnet et al. 2011, Sterken et al. 2012).

As discussed before concerning the dynamics of whole leaf expansion, dynamics of cell division and cell expansion are more pertinent phenotyping strategies to describe several cellular traits susceptible to the effects of mutations during leaf development. Protocols for kinematic analysis of cell division and cell expansion in the epidermis are well-established for the Arabidopsis leaf (Donnelly et al. 1999, De Veylder et al. 2001, Fiorani and Beemster 2006, Rymen et al. 2010). As in most dicotyledonous leaves, a cellular tip-to-base developmental gradient appears, with an increase in cell size in zones at the tip of the leaf and progressively towards the base. This gradient corresponds to a slowing down and later on cessation of cell division more or less abruptly depending on genotypes, growing conditions and the position of the leaf on the rosette (Donnelly et al. 1999, Wiese et al. 2007, Tisné et al. 2011, Andriankaja et al. 2012). Only precise spatial and temporal analyses of cell division and expansion could reveal a few contrasted phenotypes that were not visible at the macroscopic scale of the whole leaf or only considering final cell number or size (Ferjani et al. 2007, Tisné et al. 2011). Superimposed on those cellular variables is the endoreduplication process that interacts with all other variables (Cookson et al. 2006). Cells in Arabidopsis leaves can undergo a few numbers of endocycles giving rise to cells with 64C nuclei that are usually larger in area. High-throughput assessment of the extent of endoreduplication in Arabidopsis leaves gave insight into Quantitative Trait Loci (QTLs) and genes controlling this process together contributing to a better understanding of how this unfolds during leaf growth (Massonnet et al. 2011, Sterken et al. 2012).

From Floral Transition to Inflorescence Architecture

At the onset of the floral transition, the shoot apical meristem that produced leaves on its flanks during the vegetative phases turns into a reproductive meristem and starts to produce flowers while pre-existing leaf primordia that have not developed yet into rosette leaves become cauline leaves (Alvare-Buylla et al. 2010). In annual plants such as *Arabidopsis thaliana*, the date of transition from the adult vegetative phase to the reproductive one is an important trait since the leaf area produced during the vegetative phases will determine the amount of intercepted light, carbon availability, and capacity in grain production. An early transition can be unfavorable for whole plant production as it limits the vegetative resources available during the reproductive phase. However, in some cases, a late transition can be unfavorable when, for example, water in the soil becomes transient during flowering and thus, developing a high transpiring leaf area during the vegetative phase would increase water consumption and cause pronounced stress later on (Tardieu 2012).

Various studies have shown that flowering date exhibits phenotypic plasticity with respect to general resource levels of light intensity, water and nutrients (Pigliucci and Schlichting 1995, Pigliucci et al. 1995) as well as to environmental triggers such as vernalization and day length (Lempe et al. 2005). It can be detected by early events such as the initiation of the first flower primordia that have to be recorded under a microscope. But the monitoring of the appearance of flower primordia is destructive and thus requires relatively large plant populations for sequential time course records. As a consequence, in practice, the duration of the phase between seed germination and floral transition is evaluated by a number of indirect indicators or later events that are macroscopically visible. Dates of different stages such as the opening of the first floral buds, the elongation of the floral stem reaching 1-mm height, or even later the opening of the first flower on the inflorescence can be recorded in original habitats, in greenhouse or controlled growth chambers without altering plants' integrity and are thus more frequently used as surrogates for flowering time. Another indirect and widely used trait that correlates with the floral transition as shown in different mutants, accessions or populations is the total number of leaves produced during the vegetative developmental phase (Méndez-Vigo et al. 2010).

Whatever the phenotypic indicator recorded to estimate flowering date, it is a particularly well-studied character in Arabidopsis. Because of its systematic measurement in mutants, accessions and recombinant inbred lines, several genes affecting flowering date have been identified for years and flowering time genetic networks have been set up (Koornneef et al. 1991, Ehrenreich et al. 2009, Flowers et al. 2009). In contrast, less is known about the formation and patterning of internodes on the inflorescence that are key determinants of inflorescence architecture with important roles in pollination and seed dispersal. More or less compact inflorescences have been observed across the wide genetic variability available in *Arabidopsis thaliana*. Inflorescence architecture phenotypes have been mainly characterized by measuring the divergence angle and internode length between successive organs along the main inflorescence stem (Fig. 1). A spectrum of inflorescence architectures ranging from short internodes, to downward-pointing pedicels, to clusters of flowers has been reported and a few genes that controlled these patterns have been identified (Peaucelle et al. 2007, Ragni et al. 2008, Li et al. 2012). Using loss of function, double and triple mutants, genetic action models for these genes have been proposed (Li et al. 2012). All phenotypic variables discussed here and used as surrogate parameters for transition to flowering or describing phenotypic differences in inflorescence architecture can be recorded on RGB pictures taken in automated phenotyping platforms at least when two cameras are set up on the automaton, one to take pictures from top and another for lateral view angles (Vasseur et al. 2011). However, both the automatic recognition of shoot organs and their automatic counting are still not possible and would require some further developments in image analysis for high-throughput screens.

Systematic Root Growth Phenotyping in Arabidopsis thaliana

A hallmark of germination is the protrusion of the radicle from the seed's integuments. Arabidopsis root development continues from the axis of the primary root formed at the embryonic stages. Root branches of different orders arise from the main root axis and lateral root development takes place post-embryonically (Dolan et al. 1993). The radial root tissue anatomy in Arabidopsis displays a concentric organization and can be separated into several distinct functional types encompassing an epidermal layer, the endodermis, and the pericycle surrounding the vascular tissue (xylem and phloem). The number of cortical and endodermal cell files is more variable in lateral roots than in primary roots (Dolan et al. 1993). To our knowledge, a systematic attempt to screen Arabidopsis variants for variability and its potential functional implications has not been attempted so far. Primary and lateral root axes grow at their tip and display a typical organization of cellular activities with predominant gradients in cell development from the root tip to more distal regions of the root (Beemster et al. 2002, see also Baskin et al. 2010 for revised terminology of root and shoot polarity). A quiescent center ensures the maintenance of a stem cell niche. Both maintenance of this cell niche and cell differentiation are regulated by local auxin levels and depend on the biosynthetic rates of this plant hormone and on its intercellular transport kinetics (Ding and Friml 2010). Whereas auxins generally have a stimulating effect on root growth, there is hormonal cross-talk, in particular with cytokinins which have an inhibitory effect on root development (Bielach et al. 2012, Ubeda-Tomas et al. 2012). Lateral roots and root branches arise from a specialized layer of cells in the root pericycle (Dolan et al. 1993). Root phenotyping is currently an active field of research mainly driven by the need of selecting crop root ideotypes with increased ability to cope with limited resources (water and nutrients) (Herder et al. 2010). In this section we will first synthetically highlight research in the regulation of root tip growth in Arabidopsis and second describe which technical options and setups are currently being developed to enable the phenotypic characterization of global root system's architecture parameters.

Studies of Lateral Root Initiation

Because of the potential value of manipulating regulatory pathways for translational research (transfer of traits in crops, application to crop breeding), developmental biologists have dedicated considerable effort in investigating the dynamics of emergence of lateral roots. Lateral roots arise from sub-populations of cells located in the pericycle. Their appearance, observed by microscopy in transparent growth media, is accompanied by regular patterns of asymmetric cell divisions (Parizot et al. 2012). Based on a large body of literature, it can be concluded that auxin distribution along the root axis plays a major role in controlling cell cycle reactivation and patterning of lateral root formation in Arabidopsis (Parizot et al. 2012, and references therein). In addition, it is emerging that only sub-populations of pericycle cells located at the xylem pole give rise to lateral root founder cells and that these 'hot spots' show specific associations with the adjacent vascular tissue (Parizot et al. 2012).

Phenotyping of the Cellular Mechanisms Underlying Root Tips Growth

Studies of the root apical meristem dynamics in Arabidopsis have resulted in a relatively limited number of publications. One common theme has been to establish non-invasive optical methods and develop software and computational approaches to study cell dynamics at the root tip growth zone of plants grown in artificial media based on time-lapse photography (Beemster and Baskin 1998, French et al. 2009, Walter et al. 2009). These methodologies have been applied primarily to address different research questions within the main aim of explaining root tip growth as a function of the underlying cell dynamics. Summarizing the available experiments, we highlight selected studies that have addressed: (a) the link between cell production and cell expansion activities at the root tip level with overall rates of root axis elongation (Beemster et al. 2002), and (b) hormonal regulation of cell differentiation and effects on root axis development (Swarup et al. 2005). Typically, these studies are limited to comparisons of a relatively small number of genotypes displaying either natural or induced variability. However, the output is quantitative and often with a relatively high spatial (sub-millimeter) and temporal (minutes) resolution as applied to studies of root tip growth and curvature dynamics in response to gravitropic stimuli (Chavarria-Krauser et al. 2008, Walter et al. 2009). More recently, genome-wide transcript analysis of all root cell types of Arabidopsis has been completed using a set of enhancer trap lines expressing tissue-specific fluorescent proteins and combining cell sorting protocols with molecular profiling. These experiments led to comprehensive, high-resolution expression maps and to the identification of a transcriptional network underlying developmental sequences and cell differentiation (Birnbaum et al. 2003, Petricka et al. 2012). Based on these approaches, specific studies of cell autonomous and non-cell autonomous mechanisms of root differentiation have been undertaken (e.g., Levesque et al. 2006).

Phenotyping the Architecture of Whole Root Systems

Plant roots grown in soil or in opaque media are difficult to access by means of non-destructive analyses and Arabidopsis is no exception in this respect. Simply, when plants are cultivated in pots and in soil-based systems, the only way to measure root parameters such as fresh and dry weight is to wash root systems out of the rooting medium as carefully as possible. Typically, it is impossible to preserve any original geometric arrangement of the root system. Therefore, information about root geometry (e.g., branching angles, and spatial distributions in relation to soil depth) is lost. This procedure is not only time consuming but also it is unavoidable that finer roots or a varying fraction of the root system is lost by mechanical damage, depending on the soil type, water content and mechanical compaction. This aspect is particularly relevant for Arabidopsis roots characterized by diameters of between approximately 100 and 160 µm when grown under non-limiting conditions in agar plates (80 Arabidopsis accessions from 1001 Genome Project: http://www.1001genomes.org/, Nagel K, unpublished data). This point needs to be taken into account in experimental procedures that include harvesting roots from soil. These types of measurements are usually not suitable to acquire time series of root growth and cannot address quantification of root system architecture. However, there are several techniques that are applicable to study Arabidopsis root system architecture and that are increasingly utilized in Arabidopsis phenotyping experiments. First, cultivation of Arabidopsis in transparent, agar-filled plates in vertical setups is easy and 2D imaging delivers a rich dataset including major root architecture parameters (Fig. 2, Nagel et al. 2012). Second, other soil-less cultivation methodologies involve, for instance, the use of growth pouches and could be adapted to Arabidopsis (Hund et al. 2009). Third, soil-filled rhizobox setups have been designed specifically for Arabidopsis (Devienne-Barret et al. 2006) and similar ideas recently led to fully automated systems (Nagel et al. 2012). It is interesting to notice that in a comparison of different species to validate a novel rhizobox robotic and imaging system, Nagel et al. found that the portion of the system visible through the transparent cover of rhizoboxes seems to be positively correlated with the average root diameter. Because of the relatively modest average root diameter of Arabidopsis, it is possible to visualize more than 70% of the total root system by single RGB camera shots at subsequent developmental time points during the root growth trajectory. To conclude, combining different methodologies for their respective advantages while being aware of intrinsic limitations is receiving increasing consensus (De Smet et al. 2012). In summary, similar to phenotyping procedures of root system architecture in the field (Trachsel et al. 2011), accessing root structure and geometry in soil-based assays remains very challenging. Finally, it is

ROOT SYSTEMS PARAMETERS

Primary root length (mm)
1st and 2nd order lateral root length (mm)
Total root length (mm)
Maximum depth and width root system (mm)
Number of lateral roots
Lateral root density (mm⁻¹)
Branching angle lateral roots (°)
Average diameter primary root (μm)
Average diameter lateral roots (μm)

Figure 2. Automated two-dimensional feature extraction of young Arabidopsis root systems is possible in transparent cultivation media.

Color image of this figure appears in the color plate section at the end of the book.

critical to remark that the soil environment is highly heterogeneous with distinct spatial and temporal gradients in (mobile) nutrient distributions, and temperature in particular (Nagel et al. 2012, Walter et al. 2009). Designing assays addressing at least in part this complexity should be high priority for the root research agenda. A representative image of an Arabidopsis root system grown in an agar-filled Petri dish (panel A) and a color-coded image of the same root system (panel B) are presented in Fig. 2. The primary root is colored in green and the lateral roots in red. The pictures are analyzed and the parameters indicated in panel B are extracted by the GROWSCREEN-Root software (Nagel et al. 2012). Similarly, roots can be analyzed in 2D pictures for plants growing in soil-filled rhizoboxes. However, only visible roots at the soil-transparent plate interface can be detected automatically and manual curation of individual images is currently still necessary for tracing missing parts of the root system (Nagel et al. 2012). The comparison of treatments and genotypic variation can be performed as an analysis of time series or at the same time point after germination. The *Arabidopsis thaliana* (Col-0) plant was cultivated in a system, which enables the shoot to grow outside the Petri dish while the roots grew in the agar (Nagel et al. 2006). The plant was exposed—after five days stratification—to 20°C day/ night temperature, an air humidity of 60% and a day/night length of 12 h. The image was taken 2.5 weeks after germination.

Databases, Ontologies and Data Sharing

Database Resources for Arabidopsis Phenotypes

Following the first sequencing wave of the Arabidopsis genome and the ability to characterize molecular networks by means of techniques generally referred as 'omics' technologies (genome-wide transcriptomics, metabolomics, and proteomics), plant biologists and bio-informaticians have generated database resources and analytical tools at a steady pace. In parallel, thousands of experiments have been performed recording a large variety of phenotypic traits in Arabidopsis genotypes including transgenic lines, accessions, and progeny resulting from crosses between different genotypes. The performance of assays to phenotype shoot and root traits towards increased throughput with non-invasive technologies such as image analysis (i.e., up to a few hundreds of genotypes imaged simultaneously during long periods with a relatively fine time resolution) requires the development of appropriate data management schemas (Kattge et al. 2011). Some comprehensive database resources and data sharing frameworks have been developed to organize this amount of phenotypic data for an efficient sharing of Arabidopsis phenotype knowledge (Table 1). Most of the available Arabidopsis phenotypic data available in these public databases are qualitative and limited to free text descriptions (Table 1, e.g., TAIR) and only a few public resources provide quantitative description of various Arabidopsis genotypes (Shinozaki 2010, Fabre et al. 2011). Well-structured formalisms would allow meta-analyses of available datasets (Poorter et al. 2010, Parent and Tardieu 2012). Whereas these meta-analyses have been undertaken for a certain number of crops and for naturally occurring species, the same type of approaches is, to our knowledge, still lacking in Arabidopsis. Unraveling gene function by large-scale genotype screening has been mainly based on the characterization of a genotype by an 'average phenotype' measured under a given environmental condition without information on the variability structure. In most databases, it is impossible to find in which conditions plants with this given 'average phenotype' were grown. However, it is well known that even small differences in environmental conditions can substantially affect plant phenotypic traits at different levels of organization from cells to tissue and organs (Massonnet et al. 2010, Poorter et al. 2012a). Even if recent automated platforms developed in many groups allow now a precise record of plant environmental conditions, including the soil humidity, together with the growth phenotypes (i.e., Traitmill, PHENOSCOPE, WIWAM, GROWFLUOSCREEN), these data are still not available through repository databases.

Table 1. List of public databases containing Arabidopsis phenotypic data resources.

Database	Website	Description	References
RAPID, Riken Arabidopsis Phenome Information Database	http://rarge.gsc.riken.jp/phenome/	Mutants with phenotypic information	(Kuromori et al. 2006)
TrapperDB	http://genetrap.cshl.org/TrHome.html		(Liu et al. 1995, Martienssen 1998)
FioreDB	http://www.cres-t.org/fiore/public_db/index.shtml		
BAP DB	http://bioweb.ucr.edu/bapdb/		
SeedGenesProject	http://www.seedgenes.org/index.html		(Meinke et al. 2008)
Chloroplast 2010 Project	http://bioinfo.bch.msu.edu/2010_LIMS		(Ajjawi et al. 2010)
Chloroplast Function Database	http://rarge.psc.riken.jp/chloroplast/		
AtPolyDB	http://arabidopsis.usc.edu/	Data on large collection of accessions characterised by 107 phenotypes	(Huang et al. 2011, Atwell et al. 2010)
TAIR	http://www.arabidopsis.org/	Phenotype information from the Arabidopsis literature	(Huala et al. 2001)
PHENOPSIS_DB	http://bioweb.supagro.inr.fr/phenopsis	Quantitative phenotypic data on hundreds of genotypes with meta-data	(Fabre et al. 2011)

Ontologies to Describe Plant Growth Phenotypes

In many organisms—*Drosophila melanogaster, Mus musculus*, and *Saccharomyces cerevisiae*—coordinated efforts to adopt common controlled vocabularies are ongoing, mainly in the frame of literature annotation to facilitate data reuse and meta-analyses (Ashburner et al. 2000). In genomics, the Gene Ontology project has been a successful bioinformatics initiative to standardize the representation of gene and gene product attributes in various species and genomics resources (Ashburner et al. 2000). An ontology is a formal representation of a set of concepts within a specific discipline or domain and of the relationship between those concepts. The adoption of ontologies standardizes description of the phenotypic information. It is useful for biologists as it reduces the source of misunderstanding as the same concept can sometimes refer to different meanings or the same terms can have different definitions. For example, the term 'emerged' leaf in the *Arabidopsis thaliana* rosette can refer to a leaf that is 'visible by the naked eye' (Méndez-Vigo et al. 2010) or a leaf 'with a length higher than 1 mm' (Boyes et al. 2001). It also appears when comparing some phenotyping datasets that there is sometimes confusion between different variables. For example, the keywords 'plastochron' and 'leaf initiation' return many papers that only contain phyllochron data or number of emerged leaves (i.e., Pérez-Pérez et al. 2002, Wang et al. 2008). However, Arabidopsis genotypes with the same number of emerged leaves can have different numbers of leaf primordia suggesting that leaf emergence rate cannot be used as a proxy of leaf initiation rate in all circumstances (Vanhaeren et al. 2010). These examples highlight the necessity of the adoption of common vocabulary shared within the whole scientific community to describe the same variable. With the aim to unify descriptors, *Arabidopsis thaliana* growth stages have been defined for comparative purposes (Boyes et al. 2001 for whole plant development, Alvare-Buylla et al. 2010 for flower development). They are used in many studies either to compare the rate of progression between different stages among genotypes or to specify the stage at which specific measurements were done (Massonnet et al. 2011, Sterken et al. 2012). These two initiatives have the merit of standardizing description of specific developmental phases, but they are restricted to *Arabidopsis thaliana* only.

Several multi-organism ontologies are relevant to the phenotyping of Arabidopsis developmental traits (Table 2). They cover anatomy descriptions (Plant Ontology), trait description (Phenotypic Quality Ontology, Trait Ontology), unit description (Unit Ontology, UO) and environmental context (XEML Environment Ontology). However, the phenotypic description of a plant, as in other organisms, must regroup several of these concepts that are spread across different ontologies. First, there is a notion of scale dimension. The growth phenotype can be observed at different levels such as whole

Table 2. Useful ontology public resources to describe phenotypes.

Ontology Title	Prefix	Website	Description
Phenotypic Quality *Ontology*	PATO	http://www.obofoundry.org/cgi-bin/detail.cgi	This ontology can be used in conjunction with other ontologies such as anatomical ontologies to refer to phenotypes
Plant Trait Ontology	TO	http://www.obofoundry.org/cgi-bin/detail.cgi?id=plant_trait	A controlled vocabulary that describes phenotypic traits in plants
Units of measurement	UO	http://www.obofoundry.org/cgi-bin/detail.cgi?id=unit	Metrical units for use in conjunction with PATO
Plant Ontology	PO	http://www.obofoundry.org/cgi-bin/detail.cgi?id=plant_ontology	A structured vocabulary that describes plant anatomy and morphology and stages of growth and development for all plants
XEML Environment Ontology	XEO	http://xeml.mpimp-golm.mpg.de/dnn/	An ontology of terms describing the abiotic environment

plant, individual organs, or specific cellular type in an organ. Then, because the term phenotype is not an object clearly defined as a gene or a protein, the domain definition has to be specified. For example, leaf growth can be characterized by many descriptors such as leaf length, area or weight. To formalize these descriptions together in the same formal model, the Entity-Quality (EQ) model (Mungall et al. 2010) has been proposed as a simple schema combining an Entity (the plant part under study, e.g., 'leaf') with a Quality attribution (the phenotypic trait under study, e.g., 'size'). The plant entity that is considered in the phenotypic screen can be described from anatomy ontology such as Plant Ontology (PO), whereas the phenotype quality (the domain definition) can be extracted from a trait ontology such as Trait Ontology (TO) or Phenotypic Quality Ontology (PATO) (Table 2). For example, with the EQ model, leaf length can be expressed as the combination of the leaf term (PO:0025034) and the length term (PATO:0000122). This EQ model is a convenient way to manage the multi-scale aspect with a precise definition of the part under study and it limits the redundancy of descriptors, as, the same trait, length, can be attributed to different plant parts.

Several tools are available to help biologists to annotate phenotypes using ontologies. The Phenex (Balhof et al. 2010) and Phenote (http://phenote.org) tools support the EQ model for representing phenotypes and could be configured for any kind of ontologies. Knowtator (http://knowtator.sourceforge.net/) is a text annotation tool integrated with the Protégé Ontology Editor (http://protege.stanford.edu). It has been used within the European project AGRON-OMICS (www.agron-omics.eu) project to annotate scientific literature associated with Arabidopsis phenotyping. The EQ model can also be completed by other ontologies for a deeper characterization of the phenotype taking into consideration the unit of the trait measured on the entity (Unit Ontology, UO) and the environmental context (XEML Environment Ontology, XEO). The OBO foundry (http://obofoundry.org/) and Bioportal (http://bioportal.bioontology.org) are useful resources to explore all the publicly available ontologies. Wider adoption and/or enrichment of publicly available ontologies is fundamental to ensure: (a) the possibility of richer data retrieval by using simple queries, (b) database interoperability across platforms and different plant species, and (c) reasoning and computing databases by means of the logical relationships between ontology terms and class branches.

Data Sharing

Data sharing can be considered following two approaches. In a first one, all data are centralized in a common database, whereas in the second one data are kept where they have been produced in distinct databases but with tools that allow their retrieval, distribution and sharing. The Phenomics Ontology Driven Data Management (PODD) is an Australian initiative to provide access to phenotype data for the broader community (Li et al. 2013). They proposed a centralized repository to collect all phenotype data. Based on an ontology-centric architecture, the data management system is flexible and supports the evolution of domain concepts. On the other hand, Web services are promising solutions to share data (or tools) and inter-connect data providers. Biomoby is an initiative which successfully integrates bioinformatics services from plant genomics database providers (Wilkinson et al. 2005). Web services are programs available through the network for web users or web-connected programs. Based on standard technologies, these services generally resolve issues of interoperability because services can easily be accessed using a variety of computer languages. This distributed approach can rely on existing infrastructures and does not involve centralizing all data at one location. Interestingly, Web services could also be used within workflows to automate data analysis. The Biocatalogue (http://www.biocatalogue.org/) (Bhagat et al. 2010) provides a web interface to register or search for web services dedicated to the life science community. The keyword 'Arabidopsis' returns 78 services from three different service providers.

Which Priorities to Enable Wider Systematic Phenotyping in Arabidopsis?

Plant phenomics arises as a novel discipline to meet the needs of rapid advances in both genomics research and molecular plant breeding (Houle et al. 2010). The case of Arabidopsis and other crop species whose genome have been sequenced is compelling because a wealth of genetic material generated by forward and reverse genetics is now available for further analyses and to identify useful traits for translation research and plant breeding. Furthermore, the analysis of natural variation in Arabidopsis to identify major genes underlying the expression of growth and development traits, for example by QTL analysis and association mapping, relies on the ability to measure a number of phenotypic parameters with relatively high precision. Importantly, any systematic efforts to deliver rich data matrices enabling plant biologists to link genome and phenome necessarily rely on the parallel development of integrated databases collating the wealth of information acquired by several technology platforms. In this concluding section we summarize and briefly comment upon the main technical and organizational factor aspects that still hamper faster progress in Arabidopsis phenotyping.

Development of Best Practices in Arabidopsis Phenotyping and Data Sharing

Large-scale researches in plant phenomics involve many laboratories that have developed independently their own protocols and procedures. It is crucial to renew efforts of documenting experiments as accurately as possible including plant cultivation conditions, experimental layouts and detailed plant selection procedures (Poorter et al. 2012a). In this respect, we consider that it would be highly beneficial to support initiatives defining and proposing to the Arabidopsis community a minimum set of information for plant phenotyping experiments. These types of efforts are analogous to similar initiatives undertaken especially for molecular profiling experiments and sequence data in model organisms (Taylor et al. 2008).

We also remark that the ability to access well-structured quantitative data would enormously facilitate modeling tasks. Even if database resources have been developed (see before), we must point out that a large part of phenotypic data is still disseminated in scientific literature or stored in personal computers. Published work provides access to a large amount of information. Advanced text mining tools are available for a large corpus comprising over 30,000 full text papers (http://www.textpresso.org/arabidopsis/). This resource should facilitate mining, phenotypic trait selection, curation and quantitative modeling in meta-analytical frameworks (Poorter et al. 2012b).

Time investment in extracting relevant data for a desired analysis is generally high. To this end, phenomics databases not limited to molecular signatures should become more widespread. As a corollary, an effective use of data repositories across labs would greatly benefit from the adoption of common vocabularies and application ontologies mostly based on already existing efforts. The developments of microarray repositories such as Gene Expression Omnibus or ArrayExpress have considerably improved data sharing and reuse in the transcriptomics domain. Scientific journals should encourage, and require deposit of phenotypic data in such public repositories similar to the submission of Arabidopsis gene-related data through the TAIR database: http://www.arabidopsis.org.

Linking Tightly with Technology Developments

Imaging methodologies based on 2D imaging have a potential application to systematic analyses of large number of diverse Arabidopsis genotype collections comprising forward and reverse genetics mutants, naturally occurring alleles, recombinant inbred lines and association mapping populations. We generally conclude that phenotyping approaches especially using non-invasive optical methods have matured to provide the required data to deliver quantitative data of shoot and, as a specific case, root phenotypic parameters through development. However, machine vision and imaging technologies are progressing rapidly opening unprecedented opportunities for non- or minimally invasive phenotyping (Fiorani et al. 2012). There is no doubt that, in the next few years, the refinement of multiple methodologies including 3D reconstruction of shoot and root architecture, but also the visualization of hidden organs will have a positive impact on the systematic phenotyping of Arabidopsis and other model species (Pierret et al. 2003, Jahnke et al. 2009, Dhondt et al. 2010). However, integrating new sensing systems to quantitatively measure plant structure and function requires a multi-disciplinary approach also including accurate understanding of the physical principles of the sensors' operation and the ability to apply automation and robotics.

Sustainable Development of High-throughput Phenotyping Infrastructure

The next three to five years will be crucial to implement systematic phenotyping approaches applied to large Arabidopsis germplasm collections in view of coordinated international efforts, in particular in Europe and in Australia. As an example of significant efforts aimed at increasing capacities of platforms for screening of collections and to make these efforts sustainable via medium-term funding we wish to mention EU-funded projects that are designed to enable these developments, in particular the European Plant Phenotyping Network addressing the need of opening existing phenotyping platforms to user groups under a transnational access scheme (http://www.plant-phenotyping-network.eu/). In addition, similar initiatives have been established in Australia (http://www.plantphenomics.org.au/). In parallel, national efforts ensuring medium-term commitment to the development of large national infrastructure networks have been established both in France (http://www.inra-transfert.fr/en/page.php?optim=phenome) and in Germany (http://www.dppn.de). We expect that similar initiatives will also be established in the near future in other countries as well.

Acknowledgements

We thank Dr. Kerstin Nagel, IBG2 Plant Sciences at Forschungszentrum Jülich for providing the illustration of Arabidopsis roots phenotyping in agar medium. Thibaut Bontpart and Frédéric Bouvery, Laboratoire d'Ecophysiologie des Plantes sous Stress Environnementaux, Montpellier are thanked for providing the illustration of Arabidopsis shoot traits phenotyping. Fabio Fiorani and Christine Granier acknowledge support by European Union projects, EPPN and Agron-Omics. EPPN is an Integrating Activity, Research Infrastructure Project Framework Program 7—Capacities (grant agreement No. 284443). Agron-Omics is a European sixth framework integrated project (LSHG-CT-2006-037704). Arabidopsis phenotyping at IBG2 Plant Sciences, Forschungszentrum Jülich is partly institutionally funded by the Helmholtz Association and by the German Plant Phenotyping Network (DPPN) which is funded by the German Federal Ministry of Education and Research (project identification number: 031A053).

References Cited

AGI. 2000. The Arabidopsis genome initiative. Analysis of genome sequence of the flowering plant *Arabidopsis thaliana*. Nature 408: 796–815.

Aguirrezabal, L., S. Bouchier-Combaud, A. Radziejwoski, M. Dauzat, S.J. Cookson and C. Granier. 2006. Plasticity to soil water deficit in *Arabidopsis thaliana*: dissection of leaf development into underlying growth dynamic and cellular variables reveals invisible phenotypes. Plant Cell Environ. 29: 2216–2227.

Ajjawi, I., Y. Lu, L.J. Savage, S.M. Bell and R.L. Last. 2010. Large-scale reverse genetics in Arabidopsis: case studies from the Chloroplast 2010 Project. Plant Physiol. 152: 529–540.

Alonso, J.M., A.N. Stepanova, T.J. Leisse, C.J. Kim, H. Chen, P. Shinn, D.K. Stevenson, J. Zimmerman, P. Barajas, R. Cheuk, C. Gadrinab, C. Heller, A. Jeske, E. Koesema, C.C. Meyers, H. Parker, L. Prednis, Y. Ansari, N. Choy, H. Deen, M. Geralt, N. Hazari, E. Hom, M. Karnes, C. Mulholland, R. Ndubaku, I. Schmidt, P. Guzman, L. Aguilar-Henonin, M. Schmid, D. Weigel, D.E. Carter, T. Marchand, E. Risseeuw, E. Brogden, A. Zeko, W.

Crosby, C. Berry and J. Ecker. 2003. Genome-wide insertional mutagenesis of *Arabidopsis thaliana*. Science 301: 653–657.

Alvarez-Buylla, E.R., M. Benítez, A. Corvera-Poiré, A. Chaos Cador, S. de Folter, A. Gamboa de Buen, A. Garay-Arroyo, B. García-Ponce, F. Jaimes-Miranda, R.V. Pérez-Ruiz, A. Piñeyro-Nelson and Y.E. Sánchez-Corrales. 2010. Flower Development. Arabidopsis book 2010. American Society of Plant Biologists Rockville e0127.

Andriankaja, M., S. Dhondt, S. De Bodt, H. Vanhaeren, F. Coppens, L. De Milde, P. Mühlenbock, A. Skirycz, N. Gonzalez, G.T. Beemster and D. Inzé. 2012. Exit from proliferation during leaf development in *Arabidopsis thaliana*: a not-so-gradual process. Dev. Cell 17: 64–78.

Arvidsson, S., P. Pérez-Rodríguez and B. Mueller-Roeber. 2011. A growth phenotyping pipeline for *Arabidopsis thaliana* integrating image analysis and rosette area modeling for robust quantification of genotype effects. New Phytol. 191: 895–907.

Ashburner, M., C.A. Ball, J.A. Blake, D. Botstein, H. Butler, J.M. Cherry, A.P. Davis, K. Dolinski, S.S. Dwight, J.T. Eppig, M.A. Harris, D.P. Hill, L. Issel-Tarver, A. Kasarskis, S. Lewis, J.C. Matese, J.E. Richardson, M. Ringwald, G.M. Rubin and G. Sherlock. 2000. Gene ontology: tool for the unification of biology. The Gene Ontology Consortium. Nat. Genet. 25: 25–29.

Atwell, S., Y.S. Huang, B.J. Vilhjálmsson, G. Willems, M. Horton, Y. Li, D. Meng, A. Platt, A.M. Tarone, T.T. Hu, R. Jiang, N.W. Muliyati, X. Zhang, M.A. Amer, I. Baxter, B. Brachi, J. Chory, C. Dean, M. Debieu, J. de Meaux, J.R. Ecker, N. Faure, J.M. Kniskern, J.D. Jones, T. Michael, A. Nemri, F. Roux, D.E. Salt, C. Tang, M. Todesco, M.B. Traw, D. Weigel, P. Marjoram, J.O. Borevitz, J. Bergelson and M. Nordborg. 2010. Genome-wide association study of 107 phenotypes in *Arabidopsis thaliana* inbred lines. Nature 465: 627–631.

Balhoff, J.P., W.M. Dahdul, C.R. Kothari, H. Lapp, J.G. Lundberg, P. Mabee, P.E. Midford, M. Westerfield and T.J. Vision. 2010. Phenex: Ontological Annotation of Phenotypic Diversity. PLoS One 5(5): e10500.

Baskin, T.I., B. Peret, F. Baluška, P.N. Benfey, M. Bennett, B.G. Forde, S. Gilroy, Y. Helariutta, P.K. Hepler, O. Leyser, P.H. Masson, G.K. Muday, A.S. Murphy, S. Poethig, A. Rahman, K. Roberts, B. Scheres, R.E. Sharp and C. Somerville. 2010. Shootward and rootward: peak terminology for plant polarity. Trends Plant Sci. 15: 593–594.

Beemster, G. and T.I. Baskin. 1998. Analysis of cell division and elongation underlying the developmental acceleration of root growth in *Arabidopsis thaliana*. Plant Physiol. 116: 1515–1526.

Beemster, G.T.S., K. De Vusser, E. De Tavernier, K. De Bock and D. Inzé. 2002. Variation in growth rate between Arabidopsis ecotypes is correlated with cell division and A-type cyclin-dependent kinase activity. Plant Physiol. 129: 854–864.

Berná, G., P. Robles and J.L. Micol. 1999. A mutational analysis of leaf morphogenesis in *Arabidopsis thaliana*. Genetics 152: 729–742.

Bhagat, J., F. Tanoh, E. Nzuobontane, T. Laurent, J. Orlowski, M. Roos, K. Wolstencroft, S. Aleksejevs, R. Stevens, S. Pettifer, R. Lopez and C.A. Goble. 2010. BioCatalogue: a universal catalogue of web services for the life sciences. Nucleic Acids Res. Web Server issue: W689–W694.

Bhatt, A.M., J.P. Etchells, C. Canales, A. Lagodienko and H. Dickinson. 2004. VAAMANA-a BEL1-like homeodomain protein, interacts with KNOX proteins BP and STM and regulates inflorescence stem growth in Arabidopsis. Gene 328: 103–111.

Bielach, A., J. Duclercq, P. Marhavý and E. Benková. 2012. Genetic approach towards the identification of auxin–cytokinin crosstalk components involved in root development. Phil. Trans. R Soc. B 367: 1469–1478.

Birnbaum, K., J.Y. Shasha, J.J. Wang, G. Lambert, D.W. Galbraith and P.N. Benfey. 2003. A gene expression map of the Arabidopsis root. Science 302: 1956–1960.

Biskup, B., H. Scharr, A. Fischbach, A. Wiese-Klinkenberg, U. Schurr and A. Walter. 2009. A diel growth cycle of isolated leaf discs analyzed with a novel, high-throughput three-dimensional imaging method is identical to that of intact leaves. Plant Physiol. 149: 1452–1461.

Boyes, D.C., A.M. Zayed, R. Ascenzi, A.J. McCaskill, N.E. Hoffman, K.R. Davis and J. Gorlach. 2001. Growth stage-based phenotypic analysis of Arabidopsis: a model for high throughput functional genomics in plants. Plant Cell 13: 1499–1510.

Bylesjö, M., V. Segura, R.Y. Soolanayakanahally, A.M. Rae, J. Trygg, P. Gustafsson, S. Jansson and N.R. Street. 2008. LAMINA: a tool for rapid quantification of leaf size and shape parameters. BMC Plant Biol. 8: 82.

Byrne, M.E., A.T. Groover, J.R. Fontana and R.A. Martienssen. 2003. Phyllotactic pattern and stem cell fate are determined by the Arabidopsis homeobox gene BELLRINGER. Development 130: 3941–3950.

Chavarria-Krauser, A., K.A. Nagel, K. Palme, U. Schurr, A. Walter and H. Scharr. 2008. Spatio-temporal quantification of differential growth processes in root growth zones based on a novel combination of image sequence processing and refined concepts describing curvature production. New Phytol. 177: 811–821.

Cockcroft, C.E., B.G. den Boer, J.M. Healy and J.A. Murray. 2000. Cyclin D control of growth rate in plants. Nature 405: 575–579.

Cookson, S.J., A. Radziejwoski and C. Granier. 2006. Cell and leaf size plasticity in Arabidopsis: what is the role of endoreduplication? Plant Cell Environ. 28: 1355–1366.

Cookson, S.J. and C. Granier. 2006. A dynamic analyses of the shaded-induced plasticity in Arabidopsis thaliana rosette leaf development reveals new components of the shade-adaptive response. Ann. Bot. 97: 443–452.

Cookson, S.J., M. Van Lijsebettens and C. Granier. 2005. Correlation between leaf growth variables suggest intrinsic and early controls of leaf size in *Arabidopsis thaliana*. Plant Cell Environ. 28: 1355–1366.

Cookson, S.J., O. Turc, C. Massonnet and C. Granier. 2010. Phenotyping the development of leaf area in *Arabidopsis thaliana*. Methods in Molecular Biology 655: 89–103.

De Smet, I., P.J. White, A.G. Bengough, B. Dupuy, B. Parizot, I. Casimiro, R. Heidstra, M. Laskowski, M. Lepetit, F. Hochholdinger, X. Draye, H. Zhang, M.R. Broadley, B. Péret, J.P. Hammond, H. Fukaki, S. Mooney, J.P. Lynch, P. Nacry, U. Schurr, L. Laplaze, P. Benfey, T. Beeckman and M. Bennett. 2012. Analyzing lateral root development: how to move forward. Plant Cell 24: 15–20.

De Veylder, L., T. Beeckman, G.T.S. Beemster, L. Krols, F. Terras, I. Landrieu, E. Van Der Schueren, S. Maes, M. Naudts and D. Inzé. 2001. Functional analysis of cyclin-dependent kinase inhibitors of Arabidopsis. Plant Cell 13: 1653–1668.

Devienne-Barret, F., C. Richard-Molard, M. Chelle, O. Maury and B. Ney. (2006). Ara-rhizotron: an effective culture system to study simultaneously root and shoot development of Arabidopsis. Plant and Soil 280: 253–266.

Dhondt, S., H. Vanhaeren, D. Van Loo, V. Cnudde and D. Inzé. 2010. Plant structure visualization by high-resolution X-ray computed tomography. Trends Plant Science 15: 419–422.

Dhondt, S., D. Van Haerenborgh, C. Van Cauwenbergh, R.M. Merks, W. Philips, G.T. Beemster and D. Inzé. 2012. Quantitative analysis of venation patterns of Arabidopsis leaves by supervised image analysis. Plant J. 69: 553–563.

Ding, Z. and J. Friml. 2010. Auxin regulates distal stem cell differentiation in Arabidopsis roots. Proc. Natl. Acad. Sci. USA 107: 12046–12051.

Dolan, L., K. Janmaat, V. Willemsen, P. Linstead, S. Poethig, K. Roberts and B. Scheres. 1993. Cellular organisation of the *Arabidopsis thaliana* root. Development 119: 71–84.

Donnelly, P.M., D. Bonetta, H. Tsukaya, R.E. Dengler and N.G. Dengler. 1999. Cell cycling and cell enlargement in developing leaves of Arabidopsis. Dev. Biol. 215: 407–419.

Ehrenreich, I.M., Y. Hanzawa, L. Chou, J.L. Roe, P.X. Kover and M.D. Purugganan. 2009. Candidate gene association mapping of Arabidopsis flowering time. Genetics 183: 325–335.

Fabre, J., M. Dauzat, V. Nègre, N. Wuyts, A. Tireau, E. Gennari, P. Neveu, S. Tisné, C. Massonnet, I. Hummel and C. Granier. 2011. PHENOPSIS DB: an information system for Arabidopsis thaliana phenotypic data in an environmental context. BMC Plant Biol. 11: 77.

Ferjani, A., G. Horiguchi, S. Yano and H. Tsukaya. 2007. Analysis of leaf development in fugu mutants of Arabidopsis reveals three compensation modes that modulate cell expansion in determinate organs. Plant Physiol. 144: 988–999.

Fiorani, F. and G.T.S. Beemster. 2006. Quantitative analyses of cell division in plants. Plant Mol. Biol. 60: 963–979.

Fiorani, F., U. Rascher, S. Jahnke and U. Schurr. 2012. Imaging plants dynamics in heterogenic environments. Cur. Op. Biotech. 23: 227–235.

Flowers, J.M., Y. Hanzawa, M.C. Hall, R.C. Moore and M.D. Purugganan. 2009. Population genomics of the *Arabidopsis thaliana* flowering time gene network. Mol. Biol. Evol. 26: 2475–2486.

French, A., S. Ubeda-Toma's, T.J. Holman, M.J. Bennett and T. Pridmore. 2009. High-throughput quantification of root growth using a novel image-analysis tool. Plant Physiol. 150: 1784–1795.

Furbank, R.T. and M. Tester. 2011. Phenomics: technologies to relieve the phenotyping bottleneck. Trends Plant Sci. 16: 635–644.

Gómez-Mena, C., M. Piñeiro, J.M. Franco-Zorrilla, J. Salinas, G. Coupland and J.M. Martínez-Zapater. 2001. Early bolting in short days: an Arabidopsis mutation that causes early flowering and partially suppresses the floral phenotype of leafy. Plant Cell 138: 1011–1024.

Gonzalez, N., H. Vanhaeren and D. Inzé. 2012. Leaf size control: complex coordination of cell division and expansion. Trends Plant Sci. PMID: 22401845.

González-Bayón, R., E.A. Kinsman, V. Quesada, A. Vera, P. Robles, M.R. Ponce, K.A. Pyke and J.L. Micol. 2006. Mutations in the RETICULATA gene dramatically alter internal architecture but have little effect on overall organ shape in Arabidopsis leaves. J. Exp. Bot. 57: 3019–3031.

Granier, C. and F. Tardieu. 1999a. Water deficit and spatial pattern of leaf development. Variability in responses can be simulated using a simple model of leaf development. Plant Physiol. 119: 609–620.

Granier, C. and F. Tardieu. 1999b. Leaf expansion and cell division are affected by reduced intercepted light but not after the decline in cell division rate in the sunflower leaf. Plant Cell Environ. 22: 1365–1376.

Granier, C., L. Aguirrezabal, K. Chenu, S.J. Cookson, M. Dauzat, P. Hamard, J.J. Thioux, G. Rolland, S. Bouchier-Combaud, A. Lebaudy, B. Muller, T. Simonneau and F. Tardieu. 2006. PHENOPSIS, an automated platform for reproducible phenotyping of plant responses to soil water deficit in *Arabidopsis thaliana* permitted the identification of an accession with low sensitivity to soil water deficit. New Phytol. 169: 623–635.

Harb, A., A. Krishnan, M.M.R. Ambavaram and A. Pereira. 2010. Molecular and physiological analysis of drought stress in Arabidopsis reveals early responses leading to acclimation in plant growth. Plant Physiol. 154: 1254–1271.

Herder, G.D., G.V. Isterdael, T. Beeckman and I. De Smet. 2010. The roots of a new green revolution. Trends Plant Sci. 15: 600–607.

Herridge, R.P., R.C. Day, S. Baldwin and R.C. Macknight. 2011. Rapid analysis of seed size in Arabidopsis for mutant and QTL discovery. Plant methods 7: 3.

Horiguchi, G., A. Ferjani, U. Fujikura and H. Tsukaya. 2006a. Coordination of cell proliferation and cell expansion in the control of leaf size in *Arabidopsis thaliana*. J. Plant Res. 119: 37–42.

Horiguchi, G., U. Fujikura, A. Ferjani, N. Ishikawa and H. Tsukaya. 2006b. Large-scale histological analysis of leaf mutants using two simple leaf observation methods: identification of novel genetic pathways governing the size and shape of leaves. Plant J. 48: 638–644.

Houle, D., D.R. Govindaraju and S. Omholt. 2010. Phenomics: the next challenge. Nature Rev. Genet. 11: 855–866.

Huala, E., A.W. Dickerman, M. Garcia-Hernandez, D. Weems, L. Reiser, F. LaFond, D. Hanley, D. Kiphart, M. Zhuang, W. Huang, L.A. Mueller, D. Bhattacharyya, D. Bhaya, B.W.

Sobral, W. Beavis, D.W. Meinke, C.D. Town, C. Somervilleand and S.Y. Rhee. 2001. The Arabidopsis Information Resource (TAIR): a comprehensive database and web-based information retrieval, analysis, and visualization system for a model plant. Nucleic Acids Res. 29: 102–105.

Huang, Y.S., M. Horton, B.J. Vilhjálmsson, U. Seren, D. Meng, C. Meyer, M. Ali Amer, J.O. Borevitz, J. Bergelsonand and M. Nordborg. 2011. Analysis and visualization of *Arabidopsis thaliana* GWAS using web 2.0 technologies. Database (Oxford). 23, 2011: bar014. Print 2011.

Hund, A., N. Ruta and M. Liedgens. 2009. Rooting depth and water use efficiency of tropical maize inbred lines, differing in drought tolerance. Plant and Soil 318: 311–325.

Irish, V.F. 2010. The flowering of Arabidopsis flower development. Plant J. 61: 1014–1028.

Jahnke, S., M.I. Menzel, D. van Dusschoten, G.W. Roeb, J. Bühler, S. Minwuyelet, P. Blümler, V.M. Temperton, T. Hombach, M. Streun, S. Beer, M. Khodaverdi, K. Ziemons, H.H. Coenen and U. Schurr. 2009. Combined MRI-PET dissects dynamic changes in plant structures and functions. Plant J. 59: 634–644.

Jansen, M., F. Gilmer, B. Biskup, K.A. Nagel, U. Rascher, A. Fischbach, S. Briem, G. Dreissen, S. Tittmann, S. Braun, I. De Jaeger, M. Metzlaff, U. Schurr, H. Scharr and A. Walter. 2009. Simultaneous phenotyping of leaf growth and chlorophyll fluorescence via GROWSCREEN FLUORO allows detection of stress tolerance in Arabidopsis thaliana and other rosette plants. Functional Plant Biol. 36: 902–914.

Jones, H.G., R. Serraj, B.R. Loveys, L. Xiong, A. Wheaton and A.H. Price. 2009. Thermal infrared imaging of crop canopies for the remote diagnosis and quantification of plant responses to water stress in the field. Functional Plant Biol. 36: 978–989.

Juenger, T.E., J.K. Mckay, N. Hausmann, J.J.B. Keurentjes, S. Sen, K. Stowe, T.E. Dawson, E.L. Simms and J.H. Richards. 2005. Identification and characterization of QTL underlying whole-plant physiology in Arabidopsis thaliana: δ13C, stomatal conductance and transpiration efficiency. Plant Cell Environ. 28: 697–708.

Karlsson, B.H., G.R. Sills and J. Nienhuis. 1993. Effects of photoperiod and vernalization on the number of leaves at flowering in 32 Arabidopsis-thaliana (brassicaceae) ecotypes. Am. J. Bot. 80: 646–648.

Kattge, J., K. Ogle, G. Bönisch, S. Díaz, S. Lavorel, J. Madin, K. Nadrowski, St. Nöllert, K. Sartor and C. Wirth. 2011. A generic structure for plant trait databases. Methods Ecol. 2: 202–213.

Koornneef, M., C.J. Hanhart and J.H. van der Veen. 1991. A genetic and physiological analysis of late flowering mutants in *Arabidopsis thaliana*. Mol. Gen. Genet. 2298: 57–66.

Koornneef, M. and D. Meinke. 2010. The development of Arabidopsis as a model plant. Plant J. 61(6): 909–921.

Krapp, A., R. Berthomé, M. Orsel, S. Mercey-Boutet, A. Yu, L. Castaings, S. Elfthieh, H. Major, J.P. Renou and F. Daniel-Vedele. 2011. Arabidopsis roots and shoots show distinct temporal adaptation pattern towards N starvation. Plant Physiol. 157: 1255–1282.

Kuromori, T., T. Wada, A. Kamiya, M. Yuguchi, T. Yokouchi, Y. Imura, H. Takabe, T. Sakurai, K. Akiyama, T. Hirayama, K. Okada and K. Shinozaki. 2006. A trial of phenome analysis using 4000 Ds-insertional mutants in gene-coding regions of Arabidopsis. Plant J. 47: 640–651.

Lempe, J., S. Balasubramanian, S. Sureshkumar, A. Singh, M. Schmid and D. Weigel. 2005. Diversity of flowering responses in wild *Arabidopsis thaliana* strains. PLoS Genet. 1: 109–118.

Levesque, M.P., T. Vernoux, W. Busch, H. Cui, J.Y. Wang, I. Blilou, H. Hassan, K. Nakajima, N. Matsumoto, J.U. Lohmann, B. Scheres and P.N. Benfey. 2006. Whole-genome analysis of the SHORT-ROOT developmental pathway in Arabidopsis. PLoS Biol. 4e: 143.

Li, Y., L. Pi, H. Huang and L. Xu. 2012. ATH1 and KNAT2 proteins act together in regulation of plant inflorescence architecture. J. Exp. Bot. 63: 1423–1433.

Li, Y.F., G. Kennedy, F. Ngoran, P. Wu and J. Hunter. 2013. An ontology-centric architecture for extensible scientific data management systems. Future Gener. Comp. Syst. 29: 641–653.

Lièvre, M., N. Wuyts, S.J. Cookson, J . Bresson, M. Dapp, F. Vasseur, C. Massonnet, S. Tisne, M. Bettembourg, C. Balsera, A. Bédiée, F. Bouvery, M. Dauzat, G. Rolland, D. Vile and C. Granier. 2013. Phenotyping the kinematics of leaf development in flowering plants: recommendations and pitfalls. WIREs Dev. Biol. doi: 10.1002/wdev.119.

Liu, Y.G., N. Mitsukawa, T. Oosumi and R.F. Whittier. 1995. Efficient isolation and mapping of *Arabidopsis thaliana* T-DNA insert junctions by thermal asymmetric interlaced PCR. Plant J. 8: 457–463.

Martienssen, R.A. 1998. Functional genomics: probing plant gene function and expression with transposons. Proc. Natl. Acad. Sci. USA 95: 2021–2026.

Martinez-Zapater, J.M., G. Coupland, C. Dean and M. Koomneef. 1994. The transition to flowering in Arabidopsis. *In:* Arabidopsis, E.M. Meyerowitz and C.R. Somerville (eds). Cold Spring Harbor, NY: Cold Spring Harbor Press, pp. 403–433.

Massonnet, C., D. Vile, J. Fabre, M.A. Hannah, C. Caldana, J. Lisec, G.T.S. Beemster, R.C. Meyer, G Messerli, J.T. Gronlund, J. Perkovic, E. Wigmore, S. May, M.W. Bevan, C. Meyer, S. Rubio-Díaz, D. Weigel, J.L. Micol, V. Buchanan-Wollaston, F. Fiorani, S. Walsh, R. Rinn, W. Gruissem, P. Hilson, L. Hennig, L. Willmitzer and C. Granier. 2010. Probing the reproducibility of leaf growth and molecular phenotypes: a comparison of three Arabidopsis accessions cultivated in ten laboratories. Plant Physiol. 152: 2142–2157.

Massonnet, C., S. Tisné, A. Radziejwoski, D. Vile, L. de Veylder, M. Dauzat and C. Granier. 2011. New insights into the control of endoreduplication: endoreduplication is driven by organ growth in Arabidopsis leaves. Plant Physiol. 157: 2044–2055.

Meinke, D., R. Muralla, C. Sweeney and A. Dickerman. 2008. Identifying essential genes in *Arabidopsis thaliana*. Trends Plant Sci. 13: 483–491.

Meinke, D.W., J.M. Cherry, C. Dean, S.D. Rounsley and M. Koornneef. 1998. *Arabidopsis thaliana*: a model plant for genome analysis. Science 282: 662, 679–682.

Méndez-Vigo, B., M. Teresa de Andres, M. Ramiro, J.M. Martinez-Zapater and C. Alonso-Blanco. 2010. Temporal analysis of natural variation for the rate of leaf production and its relationship with flowering initiation in *Arabidopsis thaliana*. J. Exp. Bot. 61: 1611–1623.

Meyerowitz, E.M. 1989. Arabidopsis, a useful weed. Cell 56: 263–269.

Micol, J.L. 2006. Mutations in the RETICULATA gene dramatically alter internal architecture but have little effect on overall organ shape in Arabidopsis leaves. J. Exp. Bot. 57: 3019–3031.

Micol, J.L. 2009. Leaf development: time to turn over a new leaf? Curr. Opin. Plant Biol. 12: 9–16.

Mungall, C.J., G.V. Gkoutos, C.L. Smith, M.A. Haendel, S.E. Lewis and M. Ashburner. 2010. Integrating phenotype ontologies across multiple species. Genome Biol. 11: R2.

Nagel, K.A., B. Kastenholz, S. Jahnke, D. van Dusschoten, T. Aach, M. Mühlich, D. Truhn, H. Scharr, St. Terjung, A. Walter and U. Schurr. 2009. Temperature responses of roots: impact on growth, root system architecture and implications for phenotyping. Functional Plant Biol. 11: 947–959.

Nagel, K.A., A. Putz, F. Gilmer, K. Heinz, A. Fischbach, J. Pfeifer, M. Faget, S. Bloßfeld, M. Ernst, C. Dimaki, B. Kastenholz, A.K. Kleinert, A. Galinski, H. Scharr, F. Fiorani and U. Schurr. 2012. GROWSCREEN-Rhizo is a novel phenotyping robot enabling simultaneous measurements of root and shoot growth for plants grown in soil-filled rhizotrons. Functional Plant Biol. 39: 891–904.

Nagel, K.A., U. Schurr and A. Walter. 2006. Dynamics of root growth stimulation in *Nicotiana tabacum* in increasing light intensity. Plant Cell Environ. 29: 1936–1945.

O'Malley, R. and J. Ecker. 2010. Linking genotype to phenotype using the Arabidopsis unimutant collection. Plant J. 61: 928–940.

Parent, B. and F. Tardieu. 2012. Temperature responses of developmental processes have not been affected by breeding in different ecological areas for 17 crop species. New Phytol. doi: 10.1111/j.1469-8137.2012.04086.x. (Epub ahead of print) PubMed PMID: 22390357.

Parizot, B., I. Roberts, J. Raes, T. Beeckman and I. De Smet. 2012. In silico analyses of pericycle cell populations reinforce their relation with associated vasculature in *Arabidopsis*. Phil. Trans. R Soc. B 367: 1479–1488.

Peaucelle, A., H. Morin, J. Traas and P. Laufs. 2007. Plants expressing a miR164-resistant CUC2 gene reveal the importance of post-meristematic maintenance of phyllotaxy in Arabidopsis. Development 134: 1045–1050.

Pérez-Pérez, J.M., J. Serrano-Cartagena and J.L. Micol. 2002. Genetic analysis of natural variations in the architecture of *Arabidopsis thaliana* vegetative leaves. Genetics 162: 893–915.

Pérez-Pérez, J.M., S. Rubio-Díaz, S. Dhondt, D. Hernández-Romero, J. Sánchez-Soriano, G.T. Beemster, M.R. Ponce and J.L. Micol. 2011. Whole organ, venation and epidermal cell morphological variations are correlated in the leaves of Arabidopsis mutants. Plant Cell Environ. 34: 2200–2211.

Petricka, J.J., C.M. Winter and P.N. Benfey. 2012. Control of Arabidopsis root development. Annu. Rev. Plant Biol. 63: 563–590.

PHENOSCOPE : IJPB, Phénotypage haut débit chez *Arabidopsis thaliana*. (http://www.versailles.inra.fr/ijpb2/english/ppa/ppa_accueil.html).

Pierret, A., M. Kirby and C. Moran. 2003. Simultaneous X-ray imaging of plant root growth and water uptake in thin-slab systems. Plant Soil 255: 361–373.

Pigliucci, M. and C.D. Schlichting. 1995. Ontogenetic reaction norms in *Lobelia siphilitica* (Lobeliaceae): response to shading. Ecology 76: 2134–2144.

Pigliucci, M., J. Whitton and C.D. Schlichting. 1995. Reaction norms of Arabidopsis. I. Plasticity of characters and correlations across water, nutrient and light gradients. J. Evol. Biol. 8: 421–438.

Poethig, R.S. 2003. Phase change and the regulation of developmental timing in plants. Science 301: 334–336.

Poorter, H., F. Fiorani, M. Stitt, U. Schurr, A. Finck, Y. Gibon, B. Usadel, R. Munns, O.K. Atkin, F. Tardieu and T.L. Pons. 2012a. The art of growing plants for experimental purposes, a practical guide for the plant biologist. Functional Plant Biol. 39: 821–838.

Poorter, H., K.J. Niklas, P. Reich, J. Oleksyn, P. Poot and L. Mommer. 2012b. Biomass allocation to leaves, stems and roots: meta-analyses of interspecific variation and environmental control. New Phytol. 193: 30–50.

Poorter, H., U. Niinemets, A. Walter, F. Fiorani and U. Schurr. 2010. A method to construct dose-response curves for a wide range of environmental factors and plant traits by means of a meta-analysis of phenotypic data. J. Exp. Bot. 61: 2043–2055.

Pouteau, S. and C. Albertini. 2009. The significance of bolting and floral transitions as indicators of reproductive phase change in Arabidopsis. J. Exp. Bot. 60: 3367–3377.

Ragni, L., E. Belles-Boix, M. Gunl and V. Pautot. 2008. Interaction of KNAT6 and KNAT2 with BREVIPEDICELLUS and PENNYWISE in Arabidopsis inflorescences. Plant Cell 20: 888–900.

Reinhardt, D., E.R. Pesce, P. Stieger, T. Mandel, K. Baltensperger, M. Bennett, J. Traas, J. Friml and C. Kuhlemeier. 2003. Regulation of phyllotaxis by polar auxin transport. Nature 4268: 255–260.

Rymen, B., F. Coppens, S. Dhondt, F. Fiorani and G.T.S. Beemster. 2010. Kinematic analysis of cell division and expansion. Methods in Molecular Biology 655: 203–227.

Serrano-Cartagena, J., P. Robles, M.R. Ponce and J.L. Micol. 1999. Genetic analysis of leaf form mutants from the Arabidopsis Information Service collection. Mol. Gen. Genet. 261: 725–739.

Shinozaki, K. 2010. The Chloroplast Function Database: a large-scale collection of Arabidopsis Ds/Spm- or T-DNA-tagged homozygous lines for nuclear-encoded chloroplast proteins, and their systematic phenotype analysis. Plant J. 61: 529–542.

Skirycz, A., S.D. Bodt, T. Obata, I.D. Clercq, H. Claeys, R.D. Rycke, M. Andriankaja, O.V. Aken, F.V. Breusegem, A.R. Fernie and D. Inzé. 2010. Developmental stage specificity and the role of mitochondrial metabolism in the response of Arabidopsis leaves to prolonged mild osmotic stress. Plant Physiol. 152: 226–244.

Smith, H.M. and S. Hake. 2003. The interaction of two homeobox genes, BREVIPEDICELLUS and PENNYWISE, regulates internode patterning in the Arabidopsis inflorescence. Plant Cell. 15: 1717–1727.

Sterken, R., R. Kiekens, J. Boruc, F. Zhang, A. Vercauteren, I. Vercauteren, L. De Smet, S. Dhondt, D. Inzé, L. De Veylder, E. Russinova and M. Vuylsteke. 2012. Combined linkage and association mapping reveals CYCD5;1 as a quantitative trait gene for endoreduplication in *Arabidopsis*. Proc. Natl. Acad. Sci. USA 109: 4678–4683.

Swarup, K., E.M. Kramer, P. Perry, K. Knox, H.M. Leyser, J. Haseloff, G.T.S. Beemster, R. Bhalerao and M.J. Bennett. 2005. Root gravitropism requires lateral root cap and epidermal cells for transport and response to a mobile auxin signal. Nat. Cell Biol. 7: 1057–1065.

Tardieu, F. 2012. Any trait or trait-related allele can confer drought tolerance: just design the right drought scenario. J. Exp. Bot. 63: 25–31.

Taylor, C.F., D. Field, S.A. Sansone, J. Aerts, R. Apweiler, M. Ashburner, C.A. Ball, P.A. Binz, M. Bogue, T. Booth, A. Brazma, R.R. Brinkman, A. Michael Clark, E.W. Deutsch, O. Fiehn, J. Fostel, P. Ghazal, F. Gibson, T. Gray, G. Grimes, J.M. Hancock, N.W. Hardy, H. Hermjakob, R.K. Julian Jr., M. Kane, C. Kettner, C. Kinsinger, E. Kolker, M. Kuiper, N. Le Novère, J. Leebens-Mack, S.E. Lewis, P. Lord, A.M. Mallon, N. Marthandan, H. Masuya, R. McNally, A. Mehrle, N. Morrison, S. Orchard, J. Quackenbush, J.M. Reecy, D.G. Robertson, P. Rocca-Serra, H. Rodriguez, H. Rosenfelder, J. Santoyo-Lopez, R.H. Scheuermann, D. Schober, B. Smith, J. Snape, C.J. Stoeckert, K. Tipton, P. Sterk, A. Untergasser, J. Vandesompele and S. Wiemann. 2008. Promoting coherent minimum reporting guidelines for biological and biomedical investigations: the MIBBI project. Nat. Biotech. 26: 889–896.

Tisné, S., F. Barbier and C. Granier. 2011. The ERECTA gene controls spatial and temporal patterns of epidermal cell number and size in successive developing leaves of *Arabidopsis thaliana*. Ann. Bot. 108: 159–168.

Tisné, S., M. Reymond, D. Vile, J. Fabre, M. Dauzat, M. Koornneef and C. Granier. 2008. Combined genetic and modeling approaches reveal that epidermal cell area and number in leaves are controlled by leaf and plant developmental processes in Arabidopsis. Plant Physiol. 148: 1117–1127.

Trachsel, S., S.M. Kaeppler, K.M. Brown and J.P. Lynch. 2011. Shovelomics: high throughput phenotyping of maize (*Zea mays* L.) root architecture in the field. Plant and Soil 341: 75–87.

Traitmill: Cropdesign, Platform and Process (http://www.cropdesign.com/tech_traitmill.php).

Tsuge, T., H. Tsukaya and H. Uchimiya. 1996. Two independent and polarized processes of cell elongation regulate leaf blade expansion in *Arabidopsis thaliana* (L.) Heynh. Development 122: 1589–1600.

Ubeda-Tomás, S., G.T. Beemster and M.J. Bennett. 2012. Hormonal regulation of root growth: integrating local activities into global behaviour. Trends in Plant Science. epub ahead of print.

Vanhaeren, H., N. Gonzalez and D. Inzé. 2010. Hide and seek: uncloaking the vegetative shoot apex of *Arabidopsis thaliana*. Plant J. 63: 541–548.

Vasseur, F., F. Pantin and D. Vile. 2011. Changes in light intensity reveal a major role for carbon balance in Arabidopsis responses to high temperature. Plant Cell Environ. 34: 1563–1579.

Walter, A., H. Scharr, F. Gilmer, R. Zierer, K.A. Nagel, M. Ernst, A. Wiese, O. Virnich, M.M. Christ, B. Uhlig, S. Jünger and U. Schurr. 2007. Dynamics of seedling growth acclimation towards altered light conditions can be quantified via GROWSCREEN: a setup and procedure designed for rapid optical phenotyping of different plant species. New Phytol. 174: 447–455.

Walter, A., W.K. Silk and U. Schurr. 2009. Environmental effects on spatial and temporal patterns of leaf and root growth. Ann. Rev. Plant Biol. 60: 279–304.

Wang, J.W., R. Schwab, B. Czech, E. Mica and D. Weigel. 2008. Dual effects of miR156-targeted SPL genes and CYP78A5/KLUH on plastochron length and organ size in *Arabidopsis thaliana*. Plant Cell 20: 1231–1243.

Weight, C. and D.R.W. Parnham. 2008. LeafAnalyser: a computational method for rapid and large-scale analyses of leaf shape variation. Plant J. 53: 578–586.

Wiese, A., M.M. Christ, O. Virnich, U. Schurr and A. Walter. 2007. Spatio-temporal leaf growth patterns of Arabidopsis thaliana and evidence for sugar control of the diel leaf growth cycle. New Phytol. 174: 752–761.

Wilkinson, M., H. Schoof, R. Ernst and D. Haase. 2005. BioMOBY successfully integrates distributed heterogeneous bioinformatics Web Services the PlaNet exemplar case. Plant Physiol. 138: 5–17.

WIWAM: Plant Systems Biology, Systems biology of drought tolerance in Arabidopsis. (http://www.psb.ugent.be/yield research/166 projects2).

Zeller, G., S.R. Henz, C.K. Widmer, T. Sachsenberg, G. Rätsch, D. Weigel and S. Laubinger. 2009. Stress-induced changes in the *Arabidopsis thaliana* transcriptome analyzed using whole-genome tiling arrays. Plant J. 58: 1068–1082.

Zhang, X., R.J. Hause and J.O. Borevitz. 2012. Natural genetic variation for growth and development revealed by high-throughput phenotyping in *Arabidopsis thaliana*. G3 (Bethesda) 2: 29–34.

7

Challenges of Crop Phenomics in the Post-genomic Era

Vasilis C. Gegas,[1], Alan Gay,[2] Anyela Camargo[2]
and John H. Doonan[2]*

Introduction

The idea of linking genes to phenotypes, with the aim of manipulating those phenotypes, has been one of the main drivers behind the huge investment in developing genetic and genomic platforms in both model plants (such as Arabidopsis (http://signal.salk.edu/; http://arabidopsis.info/) and Brachypodium (http://www.brachypodium.org/)) and crops, of which rice and maize present perhaps the best examples (http://signal.salk.edu/cgi-bin/RiceGE; http://rmd.ncpgr.cn/; http://tos.nias.affrc.go.jp/; http://maizecoop.cropsci.uiuc.edu/mgc-info.php). Indeed, modern genetics has been instrumental in discovering the primary genetic determinants of very many phenotypes (or traits), such as resistance to diseases, development and response to defined environmental variables to mention just but a few. Most attention has focused on single gene knockouts and on the resultant simple, relatively discreet, traits. These traits are much easier to analyse than the polygenic continuous traits that are of general importance in agriculture.

---

[1] Limagrain UK Ltd., Joseph Nickerson Research Centre, Rothwell, LN7 6DT, Lincs, UK.
[2] National Plant Phenomics Centre, IBERS, Aberystwyth University Aberystwyth, Ceredigion, SY23 3EB, UK.
* Corresponding author

In the face of global grand challenges for sustainable food production and safety, the emphasis has shifted to more complex traits (e.g., yield) that may require different approaches (such as Marker-assisted selection) for their analysis and manipulation (Pérez-de-Castro et al. 2012). Next-generation genomic technologies give increasingly ready access to the genomic variation available in our major crop plants, but similar access to the phenotypic variation remains relatively expensive and slow by comparison (see Table 1), and is referred to as the "phenotyping bottleneck" (Furbank and Tester 2011).

With this problem in mind, the research community has begun to consider: what is a useful phenotypic equivalent of the genome. The phenome can be defined as the physical manifestation of the complex interaction between genome and environment and is manifest as traits at the chemical, cellular, developmental, and physiological levels (Fig. 1; Bilder et al. 2010). Phenotyping aims at the quantitative measurement of a range of traits, and in the case of crops these include such diverse features as photosynthesis, growth or biomass accumulation, architecture and development of both root and shoot systems, as well as molecular and cellular characteristics that are important for quality-related traits. The very high dimensionality of the phenome precludes a complete description across all the possible

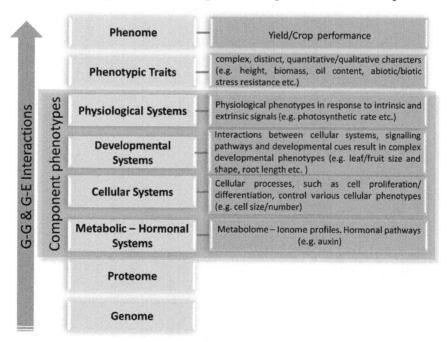

Figure 1. Schematic illustration of an eight-step "genome to phenome" mechanistic model for crop phenomics research.

Table 1. Open access databases of phenomics information for various crop species.

Species	Database	Website	Reference
Barley	SCRI Barley Mutants	http://bioinf.scri.ac.uk/barley/	(Caldwell et al. 2004)
	GeneNetwork	http://genenetwork.org/dbdoc/SXMPublish.html	(Druka et al. 2008)
Maize	MaizeGDB OPTIMAS-DW	http://www.maizegdb.org/rescuemu-phenotype.php http://www.optimas-bioenergy.org/optimas_dw	(Lawrence et al. 2007) (Colmsee et al. 2012)
Rice	Oryza Tag Line (OTL)	http://urgi.versailles.inra.fr/OryzaTagLine	(Larmande et al. 2008)
	Rice Mutant Database (RMD)	http://rmd.ncpgr.cn	(Zhang et al. 2006)
	Tos17	http://pc7080.abr.affrc.go.jp/phenotype	(Miyao et al. 2007)
	OryGenesDB	http://orygenesdb.cirad.fr/index.html	
Sorghum	Sorghum Research	http://www.lbk.ars.usda.gov/psgd/index	(Xin et al. 2008)
Brassicas	Shanghai RAPESEED Database	http://rapeseed.plantsignal.cn/	(Wu et al. 2008)
	CropStoreDB	http://www.cropstoredb.org/	(Love et al. 2012)
Soybean	Soybean Mutation Database	http://www.soybeantilling.org/psearch.jsp	N/A
Lotus	Legume Base	http://www.shigen.nig.ac.jp lotusjaponicus/top/top.jsp /bean/	N/A
Tomato	Tomato Mutant Database	http://zamir.sgn.cornell.edu/mutants	(Menda et al. 2004)
	LycoTILL	http://www.agrobios.it/tilling/index.html	N/A

levels of granularity, so most studies focus on the immediately relevant and approachable aspects. The challenge of capturing even a selective part of this increasingly complex, multivariate (or high-dimensional) phenome space has led to the emergence of phenomics as a multidisciplinary science dedicated to the systematic study and objective description of phenomes. Phenomics, therefore, aims to address the "genome-to-phenome" problem through a vertically integrated approach in which relevant *component* phenotypes are initially defined and quantified and subsequently the multi-level relationship between those *component phenotypes* and the phenome is established (Fig. 1). The term *component* phenotype is used here in the context of crop phenomics, where the phenome in question is extremely complex, i.e., yield or crop performance, and is the sum of component phenotypes that are characteristic of a particular organizational level (e.g., cellular systems) and are integral components of the model(s) defining the whole organism or crop phenome.

Parameters that are particularly crucial for crop phenomics (and have perhaps received little formal consideration in model species) include the community level (is the phenotype being described at individual or population level, for example) and a formalized and widely accepted method to capture environmental variables. Many of our key crop species are rarely grown as individuals and the community in which they grow can have dramatic effects on plant development and yield. Plants grown in a monoculture under good agronomic practice can appear and perform very differently from the same genotype in a mixed community (for example, competing with weeds).

The Need for Scaling in Phenotyping

"Scaling" in this context refers to the increase (orders of magnitude) in both depth and throughput of phenotyping. It can include both the number of different phenotypes recorded for an individual as well as the number of plants or organs analysed. In the context of crops, moving from a lab to a field-based phenomics strategy often necessitates scaling in addition to handling the increased complexity due to variable (and often poorly defined or recorded) environments (i.e., locations) or treatments. The current demand for increased throughput is driven by a number of factors including the recent revolution in genotyping, cost-effective automation and the increased cost of traditional phenotyping, not to mention its somewhat variable implementation.

Genomic information is essentially a linear code that is relatively simple to acquire and the technology lends itself to scaling. It is also independent (largely) of the environment in which the subject is grown and, although the environment can modify the basic code, there is little requirement for different platforms or strategies for genomics on plants grown in different environments. Phenotyping, however, often needs tailor-made strategies that depend on the species and the trait of interest. The major limitation in most genetic studies, therefore, is no longer the acquisition of genomic information but rather the difficulty and expense of acquiring sufficient quantitative phenotypic data. Additional complexity occurs as the environment can radically alter the observed phenotype. Unfortunately, in very many cases, the environment has been poorly characterized and often reported only at a very superficial level that makes it impossible to integrate results between different investigations.

Genomic and genetic resources for many species are exponentially increasing. Indeed comprehensive collections of mutant lines and natural accessions are available for Arabidopsis (Cao et al. 2011, O'Malley and Ecker 2010, Weigel and Mott 2009), *Medicago* (Cannon et al. 2009), rice (An et al. 2005, Krishnan et al. 2009), *Brachypodium* (Vogel et al. 2010) and maize (Settles et al. 2007). Even as genetic diversity is being lost, germplasm is increasingly being sampled and documented on a global scale and made available to researchers and breeders. Many new genetically diverse populations are being created both by traditional means—crosses between members of the same species, wide crosses between related but previously geographically dispersed species—and via biotechnological approaches. Superimposed on this explosion of diversity are the recent advances in genomics and other molecular analyses that allow whole populations of genetically distinct lines to be characterized down to the molecular scale, at a relatively modest and decreasing cost. However, utilizing these genetic resources in order to understand genome function and its response to environment and accelerate gene discovery, genomic information needs to be coupled with equally deep and extensive phenotypic information.

The magnitude and ease with which genomic information can now become available has allowed the development of sophisticated approaches such as genomic selection for predicting complex traits and therefore facilitating selection in breeding programmes especially under diverse or challenging environments. However, the successful implementation of this approach in breeding relies heavily on acquiring reliable phenotypic data on a scale similar to that of the genotyping data (Cabrera-Bosquet et al. 2012).

The problem is given some considerable degree of urgency with rapidly decreasing natural diversity due to habitat loss and other causes and the demand for more productive crop varieties. A step-change is required to reduce phenotyping costs and increase throughput, while capturing objective data under well-defined conditions and rendering it in a useful format that can be easily integrated with other large-scale datasets.

Biotechnological applications have also created a demand for objective systematic phenotyping: large-scale shotgun screens aimed at candidate gene selection use high-throughput techniques to assess transgenic plant performance and select the rare lines with potential agricultural merit (Reuzeau et al. 2006). Regulatory approval of transgenic plants for use in agriculture, at least in Europe, may demand a high level of mechanistic understanding and a demonstration of substantive equivalence, although this is likely to decrease as the public becomes more comfortable with the technology (Lusk and Rozan 2006). Indeed, in many countries, regulation has been relaxed or never imposed and the success of a transgenic variety depends on plant performance in the field.

Many key traits lend themselves to direct visual assessment—the combination of plant height and response to added nitrogen that gave rise to the Green Revolution, for example—and large gains in performance have been achieved using "low-tech" approaches. Scaling for many end-point traits (i.e., yield) requires little in the way of further technical innovation—adapted combines have long been used to measure yield on large breeding populations, and replicated over many sites and years and this remains a highly effective, though very expensive, means of phenotyping in most commercially important crops.

However, yield is a good example of an agronomic trait that, from a mechanistic (or reductionist) point of view, can be considered as "composite"—that is to say it represents the integration of many different biological processes whose importance can depend on many factors, including species, strain, environment and agronomy. Understanding these component processes and their contribution, particularly where there are innate functional conflicts, may allow us to devise more efficient breeding strategies.

Automation can be highly effective for both destructive and non-destructive analyses of large numbers of individual plants grown under controlled environment conditions. Non-destructive analysis can provide dynamic information on plant growth, development and response to changes in the environment and, in principle, is an attractive means of phenotyping large populations. However, applying these approaches in the field can be problematical. Many of the techniques employed depend on precise quantitative imaging to reveal differences in plant performance and the variable environments found in most field conditions act to frustrate such measurements (reviewed by Walter et al. (2012)).

The Lab-to-field "Valley of Death"[1]

Technology transfer between Controlled Environment Rooms (CERs, i.e., growth rooms where light, temperature and air composition can be precisely controlled and monitored) and the field has never been a serious issue for genomics since the genetic profile of a plant remains largely unchanged regardless of the environment (for SNP discovery, genome sequencing, etc.). Platforms and methodologies, initially developed in model species, are equally effective at collecting genomic information from diverse crop species. Indeed the genomes of rice (International Rice Genome Sequencing Project 2005), maize (Schnable et al. 2009) and sorghum (Bedell et al. 2005) are already sequenced and other crop species such as wheat (Brenchley et al. 2012) and barley (International Barley Genome Sequencing Consortium et al. 2012) are following soon and, as the cost of sequencing rapidly reduces, the mining of extensive germplasm collections for allelic variants using genome re-sequencing will soon become routine (Pérez-de-Castro et al. 2012).

In the case of phenomics, however, phenome plasticity renders the lab-to-field transition particularly challenging and is often considered as one of the main obstacles for phenomics crop research. Indeed in the nascent field of plant phenomics, there is a constant debate regarding the merits (and shortcomings) of lab versus field phenotyping, with those coming from an academic background tending to favour controlled conditions, whereas those from an agronomic or breeding background favour field phenotyping. Both undoubtedly have their place and purpose but it is necessary to be clear about the limitations of each. One of the main arguments in favour of lab-based phenotyping is that it allows the acquisition of high-quality data in a high-throughput and highly reproducible way, whereas field-based phenotyping is traditionally seen as a simplistic, low-resolution and low-throughput approach. It is often argued that CER experiments are relatively ineffective in exploring the intricate environmental parameter space and its effect on physiological traits and complex phenomes such as yield. On the other hand, field experiments and trials usually include the effects that environmental or population stresses have on traits, although in a manner that can differ for successive crops. Field-based phenotyping is therefore considered irreplaceable when screening for environmentally sensitive traits or selecting for specific target environments. Thus, we now need to consider how we may best prioritize our efforts to improve crop phenotyping.

---

[1] The expression "Valley of Death" is often used in the context of venture capital to describe the period in the life of a start-up company between R&D or innovation phase and commercialization (Manheim 2011). The term is also used to describe the gap between academic research and industrial application (Butler 2008).

Phenotype Prioritization for Crop Phenomics Research

A crucial step towards bridging the "valley of death" is a comprehensive understanding of those properties and qualities that make a phenotype (or trait/character) suitable for a phenomics strategy (as opposed to a component phenotype approach) both under lab and field conditions. Prioritizing suitable phenotypes, therefore, becomes even more important. Some of the criteria for such prioritization might include:

The Properties and Cost of Trait Measurements

To date, the main efforts have focused on the automation of trait measurement, both at analytical and instrumentation level. This is hardly a surprise, given the significant financial and operational costs of measurements, especially as they scale up from the lab to the field level and as throughput increases.

Direct quantification of certain phenotypes can be very expensive or impractical to scale. For this reason, the use of "surrogate" or "proxy" measures to describe traits and capture phenotypic variation is an attractive approach. Proxy measures can provide an effective way to capture phenotypic variation, given a careful calibration using manually collected data, and they can be well suited to high-throughput platforms for field applications.

Central to any crop phenomics strategy is that the phenotypic measurements made capture the key aspects of the contributing component that allow the breeder to dissect the trait sufficiently to identify strategies that could lead to crop improvement. Phenotypic "space" is highly dimensional and the exhaustive analysis of a phenotype is usually unattainable. This raises questions as to how extensively and intensively we need to measure a particular phenotype (Houle 2010). Failure to address this issue could result in a phenomics strategy that is too intensive, measuring aspects of the phenotype that might be irrelevant to the goal (Houle 2010). Conversely, a strategy that measures only a limited number of the phenotype's attributes will have a limited ability to capture the existing phenotypic variation, especially in cases of complex pleiotropic effects.

Any phenotypic measure or screen should also have **sufficient reliability**, expressed as both **internal consistency** (reduces sample-to-sample experimental error) and **discriminative power** (captures a wide spectrum of the trait variation (Bilder et al. 2010)). In the latter case, measurements with a restricted range will fail to capture subtle genetic effects due to lack of sensitivity. Robustness, which refers to the stability of the phenotypic measurements either spatially, temporarily or developmentally, also impacts on reliability and could significantly weaken the effectiveness of the screen simply because an important attribute of the phenotype, i.e., developmental plasticity (at tissue or organ level) or stage of crop development, is disregarded.

Currently phenomics is almost synonymous to high-throughput phenotyping approaches. Therefore, a critical property of any measure or screen is how amenable it is to a high-throughput measurement. This property becomes particularly important in the field where unstable and unpredictable conditions and the physical scale of the experimental site could hinder phenotypic methods that appear functional and robust in the lab. This is particularly evident in root architecture phenotyping. Although lab-based phenotyping platforms have now been developed for high-throughput imaging for the study of root architecture and the kinetics of root development (Clark et al. 2011, Iyer-Pascuzzi et al. 2010, Mooney et al. 2011), a high-throughput method for field-based root phenotyping remains elusive.

In recent years, various state-of-the-art phenomics centres have been established aimed at integrating sensor platforms with precisely controlled conditions for high-throughput and deep plant phenotyping in a cost-effective manner. Examples of such centres include the Plant Accelerator (http://www.plantaccelerator.org.au/) and the High Resolution Plant Phenomics Centre (http://www.plantphenomics.org/HRPPC) in Australia, the Jülich Plant Phenotyping Centre (http://www.fz-juelich.de/ibg/ibg-2/EN/methods_jppc/methods_node.html) in Germany and the National Plant Phenomics Centre (http://www.phenomics.org.uk/temp-site/about.html) in the UK. Although these centres utilize platforms designed primarily for lab-based phenotyping, they also seek to apply relevant technologies under field conditions at experimental (e.g., mapping and experimental populations) and industrial scale (e.g., breeding and precision farming) and they therefore have the potential to provide an effective bridge for lab-to-field phenomics.

The properties discussed above can significantly affect the suitability and cost-effectiveness (some more than others) of a particular screen for crop phenotyping and determine its translational potential as a selection tool in commercial breeding. It is difficult to assess the financial implications (e.g., returns to investment) of adopting new technologies in breeding programmes, for this type of information is usually commercially sensitive and therefore non-publicly available. However, some of the existing literature does give an idea of the potential economic gains for using the appropriate phenotyping tools in the context of a breeding strategy (Brennan and Morris 2001, Brennan et al. 2007, Brennan and Martin 2007). For example, an economic assessment of the use of indirect physiological measurements for stomatal aperture-related traits (e.g., canopy temperature, carbon isotope decomposition, etc.), as selection criteria for yield potential in wheat, indicated that such approaches can reduce the cost of developing new varieties by enabling the early identification of the least favourable lines, and therefore, reducing the cost of yield trials at later generations (i.e., "no-hopers" are discarded early), and increasing the probability for selecting successful lines (i.e., higher frequency of lines with high yield potential) (Brennan et al. 2007, Brennan and Martin 2007). In the same analysis, the costs associated with the different physiological measurements used for stomatal aperture traits seem to have a significant financial bearing on the potential benefits of a particular screen. For example, the total costs per plot for the leaf porosity measure (measured with a hand-held porometer) were almost double as compared with the canopy temperature depression screen (measured with a hand-held IR thermometer) with labor cost being the main contributor to this (Brennan and Martin 2007).

Several criteria, underpinning the successful application of indirect measures as selection tools in breeding programmes, were proposed previously (Araus et al. 2002, Bänziger et al. 2000, Jackson 2001, Reynolds et al. 2001) and can be summarized as follows: (a) high heritability, (b) high phenotypic and genetic correlation with performance, (c) significant genetic variation (ideally in agronomically acceptable backgrounds), and (d) low cost.

Knowledge of Environmental Effects

Plants, as sessile organisms, are highly adapted to coping with a variable environment and, as a result, traits can be strongly influenced by environmental conditions. Although breeding aims at producing varieties that perform well under diverse environments, not all traits are equally "resilient". Therefore, appropriate documentation of the environment is essential as part of any crop phenomics strategy, as is an appreciation of the salient environmental variables affecting particular traits. This becomes particularly important in trait-based breeding and when specific environments are being targeted. Breeding strategies usually try to account for confounding effects associated with environmental conditions by performing replicated trials across as many diverse environments as possible over several years. This approach is pragmatic in that it provides a platform for phenotyping under realistic conditions, but it requires an extensive trialing infrastructure, which can span across international boundaries, at considerable cost. A good knowledge of the different environmental attributes (seasonal fluctuations in rainfall and temperature) of the trial locations is essential (Hobbs and Sayre 2001, Masuka et al. 2012). In drought breeding, for example, the phenotyping environment should allow the stress to occur during the developmental stages that are most critical for reduction in yield potential (Bänziger et al. 2000, Grant et al. 1989). Detailed climatic metadata of the testing environments is important not only for experimental design but also for interpretation and integration across sites. For breeding purposes, the expense of collecting and curating extensive sets of metadata has to be balanced with the economic benefit. Unfortunately, the metadata associated with many field and CER experiments has not progressed, it sometimes seems, much since the days of Captain Cook, if it actually equals it, when Royal Navy ships began the tradition of systematically recording several meteorological parameters several times per day. This archive now underpins diverse research projects on climate change (García-Herrera et al. 2005, Wheeler et al. 2010). Lacking naval discipline and driven by other priorities, the environmental metadata in bio-science publications varies from the arbitrary to woefully absent and, even if referred to, is rarely easily accessible. A multisite experiment using Arabidopsis grown under CER has demonstrated the desirability of detailed environmental metadata (Massonnet et al. 2010), and suggests that minimal common standards need to be established, even for so-called controlled environments. There is, therefore, a desperate need to establish and enforce standards for metadata collection, both in academia and in industry, if we are to realize the full value of field experiments.

Soil is another, often neglected, aspect of the plants' environment that has major effects on crop phenotypes. Soil heterogeneity can introduce significant residual variation in field experiments especially when screening for drought tolerance, nutrient deficiencies or toxicities (Bänziger et al. 2000, Hayes 1925). However, selection of phenotyping sites for field experiments or breeding trials is usually based on empirical assessment of the field (or a routine soil analysis at best) and occasionally on detailed cartography of soil variability. Interestingly, in precision agriculture, mapping of soil variability across fields based on either "on-the-go" or remote sensors is now used for managing nitrogen application and other crop inputs (Adamchuk et al. 2004, Bah et al. 2012, Hively et al. 2011). Variability in the soil physical and chemical properties across the experimental fields can significantly affect the efficiency as well as the reproducibility of the screen (Brownie et al. 1993, Hayes 1925). Soil texture (the relative amount of sand, silt and clay) and parent rock have a direct effect on parameters, such as nutrient dynamics, water holding capacity and soil penetration resistance (Hillel 1980), which in turn affect root development and function and the ability of plants to access water (Bänziger et al. 2000). A heterogeneous soil texture across an experimental field could prevent the uniform application of a stress, such as drought, and introduce additional experimental error. In low nitrogen experiments, soil properties and soil variability is even more important (Godwin and Miller 2003). Screening for yield potential under low N input requires, firstly, that the experimental site is sufficiently depleted of residual nitrogen so the minimum N treatment can be achieved and, secondly, that soil properties that affect N supply and mineralization rate, such as acidity, texture and organic matter (Hillel 1980), are similar across the field (Bänziger et al. 2000). Failure to apply those two principles could significantly frustrate the phenotyping efforts, either because a clear difference in yield between the low and high N treatment cannot be observed (low N treatment is not discriminative enough) or because of inflated experimental error due to heterogeneous availability of the nutrient. As in the case of drought breeding, it is important to determine the developmental stage at which the low N stress has the most detrimental effects on yield. Ensuring N availability at the correct developmental stage depends not only on soil N supply, but also on intrinsic biological processes such as N assimilation and remobilization that determine the amount of available nitrogen in the actual crop (Bänziger et al. 2000, Hirel et al. 2007).

Controlled environment phenotyping allows, in theory, precise control of the growing conditions, thus removing some of the intrinsic variation found under field conditions. It reduces some of the imaging problems occurring in the field as many imaging techniques are very sensitive to environmental conditions. Even relatively robust techniques, such as RGB imaging (photography), are remarkably sensitive to the ambient lighting environment. This general principle extends to other forms of imaging and to other sensors and monitors. Often this environmental noise can completely obscure and confuse phenotypic differences.

For relatively simple and well-defined traits, such as leaf or flower shape and size, shoot growth, etc., lab-based phenotyping has been proven very effective. Most CER facilities are now able to provide a number of different, though often simplistic, growth conditions (light and dark, high and low temp, drought, salinity, etc.) that mimic aspects of the environment and can be monitored very precisely throughout the life cycle of the plant. Moreover, controlled conditions allow the use and evaluation of very sensitive imaging technology, such as hyperspectral cameras and laser scanning sensors, enabling the harvest of meaningful physiological data by minimizing environmental noise.

Root imaging is still undergoing active evaluation, mainly in a CER/ lab setting. Soil, apart from being heterogeneous, is also opaque to most of the easily manipulated wavelengths of electromagnetic radiation. Those wavelengths that do penetrate do not easily discriminate plant tissue (root) from soil and water although computational approaches show promise in overcoming this limitation (Flavel et al. 2012, Mooney et al. 2011).

Other solutions include rhizotrons and soil substitutes. Rhizotrons are 2-D cassettes that restrain root growth into a single plane, generally immediately next to a transparent viewing plate that allows reiterative assessment of growth. Their general applicability to crop phenomics remains to be seen as they would fail to capture the 3-D complexity of a typical field. However, the approach provides a direct assessment of root growth in response to various stress environments (Laperche et al. 2006). Plants can adapt to non-soil-based growing media that, if optically clear, can be coupled with sophisticated image segmentation software, allowing high resolution dynamic 2-D and 3-D imaging of developing root systems (Clark et al. 2011, Downie et al. 2012, Iyer-Pascuzzi et al. 2010). One of the criticisms against using artificial liquid or gel substrates or restricted root runs is that they under-represent the complexity and dynamics of the real soil conditions. Poorter et al. (2012) reported that pot size (as used in CER experiments) can have a significant effect on plant growth and therefore influence the differences between genotypes and treatments: a meta-analysis on 65 studies revealed that a two-fold increase in the pot size caused a 43% increase in biomass production. A reduction in net photosynthesis, rather than leaf morphology or biomass allocation, appears to be the major explanatory factor for the reduced growth in small pots (Poorter et al. 2012).

Bridging the "Valley of Death"

From the discussion above, careful evaluation of the particular properties of the phenome and its constituent phenotypes is an important step towards designing a successful phenomics strategy for crop species. Developing cost-effective phenotyping methodologies that can be scaled and automated is an urgent requirement to accelerate plant breeding.

Sensor- and Imaging-based Strategies for Crop Phenotyping

Non-destructive sensor-based methods for evaluating physiological status of crops have been available for decades. These technologies have been refined and underpin many aspects of precision farming and crop management (Pinter et al. 2003). Sensor-based phenotyping is a good exemplar for a phenomics approach that is scalable and bridges the gap between CER and field.

One broad class of sensors operates on the principle that the amount and spectrum of electromagnetic radiation reflected (reflectance sensors) or emitted (fluorescence sensors) from a plant or a crop canopy can be often associated with its physiology. Such sensor-acquired data, used to calculate appropriate vegetation indices, provide rapid (dynamic) non-destructive proxy measurements that can be calibrated against traits such as canopy biomass, senescence, plant nitrogen and water status, chlorophyll content and biotic stresses. In principle, these sensor-based technologies are inherently scalable and can operate on terrestrial (e.g., hand-held, tractor- or cherry picker-mounted), semi-remote (e.g., unmanned aerial vehicles) or even orbital platforms (reviewed in White et al. (2012)).

Spectral measurements of canopy reflectance are perhaps the most common method for estimating physiological parameters of crop canopies. Those relying on natural light are referred to as passive sensors, while those using artificial light are active sensors. Reflectance sensors are also distinguished on the basis of the wavebands they record, into broadband (>50 nm intervals) and narrowband (4–10 nm intervals) spectral sensors (Muller et al. 2008). The former are the most commonly used because of the simplicity of operation and the lower cost (but the differences in cost are reducing) but the restricted spectral discrimination often presents a limitation. These sensors provide useful information and a number of commercial versions are widely used in field operations, including the Yara-N-Sensor (450–900 nm), GreenSeeker (650 nm, 770 nm) and CropCircle (420–800 nm) (Samborski et al. 2009). All three of these sensors can operate with tractor-mounted platforms and provide direct calculations of vegetation indices (e.g., NDVI, NWI, etc.) that are used to estimate crop N status. Although these technologies are designed primarily for precision farming (e.g., variable-rate N application), the hand-held versions of GreenSeeker and CropCircle have the potential for use in experimental applications, such as selection for yield potential under N or drought stress in experimental populations or breeding material. GreenSeeker has been used to estimate NDVI of durum wheat genotypes grown in pots under different water and N regimes (Cabrera-Bosquet et al. 2011). These NDVI measurements were found to correlate strongly with biomass, total green area and above-ground N content. However, the limited number of wavebands (2 to 3 wavebands at any one time) restricts calculation to a small set of vegetative indices and simple ratios. CropScan (CropScan Inc., Rochester, USA (Jørgensen et al. 2006, Nilsson 1991)) is a hand-held multispectral radiometer that improves on this limitation by accommodating up to 16 wavebands at a time (in the 450–1750 nm region). The instrument measures both incident and reflected irradiation (almost) simultaneously and, therefore, the canopy reflectance measurements are corrected for differences in sun angle and light conditions. However, CropScan uses a passive sensor that requires a certain minimum level of incident irradiation to operate accurately. This hinders the function of the instrument under low light conditions (e.g., low or thick clouds).

Hyper-spectral sensors record hundreds of continuous narrow wavebands in the visible and near-infrared (NIR) regions and provide an alternative screening method when higher spectral resolution is required, or when the informative wavelengths are unknown (Blackburn 2007). Narrowband-derived vegetation indices were shown to have increased predictive power for various canopy parameters, such as green area index (GAI), shoot dry matter and total amount of shoot nitrogen, in wheat (Aparicio et al. 2000, Ferrio et al. 2005) and oilseed rape (Muller et al. 2008), and also to provide accurate estimates of nitrogen uptake under different drought stress levels in maize (Winterhalter et al. 2011). Hyper-spectral reflectance spectroscopy has also been used in estimating leaf chlorophyll content for stress-related phenotyping and precision N applications (Delegido et al. 2010, Sims and Gamon 2002).

Chlorophyll measurements, using fluorescence-based techniques (sensors and imaging), have been used extensively as a means for assessing photosynthetic performance in response to various environmental stresses (Baker and Rosenqvist 2004). There are four main sensing approaches, namely variable chlorophyll fluorescence, chlorophyll emission ratio, blue-green fluorescence and screening of chlorophyll fluorescence by phenolic compounds (Tremblay et al. 2011). Detailed physiological and technical aspects of these approaches and their application as screening tools have been reviewed elsewhere (Baker and Rosenqvist 2004, Berger et al. 2010, Furbank and Tester 2011, Tremblay et al. 2011) so we will restrict discussion to recent advances in fluorescence sensing that have shown potential for field high-throughput phenotyping, especially for nitrogen use efficiency-related traits. Measurement of laser-induced chlorophyll fluorescence has been used for determining plant biomass and N status in oilseed rape, wheat, maize and wood species (Bredemeier and Schmidhalter 2005, Kolber et al. 2005, Thoren and Schmidhalter 2009). The principle behind this approach is that chlorophyll molecules, when excited by a laser beam (630 nm), emit at two fluorescence maxima (690 nm and ~730 nm) the ratio of which was found to be correlated with above-ground N content, N uptake and biomass (Bredemeier and Schmidhalter 2005, Thoren and Schmidhalter 2009). Chlorophyll fluorescence estimated with this method is not influenced by soil interference and the platforms used are compatible with semi-remote setups.

Another promising apparatus for estimating various physiological parameters is the Multiplex, which can be used either as hand-held or tractor-mounded fluorescence sensor. The instrument uses four different wavelengths (UV-A, blue, green and red) to induce fluorescence signals (Tremblay et al. 2011). Various ratios between these excitation and fluorescence signals can be used to estimate chlorophyll, flavonoid, and anthocyanin in the leaves and to assess crop N status (Zhang et al. 2012).

Digital imaging as a tool for phenotyping platforms has attracted a lot of interest in recent years mainly due to the relatively low cost, high versatility, inherent scalability, and wide range of applications. Digital imaging at the visible spectrum (RGB images) represents the least complicated and probably the most commonly used type of imaging and various platforms have been developed for performing growth analysis of above- and below-ground organs under controlled conditions and different environmental stresses. For example, simple image capture and processing was used to assess salt tolerance mechanisms in cereals (Rajendran et al.

2009). The PHENOPSIS platform was used to evaluate natural variation in rosette growth in a panel of Arabidopsis accessions grown under soil water deficit (Granier et al. 2006). Similarly, the GROWSCREEN setup was used to measure leaf area and relative growth rate of tobacco plants grown under varying light regimes (Walter et al. 2007). More recently a new phenotyping platform, GROWSCREEN-Rhizo, was reported to be capable of simultaneously imaging roots and shoots of various plant species, including crop species, grown in soil-filled rhizotrons (Nagel et al. 2012). Interestingly, the authors reported that this setup was capable of assessing the effects of soil compaction in the root architecture of barley and maize plants suggesting that technologies such as this have the potential for bridging the lab-to-field gap especially for root traits (Nagel et al. 2012). Most of the platforms used for phenotyping root architecture-related traits employ digital imaging of plants grown in transparent containers in gel-based medium (e.g., gellan-gum). This setup allows the acquisition of multiple 2-D images of the growing root system and subsequent 3-D reconstruction and quantification of root features (Clark et al. 2011, Fang et al. 2009, Iyer-Pascuzzi et al. 2010, Sozzani and Benfey 2011).

Digital imaging has also been used to develop vegetation indices that are subsequently used as proxy measures for estimating nitrogen content, biomass or responses to abiotic stresses. Rorie et al. (2011) used digital imaging to quantify leaf "greenness" in maize under different N applications. By combining hue, saturation and brightness values they derived a dark green colour index (DGCI) which was highly correlated with leaf N concentration across years (Rorie et al. 2011). On another application, hue, saturation and brightness were used to estimate digital ground cover and assess the effects of ground cover on soil water evaporation in four wheat populations (Mullan and Reynolds 2010). Casadesús et al. (2007) used digital images to identify picture-derived vegetation indices (picVIs) in durum wheat trials under different water conditions (Casadesús et al. 2007).

Thermal imaging (IR thermography) has been used extensively to measure plant or canopy temperature as an indirect method for studying stomatal responses under water stress conditions (Reynolds et al. 2009). There is a strong association between stomatal conductance, transpiration rate and leaf temperature in diverse environments (Jones et al. 2009). Genotypes that are drought tolerant as a result of higher water uptake and/or access to deep soil water are generally able to sustain higher stomatal conductance and higher transpiration rate under water deficit and, therefore, can be selected, based on their lower canopy temperature (Jones 2007, Leinonen et al. 2006). Thermal IR measurements have been used to screen wheat and barley for stomatal behaviour under saline conditions (Sirault et al. 2009). As salt concentration increases, stomatal conductance decreases causing the leaf temperature to increase. Based on this observation, a high-throughput and non-destructive method was devised to allow homogeneous stress application, and consistent IR image capture and processing for screening for salinity tolerance under controlled conditions (James and Sirault 2012, Sirault et al. 2009). Despite some successful applications (Reynolds et al. 2009), thermography under field conditions remains a challenge. The main problems in scaling up from the lab to the field include: (1) the temporal variation in canopy temperature due to constant environmental changes (e.g., wind speed and illumination), (2) spatial variation in leaf temperature as a result of leaf and sun angle (i.e., changes in the fraction of sunlit to shaded leaves), and (3) the distinction between the canopy temperature and that of the background soil and crop residue (Jones et al. 2009). Some of these issues could be addressed by using either elevated-view (e.g., from a cherry-picker) or air-borne imaging (e.g., balloon or AUV) that allows the capture of numerous plots per image and rapid, almost simultaneous, coverage of the entire experimental field (reviewed in Jones et al. (2009)).

Near-infrared (NIR) and far-infrared (FIR) imaging have also been proposed as tools for high-throughput phenotyping, especially for drought tolerance-related traits. NIR imaging seems a promising phenotyping tool for determining plant and soil water content (Kobori and Tsuchikawa 2009, Thiel et al. 2010).

Integrating Field and CER Phenotyping in Crop Research

Within commercial plant breeding, phenotyping must be cost-effective and appropriate to the stage of the programme. A typical molecular-assisted breeding programme might consists of a number of distinct stages: for example, (1) screening and selecting candidate parents, (2) early generation selection for target traits, (3) development of markers that facilitate the tracing of trait-related QTL/genes and (4) the evaluation of genotypes in the field (Reynolds et al. 2009). Successful strategies need to fulfill an array of requirements at technical, biological and economic levels. Strategies combining controlled environment and field phenotyping at different stages can have advantages at a number of levels, especially for traits such as drought tolerance and N use efficiency.

A typical case study of a multi-step phenotyping programme might include an initial large-scale screen, often simple and somewhat subjective, of diverse genotypes usually under field conditions to keep costs manageable. Subsequent selection steps lead to progressively smaller numbers of individuals that can be analysed in detail (at cellular, biochemical, and metabolic level, as appropriate to the trait(s) of interest) (Mir et al. 2012, Sinclair 2011). Sinclair et al. (2000) described a three-step approach to identify soybean genotypes with increased N_2 fixation tolerance under water deficit (Sinclair et al. 2000): approximately 3000 plants were initially grown under well-watered field conditions, screened for variation in petiole ureide concentration and lines with reduced ureide subsequently subjected to a direct N accumulation screen under water deficit. Based on this screen, a further selection was made and these lines were grown under controlled conditions where an acetylene reduction assay was used to monitor the N_2 fixation activity during a soil-drying cycle. The lines identified at the end of this process were then used as genetic resources in breeding or for more sophisticated studies to elucidate the exact mechanism(s) conferring the tolerance (Sinclair et al. 2000).

Another model was used at CIMMYT for the development of advanced lines with complementary physiological traits for drought tolerance (Reynolds et al. 2009). In this system, candidate parents were phenotyped, using direct and proxy screens, for an array of different characteristics that may have an impact on the target trait. Parental lines, possessing as many of these characters as possible, were then combined through two- and three-way crosses. For example, in a three-way cross, the first parent contributed high biomass, a low value for carbon isotope discrimination (proxy screen for transpiration efficiency) and high chlorophyll concentration (proxy for the stay-green trait). The second parent contributed low canopy temperature (indicative of access to deep water), while the third parent contributed high early vigor and high leaf wax, associated with reduced water losses and photo-protection, respectively (Reynolds et al. 2009 and references therein). F2 progeny of these crosses were then grown under well-watered conditions and selection was performed on the basis of simple characters, such as height and general phenology. A fourth step involved early generation selection on bulk F3 and F4 material grown under drought-stress conditions using canopy temperature as proxy screen; bulks with "warmer" canopies were discarded (Reynolds et al. 2009). Spectral reflectance indices can also have an application during the early selection stages: normalized water indices based on NIR wavelengths were associated with soil water potential (ψ_{soil}) and available volumetric soil water (AVSW) in wheat under water-deficit conditions (Gutierrez et al. 2010).

A model that integrates CER and field phenotyping at the earliest stages of selection and feeds into both breeding and genetic characterization pipelines has been proposed recently (Wasson et al. 2012). Although this model was described in the context of improving root system and water uptake, its general principles could find wider applications in breeding, especially for novel traits for which high-throughput and reliable screens are not available. In such a model, the first step is to identify the proxy traits that constitute the plant ideotype for the trait of interest and for the target environment. Effective screening procedures may have to be developed and assessed under field and laboratory conditions in parallel, ensuring that any laboratory screen captures useful aspects of the trait under field conditions in a cost-effective manner and so confers an advantage upon return to the field. Successful screens would rapidly evaluate diverse germplasm and identify candidate donors that can be used as parents for the introgression of the desired traits into elite material through a backcrossing breeding scheme, and for the development of mapping populations for QTL, molecular marker and gene discovery (Wasson et al. 2012). This approach depends on suitable laboratory-based screens but holds the promise of accelerating generational turnover and reducing the time to produce new varieties.

Data Handling

Metadata and Data Integration. The concepts of genotype and phenotype are fundamental in all of genetics, developmental and evolutionary biology. Plant breeding requires the integration of these concepts to understand how and why phenotypic expression varies with the environment. Traditional plant breeding is based on phenotypic selection. Selection programmes frequently have focused on genes of major effects that give a clear phenotype. The rate of genetic gain in breeding programmes can increase either by extending the amount or nature of variation available for selection, or by accelerating the selection process to produce varieties more rapidly. The various omics platforms increase the ability to discover genes and pathways that control specific traits and provide screening and analysis platforms to support selection strategies. A significant challenge in gene discovery based on genetic studies is the final identification of the gene or regulatory sequence responsible for the phenotype. Omics technology can help to identify likely candidates that underlie genetic position and to elucidate the biological role or process that determines the gene effect. The challenge is then the integration of the full omics datasets into the modelling, population structure and selection strategies (Langridge and Fleury 2011). To perform genetic analysis at a systems level, several research groups have been doing this in the past couple of years and more recently the phenomics component has been added to the pipeline. Druka et al. (2008) used GeneNetwork to integrate barley genotypic, phenotypic and mRNA abundance datasets and used a combination of correlation analysis and linkage mapping to identify and substantiate gene targets for saturation mapping and positional cloning. Acharjee et al. (2011) integrated omics datasets from tubers of a diploid potato population to construct networks that led to the association of known and uncharacterized metabolites with genes associated with the carotenoid biosynthesis pathway (Acharjee et al. 2011). Zamboni et al. (2010) identified putative stage-specific biomarkers for berry development and withering through an integrative omics approach using two different strategies, one hypothesis free and the other hypothesis driven. Osorio et al. (2012) conducted an integrative comparative analysis of transcriptomics and metabolomics data from tomato and pepper fruits to gain a broader systems perspective and to identify additional common and distinct molecular regulatory events during development and ripening.

Currently, experimental genotypic data can be lifted using high-throughput technologies. To fully understand the context, methods, data and conclusions that pertain to a biological experiment, it is necessary to access a range of background information (commonly referred to as metadata), at an appropriate level of granularity (Taylor et al. 2008). This applies equally to crop biology and breeding. Appropriate metadata standards facilitate interpretation and use of the resultant complex datasets as well as prevent unnecessary repetition of work (Sansone et al. 2012). Standards have been defined for reporting the metadata from certain types of experiment (e.g., MIAME and ISA frameworks (Brazma et al. 2001, Haug et al. 2012)) and these minimal reporting guidelines enable data sharing and re-use. High-quality digital phenotypic data also provides opportunities for analysis, re-analysis and integration with other 'omics datasets, but only if it has explicit metadata, ultimately enabling the mathematical modelling of molecular networks controlling complex traits such as development, stress tolerance and metabolism or even the interactions of organisms within a community.

Linking phenotypic to other information sets generated by 'omics technologies can reveal insight into the cellular mechanisms underlying phenotypic variation (Bayjanov et al. 2012). However, such integration is in its infancy and would be greatly facilitated by the use of commonly accepted standards in plant biology to enable cross-talk between heterogeneous information sources. Efforts have been initiated to create pipelines that allow the sharing and integration of phenotypic data with other omics data, including other phenotypic data sources (Barrett et al. 2009). The plant phenomics community is currently discussing the implementation of integrative mechanisms (a commonly accepted ontology for annotation, standards to describe environmental metadata, experimental designs, wetlab procedures, as well as user-friendly interfaces).

As genomics and phenomics become ever more integrated into breeding programmes, there is increasing need to consider how to manipulate, share and visualize the large datasets produced. High-throughput 'omics technologies can produce overwhelming amounts of data in the quest to understand non-trivial complex questions, such as what are the effects of phenotypic plasticity as a result of genome–environment interaction (Shah et al. 2007). Large datasets can easily be generated when a biological system is measured simultaneously at different levels (mRNAs, proteins, metabolites and phenotypic traits) and although much of this data is not yet routinely used in breeding programmes, the reducing cost of such technologies may permit this in the foreseeable future. Large and diverse datasets place demands on computational tools that preclude standard approaches such as notebooks and spreadsheets. Data annotation, and the diversity of breeding/research questions, may require redesign of user interfaces. Analysis of high-throughput data is necessarily statistical to account for the experiment's design and to correctly address the large number of observed values and the small number of samples. The volume and characteristics of the data should be appropriately summarized before any sort of comprehension is attempted.

Data Storage and Sharing. Modern high-throughput phenotypic platforms often use image and other sensor data as raw material to extract and quantitate "features" of interest and associate them with meaningful traits. As with Next Generation Sequencing (NGS), the original image files can be bulky but, unlike NGS, there is an argument for retaining this "raw" data and making it available to the research community. Quantitative image analysis is still developing rapidly particularly in 3-D and 4-D and these datasets represent a largely untapped resource for further mining. The availability of microarray and other omic datasets in the public domain provides a good model for this (Barrett et al. 2009, Haug et al. 2013). The arguments in favour of open access data are now well established, and protocols and principles for data sharing are emerging. This effort laid the foundation for the release of genomic data and the development of online resources, accessible by anyone, for any purpose, that now underpin all modern biomedical research.

To be generally useful, a commonly accepted system to store and query metadata directly related to the image is urgently required (Salehi et al. 2010). For example, interpreting a time-lapse experiment with a number of treatments requires understanding which files represent which time points for which samples, and there is no standard method for organizing the data to reflect this information (Walter et al. 2010). Currently available methods either do not offer sufficient compression rates, or require a great amount of CPU time for decompression and loading every time the data is accessed. Moreover, existing hardware environments do not provide sufficient storage space and computational power to store and process the data due to their enormous size. For example, a typical sensor-based dataset—RGB, fluorescent, NIR, IR—for a drought experiment with 100 lines, four treatments and six replicated could exceed 0.07 Tb. To make the best use of these increasingly complex and large image-based data resources, the scientific community must develop appropriate methodology (Qiao et al. 2012). This requires the involvement of multidisciplinary research groups that can look at the problem from both the biological and engineering point of view.

Future Prospects

One can imagine that many existing crop varieties occupy zones of phenotypic optima or near-optima (peaks)—for the current enviro-agronomic conditions—and it may be difficult to achieve further striking improvements. However, other crops may occupy localized optima that are constrained for example by a restricted genetic base, sub-optimal agronomic regimes, or even socio-cultural practices (for example, local agricultural shows in the 19th/early 20th centuries were essentially maize "beauty contests" and are reputed to have constrained yield improvements for decades). As in hill-walking, localized peaks of performance often give the impression that they represent the summit and the journey across valleys can restrict further upward progress. To carry this analogy further, these valleys can be significant barriers to further progress especially if there is no hint that higher land exists. A map is useful to identify these other higher peaks and exploration of the pheno-genotypic landscape is only just beginning. This holds the promise of providing a means to exploit genetic diversity in an effective and systematic manner.

References Cited

Acharjee, A., B. Kloosterman, R.C.H. de Vos, J.S. Werij, C.W.B. Bachem, R.G.F. Visser and C. Maliepaard. 2011. Data integration and network reconstruction with ~omics data using Random Forest regression in potato. Anal. Chim. Acta 705: 56–63.

Adamchuk, V.I., J.W. Hummel, M.T. Morgan and S.K. Upadhyaya. 2004. On-the-go soil sensors for precision agriculture. Computers and Electronics in Agriculture 44: 71–91.

An, G., D.H. Jeong, K.H. Jung and S. Lee. 2005. Reverse genetic approaches for functional genomics of rice. Plant Molecular Biology 59: 111–123.

Aparicio, N., D. Villegas, J. Casadesus, J.L. Araus and C. Royo. 2000. Spectral vegetation indices as nondestructive tools for determining durum wheat yield. Agronomy Journal 92: 83–91.

Araus, J.L., G.A. Slafer, M.P. Reynolds and C. Royo. 2002. Plant breeding and drought in C3 cereals: what should we breed for? Annals of Botany 89: 925–940.

Bah, A., S.K. Balasundram and M.H.A. Husni. 2012. Sensor technologies for precision soil nutrient management and monitoring. American Journal of Agriculture and Biological Sciences 7: 43–49.

Baker, N.R. and E. Rosenqvist. 2004. Applications of chlorophyll fluorescence can improve crop production strategies: an examination of future possibilities. Journal of Experimental Botany 55: 1607–1621.

Bänziger, M., G.O. Edmeades, D. Beck and M. Bellon. 2000. Breeding for Drought and Nitrogen Stress Tolerance in Maize: From Theory to Practice. CIMMYT, Mexico, D.F.

Barrett, T., D.B. Troup, S.E. Wilhite, P. Ledoux, D. Rudnev, C. Evangelista, I.F. Kim, A. Soboleva, M. Tomashevsky, K.A. Marshall, K.H. Phillippy, P.M. Sherman, R.N. Muertter and R. Edgar. 2009. NCBI GEO: archive for high-throughput functional genomic data. Nucleic Acids Research 37(Database issue): D885–D890.

Bayjanov, J.R., D. Molenaar, V. Tzeneva, R.J. Siezen and S.A.F.T. van Hijum. 2012. PhenoLink—a web-tool for linking phenotype to ~omics data for bacteria: application to gene-trait matching for Lactobacillus plantarum strains. BMC Genomics 13: 170.

Bedell, J.A., M.A. Budiman, A. Nunberg, R.W. Citek, D. Robbins, J. Jones, E. Flick, T. Rohlfing, J. Fries, K. Bradford, J. McMenamy, M. Smith, H. Holeman, B.A. Roe, G. Wiley, I.F. Korf, P.D. Rabinowicz, N. Lakey, W.R. McCombie, J.A. Jeddeloh and R.A. Martienssen. 2005. Sorghum genome sequencing by methylation filtration. PLoS Biology 3: e13.

Berger, B., B. Parent and M. Tester. 2010. High-throughput shoot imaging to study drought responses. Journal of Experimental Botany 61: 3519–3528.

Bilder, R.M., F.W. Sabb, T.D. Cannon, E.D. London, J.D. Jentsch, D.S. Parker, R.A. Poldrack, C. Evans and N.B. Freimer. 2010. Phenomics: The systematic study of phenotypes on a genomewide scale. Neuroscience 164: 30–42.

Blackburn, G.A. 2007. Hyperspectral remote sensing of plant pigments. Journal of Experimental Botany 58: 855–867.

Brazma, A., P. Hingamp, J. Quackenbush, G. Sherlock, P. Spellman, C. Stoeckert, J. Aach, W. Ansorge, C.A. Ball, H.C. Causton, T. Gaasterland, P. Glenisson, F.C. Holstege, I.F. Kim, V. Markowitz, J.C. Matese, H. Parkinson, A. Robinson, U. Sarkans, S. Schulze-Kremer, J. Stewart, R. Taylor, J. Vilo and M. Vingron. 2001. Minimum information about a microarray experiment (MIAME)-toward standards for microarray data. Nature Genetics 29: 365–371.

Bredemeier, C. and U. Schmidhalter. 2005. Laser-induced chlorophyll fluorescence sensing to determine biomass and nitrogen uptake of winter wheat under controlled environment and field conditions. In: J. Stafford [ed.]. Precision Agriculture '05. 5th European Conference on Precision Agriculture, Uppsala, Sweden. ISBN 9076998698. pp. 273–280.

Brenchley, R., M. Spannagl, M. Pfeifer, G.L.A. Barker, R. D'Amore, A.M. Allen, N. McKenzie, M. Kramer, A. Kerhornou, D. Bolser, S. Kay, D. Waite, M. Trick, I. Bancroft, Y. Gu, N. Huo, M.-C. Luo, S. Sehgal, B. Gill, S. Kianian, O. Anderson, P. Kersey, J. Dvorak, W.R. McCombie, A. Hall, K.F.X. Mayer, K.J. Edwards, M.W. Bevan and N. Hall. 2012. Analysis of the bread wheat genome using whole-genome shotgun sequencing. Nature 491: 705–710.

Brennan, J. and M. Morris. 2001. Economic issues in assessing the role of physiology in wheat breeding programs. In: M.P. Reynolds, J.I. Ortiz-Monasterio and A. McNab [eds.]. Application of Physiology in Wheat Breeding. International Maize and Wheat Improvement Center (CIMMYT), Mexico, D.F. pp. 78–86.

Brennan, J.P., A.G. Condon, M. Van Ginkel and M.P. Reynolds. 2007. An economic assessment of the use of physiological selection for stomatal aperture-related traits in the CIMMYT wheat breeding programme. Journal of Agricultural Science 145: 187–194.

Brennan, J.P. and P.J. Martin. 2007. Returns to investment in new breeding technologies. Euphytica 157: 337–349.

Brownie, C., D.T. Bowman and J.W. Burton. 1993. Estimating spatial variation in analysis of data from yield trials: a comparison of methods. Agronomy Journal 85: 1244–1253.

Butler, D. 2008. Crossing the valley of death. Nature 453: 840–842.

Cabrera-Bosquet, L., J. Crossa, J. von Zitzewitz, M.D. Serret and J.L. Araus. 2012. High-throughput phenotyping and genomic selection: the frontiers of crop breeding converge. Journal of Integrative Plant Biology 54: 312–320.

Cabrera-Bosquet, L., G. Molero, A.M. Stellacci, J. Bort, S. Nogués and J.L. Araus. 2011. NDVI as a potential tool for predicting biomass, plant nitrogen content and growth in wheat genotypes subjected to different water and nitrogen conditions. Cereal Research Communications 39: 147–159.

Caldwell, D.G., N. McCallum, P. Shaw, G.J. Muehlbauer, D.F. Marshall and R. Waugh. 2004. A structured mutant population for forward and reverse genetics in Barley (*Hordeum vulgare* L.). The Plant Journal 40: 143–150.

Cannon, S.B., G.D. May and S.A. Jackson. 2009. Three sequenced legume genomes and many crop species: rich opportunities for translational genomics. Plant Physiology 151: 970–977.

Cao, J., K. Schneeberger, S. Ossowski, T. Günther, S. Bender, J. Fitz, D. Koenig, C. Lanz, O. Stegle, C. Lippert, X. Wang, F. Ott, J. Müller, C. Alonso-Blanco, K. Borgwardt, K.J. Schmid and D.

Weigel. 2011. Whole-genome sequencing of multiple Arabidopsis thaliana populations. Nature Genetics 43: 956–963.

Casadesús, J., Y. Kaya, J. Bort, M.M. Nachit, J.L. Araus, S. Amor, G. Ferrazzano, F. Maalouf, M. Maccaferri, V. Martos, H. Ouabbou and D. Villegas. 2007. Using vegetation indices derived from conventional digital cameras as selection criteria for wheat breeding in water-limited environments. Annals of Applied Biology 150: 227–236.

Clark, R.T., R.B. MacCurdy, J.K. Jung, J.E. Shaff, S.R. McCouch, D.J. Aneshansley and L.V. Kochian. 2011. Three-dimensional root phenotyping with a novel imaging and software platform. Plant Physiology 156: 455–465.

Colmsee, C., M. Mascher, T. Czauderna, A. Hartmann, U. Schlüter, N. Zellerhoff, J. Schmitz, A. Bräutigam, T. Pick, P. Alter, M. Gahrtz, S. Witt, A. Fernie, F. Börnke, H. Fahnenstich, M. Bucher, T. Dresselhaus, A. Weber, F. Schreiber, U. Scholz and U. Sonnewald. 2012. OPTIMAS-DW: a comprehensive transcriptomics, metabolomics, ionomics, proteomics and phenomics data resource for maize. BMC Plant Biol. 12: 245.

Delegido, J., L. Alonso, G. González and J. Moreno. 2010. Estimating chlorophyll content of crops from hyperspectral data using a normalized area over reflectance curve (NAOC). International Journal of Applied Earth Observation and Geoinformation 12: 165–174.

Downie, H., N. Holden, W. Otten, A.J. Spiers, T.A. Valentine and L.X. Dupuy. 2012. Transparent soil for imaging the rhizosphere. PloS One 7: e44276.

Druka, A., I. Druka, A.G. Centeno, H. Li, Z. Sun, W.T.B. Thomas, N. Bonar, B.J. Steffenson, S.E. Ullrich, A. Kleinhofs, R.P. Wise, T.J. Close, E. Potokina, Z. Luo, C. Wagner, G.F. Schweizer, D.F. Marshall, M.J. Kearsey, R.W. Williams and R. Waugh. 2008. Towards systems genetic analyses in barley: integration of phenotypic, expression and genotype data into GeneNetwork. BMC Genet. 9: 73.

Fang, S., X. Yan and H. Liao. 2009. 3D reconstruction and dynamic modeling of root architecture in situ and its application to crop phosphorus research. The Plant Journal 60: 1096–1108.

Ferrio, J.P., D. Villegas, J. Zarco, N. Aparicio, J.L. Araus and C. Royo. 2005. Assessment of durum wheat yield using visible and near-infrared reflectance spectra of canopies. Field Crops Research 94: 126–148.

Flavel, R., C. Guppy, M. Tighe, M. Watt, A. McNeill and I. Young. 2012. Non-destructive quantification of cereal roots in soil using high-resolution X-ray tomography. Journal of Experimental Botany 63: 2503–2511.

Furbank, R.T. and M. Tester. 2011. Phenomics—technologies to relieve the phenotyping bottleneck. Trends in Plant Science 16: 635–644.

García-Herrera, R., C. Wilkinson, F.B. Koek, M.R. Prieto, N. Calvo and E. Hernández. 2005. Description and general background to ships' logbooks as a source of climatic data. Climatic Change 73: 13–36.

Godwin, R.J. and P.C.H. Miller. 2003. A review of the technologies for mapping within-field variability. Biosystems Engineering 84: 393–407.

Granier, C., L. Aguirrezabal, K. Chenu, S.J. Cookson, M. Dauzat, P. Hamard, J.-J. Thioux, G. Rolland, S. Bouchier-Combaud, A. Lebaudy, B. Muller, T. Simonneau and F. Tardieu. 2006. PHENOPSIS, an automated platform for reproducible phenotyping of plant responses to soil water deficit in Arabidopsis thaliana permitted the identification of an accession with low sensitivity to soil water deficit. New Phytologist 169: 623–635.

Grant, R.F., B.S. Jackson, J.R. Kiniry and G.F. Arkin. 1989. Water deficit timing effects on yield components in maize. Agron J. 81: 61–65.

Gutierrez, M., M.P. Reynolds and A.R. Klatt. 2010. Association of water spectral indices with plant and soil water relations in contrasting wheat genotypes. Journal of Experimental Botany 61: 3291–3303.

Haug, K., R.M. Salek, P. Conesa, J. Hastings, P. de Matos, M. Rijnbeek, T. Mahendraker, M. Williams, S. Neumann, P. Rocca-Serra, E. Maguire, A. González-Beltrán, S.A. Sansone, J.L. Griffin and C. Steinbeck. 2013. MetaboLights—an open-access general-purpose repository for metabolomics studies and associated meta-data. Nucleic Acids Research 41: D781–6.

Hayes, H. 1925. Control of Soil Heterogeneity and Use of the Probable Error Concept in Plant Breeding Studies. University Farm, St. Paul.

Hillel, D. 1980. Fundamentals of Soil Physics. Academic Press, London.

Hirel, B., J. Le Gouis, B. Ney and A. Gallais. 2007. The challenge of improving nitrogen use efficiency in crop plants: towards a more central role for genetic variability and quantitative genetics within integrated approaches. Journal of Experimental Botany 58: 2369–2387.

Hively, W.D., G.W. McCarty, J.B. Reeves, III, M.W. Lang, R.A. Oesterling and S.R. Delwiche. 2011. Use of airborne hyperspectral imagery to map soil properties in tilled agricultural fields. Applied and Environmental Soil Science 2011: 358193.

Hobbs, P. and K. Sayre. 2001. Managing experimental breeding trials. *In*: M.P. Reynolds, J.I. Ortiz-Monasterio and A. McNab [eds.]. Application of Physiology in Wheat Breeding. CIMMYT, Mexico, D.F. pp. 48–58.

Houle, D. 2010. Numbering the hairs on our heads: the shared challenge and promise of phenomics. Proc. Natl. Acad. Sci. USA 107(Suppl 1): 1793–1799.

International Barley Genome Sequencing Consortium, K.F. Mayer, R. Waugh, J.W. Brown, A. Schulman, P. Langridge, M. Platzer, G.B. Fincher, G.J. Muehlbauer, K. Sato, T.J. Close, R.P. Wise and N. Stein. 2012. A physical, genetic and functional sequence assembly of the barley genome. Nature 491: 711–716.

International Rice Genome Sequencing Project. 2005. The map-based sequence of the rice genome. Nature 436: 793–800.

Iyer-Pascuzzi, A.S., O. Symonova, Y. Mileyko, Y. Hao, H. Belcher, J. Harer, J.S. Weitz and P.N. Benfey. 2010. Imaging and analysis platform for automatic phenotyping and trait ranking of plant root systems. Plant physiology 152: 1148–1157.

Jørgensen, R.N., P.M. Hansen and R. Bro. 2006. Exploratory study of winter wheat reflectance during vegetative growth using three mode component analysis. International Journal of Remote Sensing 27: 919–937.

Jackson, P. 2001. Directions for physiological research in breeding: issues from a breeding perspective. *In*: M. Reynolds, J. Ortiz-Monasterio and A. McNab [eds.]. Application of Physiology in Wheat Breeding. CIMMYT, Mexico, D.F. pp.

James, R. and X. Sirault. 2012. Infrared thermography in plant phenotyping for salinity tolerance. Methods in Mol. Biol. 913: 173–189.

Jones, H.G. 2007. Monitoring plant and soil water status: established and novel methods revisited and their relevance to studies of drought tolerance. Journal of Experimental Botany 58: 119–130.

Jones, H.G., R. Serraj, B.R. Loveys, L. Xiong, A. Wheaton and A.H. Price. 2009. Thermal infrared imaging of crop canopies for the remote diagnosis and quantification of plant responses to water stress in the field. Functional Plant Biology 36: 978.

Kobori, H. and S. Tsuchikawa. 2009. Prediction of water content in Ligustrum japonicum leaf using near infrared chemometric imaging. Journal of Near Infrared Spectroscopy 17: 151–157.

Kolber, Z., D. Klimov, G. Ananyev, U. Rascher, J. Berry and B. Osmond. 2005. Measuring photosynthetic parameters at a distance: laser induced fluorescence transient (LIFT) method for remote measurements of photosynthesis in terrestrial vegetation. Photosynthesis Research 84: 121–129.

Krishnan, A., E. Guiderdoni, G. An, Y.-I.C. Hsing, C.-D. Han, M.C. Lee, S.-M. Yu, N. Upadhyaya, S. Ramachandran, Q. Zhang, V. Sundaresan, H. Hirochika, H. Leung and A. Pereira. 2009. Mutant resources in rice for functional genomics of the grasses. Plant Physiology 149: 165–170.

Langridge, P. and D. Fleury. 2011. Making the most of "omics" for crop breeding. Trends Biotechnol. 29: 33–40.

Laperche, A., F. Devienne-Barret, O. Maury, J. Le Gouis and B. Ney. 2006. A simplified conceptual model of carbon/nitrogen functioning for QTL analysis of winter wheat adaptation to nitrogen deficiency. Theoretical and Applied Genetics 113: 1131–1146.

Larmande, P., C. Gay, M. Lorieux, C. Périn, M. Bouniol, G. Droc, C. Sallaud, P. Perez, I. Barnola, C. Biderre-Petit, J. Martin, J. Morel, A. Johnson, F. Bourgis, A. Ghesquière, M. Ruiz, B. Courtois and E. Guiderdoni. 2008. Oryza Tag Line, a phenotypic mutant database for the Genoplante rice insertion line library. Nucleic Acids Research 36(Database Issue): D1022–D1027.

Lawrence, C., M. Schaeffer, T. Seigfried, D. Campbell and L. Harper. 2007. MaizeGDB's new data types, resources and activitie. Nucleic Acids Research 35(Database Issue): D895–D900.

Leinonen, I., O.M. Grant, C.P.P. Tagliavia, M.M. Chaves and H.G. Jones. 2006. Estimating stomatal conductance with thermal imagery. Plant, Cell and Environment 29: 1508–1510.

Love, C.G., A.E. Andongabo, J. Wang, P.W.C. Carion, C.J. Rawlings and G.J. King. InterStoreDB: A Generic Integration Resource for Genetic and Genomic Data. J. Integr. Plant Biol. 2012 Apr 11.

Lusk, J.L. and A. Rozan. 2006. Consumer acceptance of ingenic foods. Biotechnol. J. 1: 1433–1434.

Manheim, A. 2011. Bridging Energy's 3rd Valley of Death. http://www.flcmidatlantic. org/2011_Commercializing-Innovation.html.

Massonnet, C., D. Vile, J. Fabre, M.A. Hannah, C. Caldana, J. Lisec, G.T.S. Beemster, R.C. Meyer, G. Messerli, J.T. Gronlund, J. Perkovic, E. Wigmore, S. May, M.W. Bevan, C. Meyer, S. Rubio-Díaz, D. Weigel, J.L. Micol, V. Buchanan-Wollaston, F. Fiorani, B.R. Sean Walsh, W. Gruissem, P. Hilson, L. Hennig, L. Willmitzer and C. Granier. 2010. Probing the reproducibility of leaf growth and molecular phenotypes: a comparison of three arabidopsis accessions cultivated in ten laboratories. Plant Physiology 152: 2142–2157.

Masuka, B., J. Araus, B. Das, K. Sonder and J. Cairns. 2012. Phenotyping for abiotic stress tolerance in maize. J. Integr. Plant Biol. 54: 238–249.

Menda, N., Y. Semel, D. Peled, Y. Eshed and D. Zamir. 2004. *In silico* screening of a saturated mutation library of tomato. Plant J. 38: 861–872.

Mir, R.R., M. Zaman-Allah, N. Sreenivasulu, R. Trethowan and R.K. Varshney. 2012. Integrated genomics, physiology and breeding approaches for improving drought tolerance in crops. Theoretical and Applied Genetics 125: 625–645.

Miyao, A., Y. Iwasaki, H. Kitano, J. Itoh, M. Maekawa, K. Murata, O. Yatou, Y. Nagato and H. Hirochika. 2007. A large-scale collection of phenotypic data describing an insertional mutant population to facilitate functional analysis of rice genes. Plant Mol. Biol. 63: 625–635.

Mooney, S.J., T.P. Pridmore, J. Helliwell and M.J. Bennett. 2011. Developing X-ray computed tomography to non-invasively image 3-D root systems architecture in soil. Plant and Soil 352: 1–22.

Mullan, D. and M. Reynolds. 2010. Quantifying genetic effects of ground cover on soil water evaporation using digital imaging. Functional Plant Biology 37: 703–712.

Muller, K., U. Bottcher, F. Meyer-Schatz and H. Kage. 2008. Analysis of vegetation indices derived from hyperspectral reflection measurements for estimating crop canopy parameters of oilseed rape (*Brassica napus* L.). Biosystems Engineering 101: 172–182.

Nagel, K.A., A. Putz, F. Gilmer, K. Heinz, A. Fischbach, J. Pfeifer, M. Faget, S. Blossfeld, M. Ernst, C. Dimaki, B. Kastenholz, A.-K. Kleinert, A. Galinski, H. Scharr, F. Fiorani and U. Schurr. 2012. GROWSCREEN-Rhizo is a novel phenotyping robot enabling simultaneous measurements of root and shoot growth for plants grown in soil-filled rhizotrons. Functional Plant Biology 39: 891–904.

Nilsson, H.-E. 1991. Hand-held radiometry and IR-thermography of plant diseases in field plot experiments. Int. J. Remote Sensing 12: 545–557.

O'Malley, R.C. and J.R. Ecker. 2010. Linking genotype to phenotype using the Arabidopsis unimutant collection. The Plant Journal 61: 928–940.

Osorio, S., R. Alba, Z. Nikoloski, A. Kochevenko, A.R. Fernie and J.J. Giovannoni. 2012. Integrative comparative analyses of transcript and metabolite profiles from pepper and tomato ripening and development stages uncovers species-specific patterns of network regulatory behavior. Plant Physiology 159: 1713–1729.

Pérez-de-Castro, A.M., S. Vilanova, J. Cañizares, L. Pascual, J.M. Blanca, M.J. Díez, J. Prohens and B. Picó. 2012. Application of genomic tools in plant breeding. Curr. Genomics 13: 179–195.

Pinter, P.J., Jr., J.L. Hatfield, J.S. Schepers, E.M. Barnes, M.S. Moran, C.S.T. Daughtry and D.R. Upchurch. 2003. Remote sensing for crop management. Photogrammetric Engineering & Remote Sensing 69: 647–664.

Poorter, H., J. Bühler, D. van Dusschoten, J. Climent and J.A. Postma. 2012. Pot size matters : a meta-analysis of the effects of rooting volume on plant growth. Functional Plant Biology 39: 839–850.

Qiao, D., W.-K. Yip and C. Lange. 2012. Handling the data management needs of high-throughput sequencing data: SpeedGene, a compression algorithm for the efficient storage of genetic data. BMC Bioinformatics 13: 100.

Rajendran, K., M. Tester and S.J. Roy. 2009. Quantifying the three main components of salinity tolerance in cereals. Plant, Cell & Environment 32: 237–249.

Reuzeau, C., V. Frankard, Y. Hatzfeld, A. Sanz, W. Van Camp, P. Lejeune, C. De Wilde, K. Lievens, J. de Wolf, E. Vranken, R. Peerbolte and W. Broekaer. 2006. Traitmill: a functional genomics platform for the phenotypic analysis of cereals. Plant Genetic Resources 4: 20–24.

Reynolds, M., Y. Manes, A. Izanloo and P. Langridge. 2009. Phenotyping approaches for physiological breeding and gene discovery in wheat. Annals of Applied Biology 155: 309–320.

Reynolds, M.P., R. Trethowan, M. van Ginkel and S. Rajaram. 2001. Application of physiology in wheat breeding. In: M. Reynolds, J. Ortiz-Monasterio and A. McNab [eds.]. Application of Physiology in Wheat Breeding. CIMMYT, Mexico, D.F. pp. 2–10.

Rorie, R.L., L.C. Purcell, M. Mozaffari, D.E. Karcher, C.A. King, M.C. Marsh and D.E. Longer. 2011. Association of "greenness" in corn with yield and leaf nitrogen concentration. Agronomy Journal 103: 529–535.

Salehi, A., M. Pathan, D. Palmer and M. Compton. 2010. SensorFeed: an architecture for model-based sensor network data enrichment. In Sixth International Conference on Intelligent Sensors, Sensor Networks and Information Processing (ISSNIP) 127–132.

Samborski, S.M., N. Tremblay and E. Fallon. 2009. Strategies to make use of plant sensors-based diagnostic information for nitrogen recommendations. Agronomy Journal 101: 800.

Sansone, S.-A., P. Rocca-Serra, D. Field, E. Maguire, C. Taylor, O. Hofmann, H. Fang, S. Neumann, W. Tong, L. Amaral-Zettler, K. Begley, T. Booth, L. Bougueleret, G. Burns, B. Chapman, T. Clark, L.-A. Coleman, J. Copeland, S. Das, A. de Daruvar, P. de Matos, I. Dix, S. Edmunds, C.T. Evelo, M.J. Forster, P. Gaudet, J. Gilbert, C. Goble, J.L. Griffin, D. Jacob, J. Kleinjans, L. Harland, K. Haug, H. Hermjakob, S.J. Ho Sui, A. Laederach, S. Liang, S. Marshall, A. McGrath, E. Merrill, D. Reilly, M. Roux, C.E. Shamu, C.A. Shang, C. Steinbeck, A. Trefethen, B. Williams-Jones, I. Xenarios and W. Hide. 2012. Toward interoperable bioscience data. Nature Genetics 44: 121–126.

Schnable, P.S., D. Ware, R.S. Fulton, J.C. Stein, F. Wei, S. Pasternak, C. Liang, J. Zhang, L. Fulton, T.A. Graves, P. Minx, A.D. Reily, L. Courtney, S.S. Kruchowski, C. Tomlinson, C. Strong, K. Delehaunty, C. Fronick, B. Courtney, S.M. Rock, E. Belter, F. Du, K. Kim, R.M. Abbott, M. Cotton, A. Levy, P. Marchetto, K. Ochoa, S.M. Jackson, B. Gillam, W. Chen, L. Yan, J. Higginbotham, M. Cardenas, J. Waligorski, E. Applebaum, L. Phelps, J. Falcone, K. Kanchi, T. Thane, A. Scimone, N. Thane, J. Henke, T. Wang, J. Ruppert, N. Shah, K. Rotter, J. Hodges, E. Ingenthron, M. Cordes, S. Kohlberg, J. Sgro, B. Delgado, K. Mead, A. Chinwalla, S. Leonard, K. Crouse, K. Collura, D. Kudrna, J. Currie, R. He, A. Angelova, S. Rajasekar, T. Mueller, R. Lomeli, G. Scara, A. Ko, K. Delaney, M. Wissotski, G. Lopez, D. Campos, M. Braidotti, E. Ashley, W. Golser, H. Kim, S. Lee, J. Lin, Z. Dujmic, W. Kim, J.

Talag, A. Zuccolo, C. Fan, A. Sebastian, M. Kramer, L. Spiegel, L. Nascimento, T. Zutavern, B. Miller, C. Ambroise, S. Muller, W. Spooner, A. Narechania, L. Ren, S. Wei, S. Kumari, B. Faga, M.J. Levy, L. McMahan, P.V. Buren, M.W. Vaughn, K. Ying, C.-T. Yeh, S.J. Emrich, Y. Jia, A. Kalyanaraman, A.-P. Hsia, W.B. Barbazuk, R.S. Baucom, T.P. Brutnell, N.C. Carpita, C. Chaparro, J.-M. Chia, J.-M. Deragon, J.C. Estill, Y. Fu, J.A. Jeddeloh, Y. Han, H. Lee, P. Li, D.R. Lisch, S. Liu, Z. Liu, D.H. Nagel, M.C. McCann, P. SanMiguel, A.M. Myers, D. Nettleton, J. Nguyen, B.W. Penning, L. Ponnala, K.L. Schneider, D.C. Schwartz, A. Sharma, C. Soderlund, N.M. Springer, Q. Sun, H. Wang, M. Waterman, R. Westerman, T.K. Wolfgruber, L. Yang, Y. Yu, L. Zhang, S. Zhou, Q. Zhu, J.L. Bennetzen, R.K. Dawe, J. Jiang, N. Jiang, G.G. Presting, S.R. Wessler, S. Aluru, R.A. Martienssen, S.W. Clifton, W.R. McCombie, R.A. Wing and R.K. Wilson. 2009. The B73 maize genome: complexity, diversity, and dynamics. Science 326: 1112–1115.

Settles, A.M., D.R. Holding, B.C. Tan, S.P. Latshaw, I. Liu, M. Suzuki, I. Li, B.A. O'Brien, D.S. Fajardo, E. Wroclawska, C.-W. Tseung, J. Lai, C.T.H. III, W.T. Avigne, J. Baier, J. Messing, L.C. Hannah, K.E. Koch, P.W. Becraft, B.A. Larkins and D.R. McCarty. 2007. Sequence-indexed mutations in maize using the UniformMu transposon-tagging population. BMC Genomics 8: 116.

Shah, A.R., M. Singhal, K.R. Klicker, E.G. Stephan, H.S. Wiley and K.M. Waters. 2007. Enabling high-throughput data management for systems biology: the bioinformatics resource manager. Bioinformatics 23: 906–909.

Sims, D.A. and J.A. Gamon. 2002. Relationships between leaf pigment content and spectral reflectance across a wide range of species, leaf structures and developmental stages. Remote Sensing of Environment 81: 337–354.

Sinclair, R. 2011. Challenges in breeding for yield increase for drought. Trends in Plant Science 16: 289–293.

Sinclair, T.R., L.C. Purcell, V. Vadez, R. Serraj, C.A. King and R. Nelson. 2000. Identification of soybean genotypes with N 2 fixation tolerance to water deficits. Crop Science 40: 1803–1809.

Sirault, X.R.R., R.A. James and R. Furbank. 2009. A new screening method for osmotic component of salinity tolerance in cereals using infrared thermography. Functional Plant Biology 36: 970–977.

Sozzani, R. and P.N. Benfey. 2011. High-throughput phenotyping of multicellular organisms: finding the link between genotype and phenotype. Genome Biology 12: 219.

Taylor, C.F., D. Field, S.-A. Sansone, J. Aerts, R. Apweiler, M. Ashburner, C.A. Ball, P.-A. Binz, M. Bogue, T. Booth, A. Brazma, R.R. Brinkman, A.M. Clark, E.W. Deutsch, O. Fiehn, J. Fostel, P. Ghazal, F. Gibson, T. Gray, G. Grimes, J.M. Hancock, N.W. Hardy, H. Hermjakob, J. Randall, K. Julian, M. Kane, C. Kettner, C. Kinsinger, E. Kolker, M. Kuiper, N. Le Novère, J. Leebens-Mack, S.E. Lewis, P. Lord, A.-M. Mallon, N. Marthandan, H. Masuya, R. McNally, A. Mehrle, N. Morrison, S. Orchard, J. Quackenbush, J.M. Reecy, D.G. Robertson, P. Rocca-Serra, H. Rodriguez, H. Rosenfelder, J. Santoyo-Lopez, R.H. Scheuermann, D. Schober, B. Smith, J. Snape, C.J.S. Jr., K. Tipton, P. Sterk, A. Untergasser, J. Vandesompele and S. Wiemann. 2008. Promoting coherent minimum reporting guidelines for biological and biomedical investigations: the MIBBI project. Nature Biotechnology 26: 889–896.

Thiel, M., T. Rath and A. Ruckelshausen. 2010. Plant moisture measurement in field trials based on NIR spectral imaging—a feasibility study. In CIGR Workshop on Image Analysis in Agriculture. August 2010, Budapest, pp. 16–29.

Thoren, D. and U. Schmidhalter. 2009. Nitrogen status and biomass determination of oilseed rape by laser-induced chlorophyll fluorescence. European Journal of Agronomy 30: 238–242.

Tremblay, N., Z. Wang and Z.G. Cerovic. 2011. Sensing crop nitrogen status with fluorescence indicators. A review. Agronomy for Sustainable Development 32: 451–464.

Vogel, J.P., D.F. Garvin, T.C. Mockler, J. Schmutz, D. Rokhsar, M.W. Bevan, K. Barry, S. Lucas, M. Harmon-Smith, K. Lail, H. Tice, J. Schmutz, J. Grimwood, N. McKenzie, M.W. Bevan, N. Huo, Y.Q. Gu, G.R. Lazo, O.D. Anderson, J.P. Vogel, F.M. You, M.C. Luo, J. Dvorak, J.

Wright, M. Febrer, M.W. Bevan, D. Idziak, R. Hasterok, D.F. Garvin, E. Lindquist, M. Wang, S.E. Fox, H.D. Priest, S.A. Filichkin, S.A. Givan, D.W. Bryant, J.H. Chang, T.C. Mockler, H. Wu, W. Wu, A.P. Hsia, P.S. Schnable, A. Kalyanaraman, B. Barbazuk, T.P. Michael, S.P. Hazen, J.N. Bragg, D. Laudencia-Chingcuanco, J.P. Vogel, D.F. Garvin, Y. Weng, N. McKenzie, M.W. Bevan, G. Haberer, M. Spannagl, K. Mayer, T. Rattei, T. Mitros, D. Rokhsar, S.J. Lee, J.K. Rose, L.A. Mueller, T.L. York, T. Wicker, J.P. Buchmann, J. Tanskanen, A.H. Schulman, H. Gundlach, J. Wright, M. Bevan, A.C. de Oliveira, C. Maia Lda, W. Belknap, Y.Q. Gu, N. Jiang, J. Lai, L. Zhu, J. Ma, C. Sun, E. Pritham, J. Salse, F. Murat, M. Abrouk, G. Haberer, M. Spannagl, K. Mayer, R. Bruggmann, J. Messing, F.M. You, M.C. Luo, J. Dvorak, N. Fahlgren, S.E. Fox, C.M. Sullivan, T.C. Mockler, J.C. Carrington, E.J. Chapman, G.D. May, J. Zhai, M. Ganssmann, S.G. Gurazada, M. German, B.C. Meyers, P.J. Green, J.N. Bragg, L. Tyler, J. Wu, Y.Q. Gu, G.R. Lazo, D. Laudencia-Chingcuanco, J. Thomson, J.P. Vogel, S.P. Hazen, S. Chen, H.V. Scheller, J. Harholt, P. Ulvskov, S.E. Fox, S.A. Filichkin, N. Fahlgren, J.A. Kimbrel, J.H. Chang, C.M. Sullivan, E.J. Chapman, J.C. Carrington, T.C. Mockler, L.E. Bartley, P. Cao, K.H. Jung, M.K. Sharma, M. Vega-Sanchez, P. Ronald, C.D. Dardick, S. De Bodt, W. Verelst, D. Inzé, M. Heese, A. Schnittger, X. Yang, U.C. Kalluri, G.A. Tuskan, Z. Hua, R.D. Vierstra, D.F. Garvin, Y. Cui, S. Ouyang, Q. Sun, Z. Liu, A. Yilmaz, E. Grotewold, R. Sibout, K. Hematy, G. Mouill.e, H. Höfte, T. Michael, J. Pelloux, D. O'Connor, J. Schnable, S. Rowe, F. Harmon, C.L. Cass, C. Sedbrook, M.E. Byrne, S. Walsh, J. Higgins, M. Bevan, P. Li, T. Brutnell, T. Unver, H. Budak, H. Belcram, M. Charles, B. Chalhoub and I. Baxter. 2010. The International BrachypodiumInitiative. 2010. Genome sequencing and analysis of the model grass Brachypodium distachyon. Nature 463: 763–768.

Walter, A., H. Scharr, F. Gilmer, R. Zierer, K.A. Nagel, M. Ernst, A. Wiese, O. Virnich, M.M. Christ, B. Uhlig, S. Jünger and U. Schurr. 2007. Dynamics of seedling growth acclimation towards altered light conditions can be quantified via GROWSCREEN: a setup and procedure designed for rapid optical phenotyping of different plant species. The New Phytologist 174: 447–455.

Walter, A., B. Studer and R. Kölliker. 2012. Advanced phenotyping offers opportunities for improved breeding of forage and turf species. Annals of Botany 110: 1271–1279.

Walter, T., D.W. Shattuck, R. Baldock, M.E. Bastin, A.E. Carpenter, S. Duce, J. Ellenberg, A. Fraser, N. Hamilton, S. Pieper, M.A. Ragan, J.E. Schneider, P. Tomancak and J.K. Hériché. 2010. Visualization of image data from cells to organisms. Nature Methods 7: S26–S41.

Wasson, A.P., R.A. Richards, R. Chatrath, S.C. Misra, S.V. Prasad, G.J. Rebetzke, J.A. Kirkegaard, J. Christopher and M. Watt. 2012. Traits and selection strategies to improve root systems and water uptake in water-limited wheat crops. Journal of Experimental Botany 63: 3485–3498.

Weigel, D. and R. Mott. 2009. The 1001 genomes project for Arabidopsis thaliana. Genome Biology 10: 107.

Wheeler, D., R. Garcia-Herrera, C.W. Wilkinson and C. Ward. 2010. Atmospheric circulation and storminess derived from Royal Navy logbooks: 1685 to 1750. Climatic Change 101: 257–280.

White, J.W., P. Andrade-Sanchez, M.A. Gore, K.F. Bronson, T.A. Coffelt, M.M. Conley, K.A. Feldmann, A.N. French, J.T. Heun, D.J. Hunsaker, M.A. Jenks, B.A. Kimball, R.L. Roth, R.J. Strand, K.R. Thorp, G.W. Wall and G. Wang. 2012. Field-based phenomics for plant genetics research. Field Crops Research 133: 101–112.

Winterhalter, L., B. Mistele, S. Jampatong and U. Schmidhalter. 2011. High-throughput sensing of aerial biomass and above-ground nitrogen uptaken in the vegetative stage of well-watered and drought stressed tropical maize hybrids. Crop Science 51: 479–489.

Wu, G., Q. Shi, Y. Niu, M. Xing and H. Xue. 2008. Shanghai RAPESEED Database: a resource for functional genomics studies of seed development and fatty acid metabolism of Brassica. Nucleic Acids Research 36(Database Issue): D1044–D1047.

Xin, Z., M. Wang, N. Barkley, G. Burow, C. Franks, G. Pederson and J. Burke. 2008. Applying genotyping (TILLING) and phenotyping analyses to elucidate gene function in a chemically induced sorghum mutant population. BMC Plant Biol. 8: 103.

Zamboni, A., M. Di Carli, F. Guzzo, M. Stocchero, S. Zenoni, A. Ferrarini, P. Tononi, K. Toffali, A. Desiderio, K.S. Lilley, M.E. Pè, E. Benvenuto, M. Delledonne and M. Pezzotti. 2010. Identification of putative stage-specific grapevine berry biomarkers and omics data integration into networks. Plant Physiology 154: 1439–1459.

Zhang, J., C. Li, C. Wu, L. Xiong, G. Chen, Q. Zhang and S. Wang. 2006. RMD: a rice mutant database for functional analysis of the rice genome. Nucleic Acids Research 34(Database Issue): D745–D748.

Zhang, Y., N. Tremblay and J. Zhu. 2012. A first comparison of Multiplex ® for the assessment of corn nitrogen status. Journal of Food, Agriculture & Environment 10: 1008–1016.

8

Yeast Phenomics— Large-scale Mapping of the Genetic Basis for Organismal Traits

Jonas Warringer and *Anders Blomberg**

Introduction

There are currently more than 1,500 yeast species described, corresponding to an estimated 1% of all species in the kingdom fungi. Yeasts represent an enormous diversity with the deepest branches stretching more than 800 million years back in time, with few unifying features other than predominantly single-celled life histories. Thus, yeast, in the broad sense use of the word, denotes a group of species as evolutionarily diverged as humans and the sea squirt *Ciona* (Dujon 2006). More often, however, yeast is used in a much more narrow sense, indicating the species *Saccharomyces cerevisiae*, or budding yeast. *S. cerevisiae* constitutes one of our oldest and economically most important domesticated organisms, with use in wine, beer, bread and drug production (Fay and Benavides 2005). Thanks to its beneficial genetics, making it ideally suited to experimentation, yeast

Department of Chemistry and Molecular Biology, University of Gothenburg, Lundberg laboratory, Medicinaregatan 9c, PO Box 462, 405 30 Gothenburg, Sweden.
Email: jonas.warringer@cmb.gu.se
* Corresponding author: anders.blomberg@cmb.gu.se

has also become a favorite model organism in genetics, molecular and cellular biology (Goffeau 2000). The yeast *S. cerevisiae* was, in 1996, the first eukaryotic organism to be sequenced (Goffeau et al. 1996). Yeast sequence diversity has continued to attract attention, leading to hemiascomycetous yeasts currently sporting the greatest number of sequenced genomes for any eukaryotic phylum. The finalized genome of *S. cerevisiae* ushered in the era of large-scale experimental biology and promoted the establishment of transcriptomics, proteomics and metabolomics as distinct fields of research. Yeast early on became a landmark-species in functional genomics, leading to the completion in yeast of the first gene-deletion collection in any organism (Giaever et al. 2002). More recently, yeast has attained a premier role in phenomics with the first use of the word in the title of a yeast research paper in 2003 (Warringer et al. 2003). We here provide a comprehensive overview of yeast phenomics, the genome-wide cataloguing of phenotypes at the organismal level, focusing specifically on traits that co-vary with the capacity to reproduce.

Strictly speaking, any characteristic at the biochemical, molecular, cellular, tissue, organ, organismal, or population level qualifies as a trait/phenotype. In yeast, traits as diverse as the germination of spores, the colony morphology on agar, changes in mRNA abundance in various genetic backgrounds, and variations in secondary metabolites that have an impact on the taste of beer would all fall under the yeast phenotype heading in the broadest sense of the word. In reality, trait variations reflecting the organism's ability to compete in ecologically relevant environments, i.e., yeast fitness, have been, and are, in the focus of yeast phenomics. Artificial gene knockouts in different organisms show little loss of fitness in standard conditions tested in the laboratory, and that also goes for yeast. The seeming unimportance of many genes that nevertheless must have been selected most likely reflects our inability to fully mimic evolutionarily/ecologically relevant conditions in the lab. This is a fundamental shortcoming of laboratory research as failure to establish associations between a gene and traits that co-vary with fitness prevents evolutionarily meaningful studies of the gene's function. A better knowledge of yeast ecology, allowing reconstruction of environments mimicking natural situations, would likely allow characterization of the many yeast genes that still are of unknown function.

Yeast Ecology and Evolution

Despite decades of intense studies, the ecology of *S. cerevisiae* remains elusive (Replansky et al. 2008). This yeast species has been encountered in a diverse set of natural conditions, such as on the surface of grapes, on oak tree bark, on damaged fruits, in forest soils, on rotten wood and on insects. In a recent study from mainland China, thousands of samples were collected from a wide array of natural sources, at remote sites. *S. cerevisiae* cultures could be established from more than 10% of these samples (Wang et al. 2012), meaning that the species is ubiquitous in the wild and frequently can be encountered in environments only marginally affected by human activity.

Many environmental factors have contributed to shape yeast phenotypes. One important selection factor, present in all niches, is competition with other microbes, often leading to a biological arms race encompassing both offensive and defensive weaponry. Microbial competition has resulted in yeast having attained a remarkably efficient capacity to ferment low complexity carbon sources (sugars) into the toxic substance ethanol, which is secreted to kill off competitors. To avoid wasting good energy and carbon, yeast has also attained the capacity to reuse this secreted ethanol, when other types of carbon resources have been depleted, and to further reduce it. As a further extension of their offensive weaponry, many yeast strains, i.e., killer strains, also produce secreted proteins that are lethal to yeast from other populations as well as bacteria, providing an additional competitive advantage (Schmitt and Breinig 2006).

Yeast has also attained a fitness edge by developing the capacity to thrive in harsh environments that are non-permissive to many other microorganisms, e.g., in acidic environments, in osmotically challenging conditions, and at low temperatures. Furthermore, anthropogenic activity has in certain locations led to yeast having evolved resistance to micro-biocides, such as copper and sulfites, compounds that are used in the industrial production of wine. Many yeast species can also tolerate niches enriched in digestive enzymes, allowing them to hitchhike with insects that, for example, have devoured fruit where yeast resides. This may facilitate dispersion over distances otherwise unattainable by the non-moving yeast. Over the years researchers have investigated the dietary effects of yeasts on the growth, fecundity and survival of a wide array of insects. A classic example of yeast-insect interactions involves cactophilic yeasts and species of the fruit fly *Drosophila* that are found in decaying cacti, with reports on non-random yeast ingestion as a result of selective feeding (Fogleman et al. 1981). Flies actively seek yeast, and the yeast metabolite acetate is a strong attractant to adult flies (Becher et al. 2010). However, insects might also affect yeast communities, e.g., as a result of transportation to new substrates, of depositing yeast-containing waste products, and by physically altering the structure of substrates. For example, larvae of *Drosophila melanogaster* have profound effects on the densities and community structure of yeasts that develop in banana substrates (Stamps et al. 2012). Future studies will tell to what extent this yeast-fly mutual relationship is the result of co-selection, and what impact it has had on the development of various yeast species. It is certainly plausible that specific physiological and mechanical properties of yeast could have been advantageous and selected for, like excretion of insect attractants, high nutritional status and resistance to passage through the insects' digestive tract.

Genomics Goes Ecology

What can we learn about yeast ecology and evolution from the current rich repertoire of yeast genome sequences? It is well established that a whole-genome duplication (WGD) occurred about 100 million years ago in the clade including *S. cerevisiae* (Wolfe and Shields 1997). Yeasts whose ancestor experienced the WGD have a number of unique characters, one of these being that they degrade hexoses to pyruvate and ethanol even in the presence of oxygen (Piskur and Langkjaer 2004). A primary outcome of the genome duplication may have been an almost immediate selective advantage to an ancestor of *S. cerevisiae*, giving it the ability to use glucose more rapidly than its ancestors and hence out-compete other yeasts when glucose was in great supply (Conant and Wolfe 2007). This argument is based on the inherent kinetics of fermentation and respiration, resulting in an increased preference for fermentation in the polyploid yeast. In this context, it should be pointed out that the WGD in the evolution of the *Saccharomyces* clade happened at the same time as the appearance of fruit- and flower-bearing angiosperms. This important evolutionary event, which resulted in rich supply of sugars in various fruits and nectars, may have opened an ecological niche to which some yeasts, such as the ancestor of *S. cerevisiae* via its rapid use of glucose through fermentation, were particularly well adapted. The *Saccharomyces* clade is still characterized by the assimilation of a rich plethora of hexoses, making various sugars an ideal test variable in phenotypic screens.

Chromosomal rearrangements occur readily in nature and are a major reshaping force during evolution. One outcome of these rearrangements in yeast is two large gene clusters: the GAL cluster (Hittinger et al. 2004) and the DAL cluster (Wong and Wolfe 2005). The GAL gene clusters appear to have originated independently in three unrelated yeast lineages, and by at least two different mechanisms: horizontal gene transfer (HGT) and gene relocation (Slot and Rokas 2010). Interestingly, species with GAL clusters have exhibited significantly higher rates of GAL pathway loss than species with un-clustered GAL genes. This suggests that clustered genes provide overall enhanced fitness by increased spread by HGT in favorable environments and increased loss in unfavorable ones, thus facilitating fungal adaptation to diverse and changing nutritional environments. The galactose content in plants varies substantially from hundreds of mg per g in legume seeds and algal mats to less than 1 mg per g in some fruits (Slot and Rokas 2010). This suggests that different yeast populations have had ample opportunity to evolve niche-dependent adaptations for galactose utilization, which is clearly evident from genome sequence analysis (Warringer et al. 2011).

The DAL gene cluster consists of six adjacent gene-encoding proteins that enable *S. cerevisiae* to use allantoin as a nitrogen source. Allantoin is converted through a series of steps into ammonia, a simpler form of nitrogen used as the internal nitrogen currency of yeast. In most species, the DAL genes are scattered around the genome, but in an ancestor of *S. cerevisiae*, they became relocated to a single subtelomeric site. These genomic rearrangements coincided with a biochemical reorganization of the purine degradation pathway, which switched to importing allantoin instead of urate. This change eliminated one of the oxygen-consuming enzymes, urate oxidase, a gene-loss found in yeasts that can grow vigorously under anaerobic conditions. An interesting mechanistic hypothesis for this linkage is based on the fact that glyoxylate, which is produced by the Dal3 reaction and removed by the Dal7 reaction, is quite toxic to yeast. Thus, there may be selection for alleles of *DAL3* and *DAL7* that interact well, are inherited together, and are strictly co-regulated to facilitate metabolic channeling. The rearrangements of the DAL cluster in the *S. cerevisiae* "*sensu stricto*" clade may have conferred some advantage in natural environments low in nitrogen. Natural sources of allantoin for yeasts are plants and insect excretion, pointing again to the close connection between yeast-plants and yeast-insects in the evolution of *S. cerevisiae*. The evolution of the DAL cluster also points to nitrogen sources as highly relevant variations to screen in yeast phenomics.

Finally, the selective pressure to economize on oxygen during evolution of the "fermentative lifestyle" in *S. cerevisiae* selected for diminished dependence on oxygen-requiring reactions. The importance of the yeast ability to grow in low oxygen is demonstrated by the way its pyrimidine synthesis pathway has been reconfigured to avoid dependence on respiration and by the existence of many duplicated genes encoding specialized hypoxic and aerobic forms of proteins (Langkjaer et al. 2003). The fermentative yeast lineages apparently had several competitive advantages, since they could grow fast with or without oxygen and they produced toxic ethanol even under aerobic conditions, the latter being important against bacterial competitors. This type of ecological warfare between yeast and bacteria through removal of oxygen and production of ethanol can be found in grape-wine ecosystems (Fleet 2003). Oxygen limitation may have been the prevalent condition during the early days of *Saccharomyces* ecology, and it is thus highly relevant to perform phenotypic screens under micro-aerobic, or anaerobic, conditions.

Yeast Fitness

Yeast fitness in a natural context is considered a composite of the ability to reproduce mitotically, survive as a vegetative cell, mate, sporulate, survive as a spore, migrate and germinate. The fitness of yeast is potentially also reflected in population level phenotypes, notably the way yeasts form colonies on solid substrates. Colony formation is the most common way of yeast growth in nature, and liquid phase cultivation might be more a spin-off of released cells from these colonies. There is clear differentiation within colonies and biofilms of *S. cerevisiae* growing on solid media, with evidence for the involvement of distinct developmental programs (Cap et al. 2012, Vachova et al. 2012). An effect of this population differentiation is that specific cells in colonies have higher survival rates than cells reproducing individually in liquid media. Dying cells of chronologically aged central colony areas may release compounds that serve as nutrients for late colony growth. Such regulated dying contributes to the longevity of the whole colony by providing nutrients released by the death of some to the benefit of others, with little net loss of assimilated nutrients to competitors.

Sporulation is another developmental program of importance to fitness. Under adverse conditions, yeast cells can form spores that are generally long-lived and resistant to external challenges. Respiratory metabolism in environments with low nitrogen content initiates yeast meiosis and formation of membrane enveloped, haploid nuclei within the anucleated mother cell, transforming it into a sporesack (Neiman 2011). Upon dissolution, the sporesack releases the spores which germinate first when encountering fermentable carbon. Consequently, yeast may spend most of its chronological life-time in nature as a spore, husbanding reserves. Spore formation was initially thought to enhance tolerance to environmental extremes (Briza et al. 1990, Dawes and Hardie 1974). However, spores appear not to be very tolerant to common ecological fluctuations, such as repeated freeze–thaw cycles or desiccation (Coluccio et al. 2008). Spore formation may instead be intended to promote dispersion via insect vectors. Digestive enzymes encountered in the insect gut often kill vegetative yeast but fail to kill yeast spores. This potentially allows yeast to hitch a ride with insects devouring nectar or pollen in which the yeast resides. Selection acting on such complex life history traits is likely to have been substantial in the evolutionary history of yeast but is, because of experimental challenges, typically disregarded in yeast phenomics.

Although the diploid state tends to dominate, yeast is found in haploid, diploid and tetraploid forms in a wide variety of ecological niches (Gerstein and Otto 2009). In nutrient-rich environments, polyploid cells reproduce clonally by asymmetric budding, producing daughter cells that are physiologically distinct but genetically identical to the mother. When encountering starvation conditions, diploids undergo meiosis. Typically, a tetrad of four haploid spores is produced, although triads, dyads or monads may emerge (Taxis et al. 2005). When returned to a nutrient-rich environment medium, spores germinate into metabolically active haploid gametes of two mating types, *MAT*a and *MAT*α. Haploid gametes may reproduce clonally for one or more generations but strive to fuse with haploids of the complementary mating type in a process akin to fertilization, producing a diploid. As spores are enclosed within a spore sack and linked by interspore bridges (Coluccio and Neiman 2004), mating tends to occur between gametes derived from the same tetrad (Ruderfer et al. 2006, Tsai et al. 2008). Enforcing the tendency towards selfing, haploids typically switch mating type after the first mitotic cell division to fuse with their own progeny (Knop 2006). Given that diploids and haploids often have radically different responses to external perturbations (Gerstein and Otto 2009, Zörgö et al. 2013), the sexual cycle certainly has an impact on yeast fitness.

Despite the complexities of yeast life history, experimental phenomics typically approximate yeast fitness to mitotic capacity in artificial laboratory environments, or even exclusively to the mitotic growth rate. Therefore, much of this chapter will focus on this type of phenotype.

Phenomics Technologies to Measure Mitotic Fitness

Growth on Agar

There are logistical advantages to growing yeast in populations, typically derived from one or a few founder individuals, on the surface of an agar matrix. Increase in the circular area, or in rare cases volume, of such a colony is considered proportional to the increase in population size and can therefore be used as a measure of population reproductive capacity or fitness. In most cases, the growth rate of each colony is not determined specifically (Baryshnikova et al. 2010). Instead, a composite of colony growth is considered as the colony size at a specified time point. Variations in initial colony size between colonies and the contribution from individual growth phases to the final colony size are typically disregarded. Several image analysis tools can perform colony segmentation and colony area quantification (Collins et al. 2006, Dittmar et al. 2010, Lamprecht et al. 2007, Lawless et al. 2010, Shah et al. 2007). However, most of these tools are similar in their fundamental design, detecting and quantifying colonies and background in similar ways. In principle, any of these could provide colony estimates at dense time intervals, allowing the generation of growth curves and the extraction of more precise information on growth rate. However, in practice, spatial and temporal bias as well as noisy data makes this task a real challenge and only a few endeavors have dared to take this step.

Phenotypic Array Analysis is such a methodology which obtains growth rates from cultivation on agar (Hartman and Tippery 2004, Shah et al. 2007). It is based on strains being diluted in liquid medium and spotted onto the agar surface in droplets that are a few microliters in volume. Following image capture at different time points, the exponential rate of growth can be obtained over 4–5 generations. More recently, an analogous system named Quantitative Fitness Analysis (QFA) has been developed (Banks et al. 2012a). Analysis of photographs from each time point produces quantitative cell density estimates and corresponding growth curves, allowing quantitative fitness measures to be extracted (Fig. 1). QFA can capture complete growth curves, including exponential and saturation phases, since cultures are heavily diluted before inoculation onto the agar. A weak point of both approaches is the rather shaky estimate of colony density at early time points (Fig. 1), and growth rate estimates have to be based on modeling of data from the whole growth experiment onto a logistic growth function. Such functions typically have difficulties handling growth curves that are atypical, such as bimodal or linear growth (Warringer et al. 2008). Another drawback of this technology is that access to nutrients is uneven over the agar plate. This creates a strong spatial bias, where cells in less competitive locations, notably colonies with few or slow growing neighbors and on the edges of the plate, enjoy substantial growth advantages and thereby tend to attain artificially high growth rates. This effect can be reduced, though

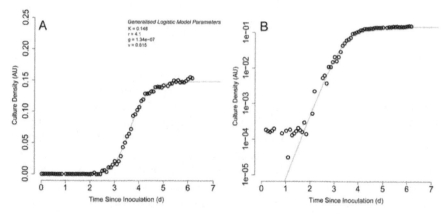

Figure 1. Growth curves from colonies growing on agar. (A) Growth curve of diluted yeast cultures growing on agar with galactose as carbon and energy source at 20°C. Images were captured robotically approximately every 2 hours. Exponential phase at this temperature was observable for approximately 1.5 days. A generalized logistic model was fit to the observed data (grey curve), and model parameters are presented: K (carrying capacity (AU)), r (growth rate $(d-1)$), g (inoculum density (AU)), v (growth symmetry). (B) As for panel A, plotted with cell density on a log scale (Banks et al. 2012b). Published with permission from the Journal of Visualized Experiments.

not eliminated, by disregarding colonies that are known to have fewer neighbors, i.e., those located close to the agar edges. An advantage of the agar system over liquid microcultivation is that cross-contamination is less of a concern because of lower risk of migration between populations through droplets. QFA has been designed for high throughput and is most useful for genome-wide genetic interaction or drug screens investigating thousands of independent cultures. The technology has recently been applied to the analysis of a point mutation in the essential telomere capping gene *CDC13* (*cdc13-1*), where it was found that Cdc13 and Yku70, believed to have complementary roles in telomere capping, displayed genetic interaction profiles that differed significantly (Addinall et al. 2011). Thus, the gene products from *CDC13* and *YKU70* might differ more in function than was earlier thought.

Growth in Liquid Cultures

The microtiterplate offers parallel liquid growth analysis in combination with ease of automation. A large number of genome-wide studies have been conducted utilizing this microcultivation technology (Liti et al. 2009a, Warringer et al. 2003, Warringer et al. 2011). Changes in optical density over time are assumed to reflect underlying variations in population density, allowing establishment of population growth curves. However, one should be aware that light absorption in microbial cultures shows strongly nonlinear characteristics at higher cell densities. This can be compensated for using a calibration function that is based on the correlation between the observed OD and the true OD, the latter obtained by sample dilution (Warringer and Blomberg 2003). If this high cell density compensation is not applied, the exponential growth phase will appear unrealistically short and the growth rate as well as the yield/efficiency value will be underestimated.

Despite its frequent use, there is a lack of knowledge about the specific growth conditions in microcultivation systems (Buchs 2001). Proper supply of oxygen, as well as removal of carbon dioxide, is hard to achieve since shaking is usually not adequate in these small volumes (Hermann et al. 2003). Below the critical shaking frequency, essentially no liquid movement inside the wells is observed and oxygen transfer is almost the same as for non-shaking conditions. At higher cell densities, it is thus to be expected that the growth on fully respiratory carbon sources would be oxygen limited. Several instruments for automated analysis during microcultivation are available on the market (see Blomberg (2011) for a detailed review). In many instances, these devices are combined incubators, shakers and readers that automatically record growth under controlled conditions (e.g., BioLector,

Bioscreen C, PowerWave). In other cases, plates are moved with robot arms between physically separated incubators, shakers and spectrophotometers. To avoid optical problems related to moisture on the inside of the lid, a common practice is to remove the lid prior to optical reading. However, this regime increases contamination and evaporation, the latter of which affects wells in different locations differently, resulting in spatial bias. One solution is to use gas- and light-transparent plastic covers that restrict evaporation. An alternative approach is used in the Bioscreen C device, where the lid is kept slightly warmer than the rest of the plate to avoid formation of moisture on the inside of the lid. Overall, liquid cultures provide high-resolution growth curves with well-separated and independent analysis of the various growth phases (Fig. 2). This is in many cases essential since different gene deletion strains display aberrations of different growth phases in different environments (Warringer et al. 2008).

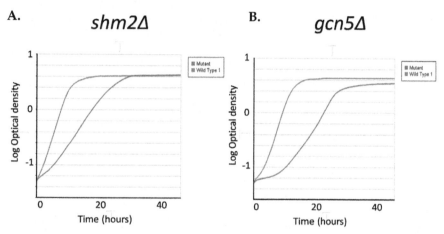

Figure 2. Growth curves of yeast cultures generated by microcultivation. Yeasts were growing in 350 μl of synthetic medium cultures with glucose as carbon and energy source. The optical density was automatically recorded every 20 minutes over two days. (A) Quantitative growth analysis of the *shm2Δ* mutant (red curve) compared to the wild type (grey curve). This mutant displays a clear change in growth rate. (B) Quantitative growth analysis of the *gcn5Δ* mutant (red curve) compared to the wild type (grey curve). This mutant displays clear changes in all three growth variables—growth lag, growth rate and growth efficiency.

Color image of this figure appears in the color plate section at the end of the book.

Growth in Competition

Many phenotypic effects of gene deletions are minor (Thatcher et al. 1998) and might escape detection in ordinary culturing systems. However, minor growth differences can be revealed as a cumulative effect after many generations of proliferation in competition with a reference strain. A powerful large-scale design is simultaneous growth competition analysis of all yeast deletion mutants. Each yeast deletion strain in the yeast deletion collection (Giaever et al. 2002) is tagged with a unique molecular bar code. Thus, mitotic fitness can indirectly be determined by monitoring the abundance of each mutant's molecular barcode over time (Giaever et al. 2002, Winzeler et al. 1999a). However, there are technical difficulties in measuring mutants with slow growth rates. Noise and bias as compared to cultivations with a single mutant and the corresponding reference strain is also elevated. Nevertheless, this methodology has been applied in an impressive study of hundreds of chemical agents, and the growth effects imposed on the whole gene deletion collection have thus been exhaustively mapped (Hillenmeyer et al. 2008). An alternative, highly sensitive growth competition assay based on co-culturing of a red fluorescent protein (RFP)-expressing wild-type strain with a mutant green fluorescent protein (GFP)-expressing strain has been developed (Breslow et al. 2008). The relative change in abundance of the two strains over time is monitored by high-throughput flow cytometry that follows the red and green fluorescence of the culture. This is used as a measure of the relative growth rate. A consequence of the reported high precision is that the authors report on detecting significant growth defects in basal medium for 45% of the deletion mutants.

Growth of Individual Cells

Microbial fitness assays have historically been limited to ensemble measurements calculating the difference in mean growth rate or the competitive fitness advantage of one population over another. However, clonal populations of cells grown in a constant environment display a striking cell-to-cell heterogeneity. Assumptions of isogenic populations are probably hugely over-optimistic, as the last common ancestor of all cells in the studied populations often lies far back in time. However, genetic variability is not deemed high enough to explain the very substantial heterogeneity observed. A crucial challenge is to understand how much of this non-genetic variability serves a biological function, e.g., does variation in itself increase population fitness? It can easily be understood that a fluctuating environment can potentially favor high variation between individuals, if selection acts on the population level rather than on the level of the individual.

To enable measurements of cell-specific growth rates, microcolony growth can be followed by time-lapse bright-field microscopy, either via seeding cells onto thin agarose slabs (Di Talia et al. 2007) or via growing cells on glass-bottomed micro-well plates (Levy et al. 2012). In the latter case, thousands of microcolony growth rates are monitored automatically, providing great population statistics. Micrographs were captured with a high-resolution microscope equipped with automated infrared high-speed focusing and a fully automated stage equipped with an environmental chamber. The authors found that cells in a presumably isogenic population display vastly different doubling times ranging from 70 to 160 minutes. A powerful aspect of microscopy-based growth rate studies is that they allow simultaneous cell-based analysis of expression/localization of GFP-tagged proteins. It was concluded that Tsl1, a protein involved in the synthesis of the disaccharide trehalose, appears to be a good molecular marker for cells both displaying slow growth in basal conditions and exhibiting increased tolerance to some stresses. This would thus constitute a bet-hedging mechanism whereby certain slow-growing cells in populations would be more robust and survive environmental change.

Phenomics on Recombinant Laboratory Strains

Starter Strain for Yeast Recombinant Collections

The genetics of *S. cerevisiae* makes it a formidable model organism. Its easily controllable life cycle, rapid mitotic reproduction in both haploid and diploid form, efficient DNA uptake, and high recombination rate facilitate efficient targeted genetic manipulation (Goffeau 2000). Together with ease of handling and access to a genome sequence of outstanding quality (Goffeau et al. 1996), these qualities have allowed creation of an unparalleled set of recombinant yeast collections. Almost exclusively, these resources have utilized the yeast universal reference strain S288C, an artificial genomic mosaic of predominantly European genetic stock (Liti et al. 2009a), as a starting point. S288C's elevated position in yeast reverse genetics derives from a natural, inactivating mutation in the mating-type switching gene *HO*, which during the 1940s enabled stable propagation of yeast lines of each mating type (Mortimer and Johnston 1986). However, the original S288C isolate also carried other detrimental alleles, notably a *GAL2* missense mutation impeding galactose uptake (Winston et al. 1995) and a transposon insertion in *HAP1*, impairing electron transfer in the respiratory chain (Gaisne et al. 1999). Repair of some of these defects and introduction of a variety of auxotrophic markers (Brachmann et al. 1998) later gave rise to the BY series of designer strains, the actual starting point for most reverse genetics collections. Most other common lab genetic backgrounds, such as sigma1278b, the YPH series and W303, share the majority of their genomes with S288C, while a few others, SK1 and Y55, primarily stem from West African stock (Liti et al. 2009a).

Yeast Recombinant Collections

Early efforts towards establishment collections of genetically modified yeast focused on transposon-mediated insertion disruption. A collection of 11,000 strains (representing 2,000 genes) was initially constructed (Ross-Macdonald et al. 1999, Ross-Macdonald et al. 1997). However, the challenges of non-random transposon insertion and variable degrees of function loss due to transposon insertion spoke in favor of targeted gene deletion approaches. A collection of designed gene deletion strains, individually replacing each yeast open reading frame with a selectable marker, was consequently constructed by an international consortium (Winzeler et al. 1999b). Encompassing haploid gene deletion strains of the two mating types, and homozygous and heterozygous diploid gene deletions, the collection contains close to 20,000 strain constructs (Giaever et al. 2002). A key feature of this genetic resource is the inclusion of flanking 20-mer DNA barcodes, uniquely identifying each deletion strain, and enabling competitive cultivation of the complete collection (see section **Growth in Competition**). An impressive number of offshoots have been spawned from the initial gene deletion collection initiative. Collections of conditionally deleterious alleles, where a controlled temperature change is used to induce partial loss of gene functionality (Ben-Aroya et al. 2008, Ben-Aroya et al. 2010, Li et al. 2011), now allows the study of essential gene functionality. Promoter replacement by titratable promoters allows the controlled reduction of gene product abundance (Mnaimneh et al. 2004), fusion of genes to degradation cassettes permits heat inducible proteins (Kanemaki et al. 2003), and insertion of foreign elements in 3′ UTRs (Yan et al. 2008) destabilizes mRNAs, also reducing gene product abundance. Extensive, but sub-genome-wide collections of such constructs are now available. At the opposite end of the spectrum, a collection that replaces native for galactose-inducible promoters enables the study of increases in gene product abundance (Osterberg et al. 2006, Sopko et al. 2006). A similar effect is achieved by placing genes on plasmids whose copy levels can be controlled by alterations of external concentrations of leucine (Krantz et al. 2009, Moriya et al. 2006, Makanae et al. 2013). Global collections of plasmid-based overexpression are also available (Ho et al. 2009, Magtanong et al. 2011), although these collections allow no control over plasmid levels and therefore may not prevent plasmid depletion in cases of severe gene toxicity. A third class of strain collections aims at quantifying molecular rather than organismal phenotypes, e.g., protein expression, localization and interaction. This is typically achieved by linking target genes to genetic markers that are traceable when expressed, allowing tracking of the composite gene products. The majority of yeast genes have been individually tagged by a traceable marker in collections based on Tandem Affinity Purification (TAP) tag (Ghaemmaghami et al. 2003) and Green Fluorescent Protein (GFP) tag, respectively (Newman et al. 2006). Although these collections primarily serve as resources for molecular phenotypes, they also inform on organismal phenotypes.

Lessons Learned from Yeast Recombinant Genetics

Extensive exploitation of the deletion collection has illuminated the gene–phenotype landscape in yeast to an unparalleled extent and fuelled the emergence of novel fields, such as chemical genomics (Lopez et al. 2008, Roemer et al. 2012) and toxico-genomics (North and Vulpe 2010). In optimal conditions, a mere 19% of yeast genes are essential for life while a further 15% contributes measurably to its mitotic growth rate, with most of these having a marginal impact (Giaever et al. 2002). Exposure to non-optimal environments, e.g., the elevation of salinity encountered in marine environments (Warringer et al. 2003), slightly increases the proportion of fitness contributing genes. Nevertheless, the majority of gene products stubbornly remain dispensable in each individual environment. This surprising observation begs explanation. First, lack of measurement precision, low degree of replication and the logistic challenge of randomizing replicates over space and time, meaning that the frequency of false negatives is likely to be high in current large-scale screens, resulting in a lack of detection power. Data from high-precision measurements also support the proposition that we often fail to detect minor contributions to fitness (Thatcher et al. 1998). Such minor effects could nevertheless have an enormous cumulative effect over longer evolutionary time spans and represent strong selection. Second, natural evolution occurs in a constantly shifting ecological setting. Thus, different selective pressures may have combined to maintain the current gene make-up of yeast, with most genes being dispensable in most single environments but indispensable in the total aggregate of environments encountered in its habitat. Indeed, considering hundreds of different environments, all but a handful of evolutionary conserved open reading frames contributed positively to fitness (Hillenmeyer et al. 2008). Finally, our tendency to approximate yeast fitness to mitotic reproduction of haploid yeasts, or even the rate of mitotic reproduction, is unlikely to reflect the true extent of historical selection acting on the yeast genome (see section **Yeast Ecology and Evolution**). Yeast maintains a complex life history, including shifts in ploidy, mating, recombination, entrance into spore form, migration and germination (Knop 2006). These shifts are likely to contribute to fitness, but are overlooked in current phenomics approaches, as is the ability to survive. Deletion of one gene copy in a diploid yeast generally fails to reduce fitness (Deutschbauer et al. 2005), not because of compensatory activation of the remaining gene copy, but because of basal protein production exceeding demands by a factor of two or more (Springer et al. 2010). Thus, although haploinsufficiency and haploproficiency, describing the inferior and superior performance of diploids retaining a single copy of a gene respectively, occur (Delneri et al. 2008), they appear to be exceptions rather than norm.

Environmental shifts typically evoke changes in the importance of several hundred genes, almost always representing a wide diversity of cellular functions. Whereas gene deletions in the absence of stress show a disproportionate tendency to impair growth, their phenotypes under various stresses, such as oxidative stress, are typically normally distributed (Fig. 3). Thus, removal of a gene is often as likely to enhance stress tolerance as impair it, suggesting that most commonly tested stresses historically have imposed at the most marginal selective pressures on the yeast genome.

Challenges thought to share mechanisms of actions sometimes show overlapping patterns of gene importance. Thus, shared patterns of gene importance may allow tentative assignments of mechanisms of actions to poorly understood challenges, such as novel drugs whose cellular targets are unknown (Parsons et al. 2004, Parsons et al. 2006). Such guilt-by-association assignments may partially be confounded by pleiotropy. For example, genes involved in intracellular transport, transcription regulation and vacuolar sorting tend to be of elevated importance in a vast range of

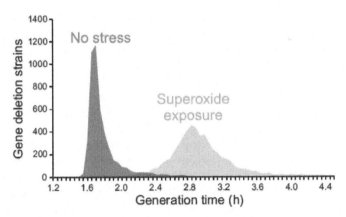

Figure 3. Frequency distribution of the generation time of yeast gene deletions in absence and presence of stress. The complete collection of haploid yeast gene deletion strains (4,700 strains) was micro-cultivated in absence and presence of stress (paraquat exposure, elevating intracellular superoxide concentrations through electron shunting to oxygen from the respiratory chain) as in Warringer et al. (2003). The mean doubling/generation time (h) of each individual population was extracted from high-density growth curves. The frequency distribution of mean population doubling times was plotted using a bin range of 0.05 h. A handful of highly superoxide sensitive strains fall outside the bin range shown, with the most superoxide sensitive strains, the superoxide dismutases *sod1Δ* and *sod2Δ*, achieving a generation time of 7.7 h and 6.0 h, respectively.

unrelated stressful environments (Dudley et al. 2005, Hillenmeyer et al. 2008). Nevertheless, pleiotropy appears less widespread than initially feared (Ericson et al. 2006). In fact, despite remarkable universality of stress responses in terms of which mRNA species increase or decrease in abundance (Causton et al. 2001, Gasch et al. 2000), gene importance patterns overall appear stress specific (Jin et al. 2008, Thorpe et al. 2004). This discrepancy, a universal stress response in terms of mRNA regulation but stress specificity in terms of which genes are important, is reflected in a complete lack of correlation between mRNA regulation and gene deletion phenotypes (Birrell et al. 2002, Giaever et al. 2002, Thorpe et al. 2004, Warringer et al. 2003). Functional redundancy of gene products, e.g., as emerging from gene duplication, may partially confound efforts to reveal gene importance through phenotyping of deletion constructs (DeLuna et al. 2008). Changes in transcript abundance may also fail to translate into protein abundance (Gygi et al. 1999), for example as a consequence of regulation of ribosomal transcript mobilization counteracting regulation of mRNA production (Warringer et al. 2010). Changes in protein abundance may finally fail to translate into changes in total protein activity, because of failure to fold, transport or modify the additional proteins correctly. Nevertheless, even taking these factors into account, the complete absence of a correlation between mRNA abundance adjustments and shifts in gene importance is troubling.

Hitherto, yeast overexpression collections have been sparingly exploited. Our phenotypic understanding of mutations enhancing protein functionality is therefore lower than our understanding of mutations impairing protein functionality. So far, the consequences of gene over expression appear strongly biased towards negative effects on fitness (Krantz et al. 2009). At least 15% of overexpressed genes impair mitotic reproduction (Sopko et al. 2006), with enrichments for cell cycle genes, signaling molecules, and transcription factors. Remarkably, in contrast to gene deletion effects which tend to be environment specific, the consequences of gene over expression appear independent of the environment (Krantz et al. 2009). Thus, the position of the upper tolerance level with regards to gene expression is constant, whereas the lower tolerance level appears to be shifting, and depends on the environment. Dosage imbalances between cell cycle components have been identified as the cause of specific gene overexpression toxicities (Kaizu et al. 2010), in agreement with the "protein complex stoichiometry hypothesis" which postulates deviations from normal component stoichiometry to disrupt protein complexes, resulting in loss of complex function (Papp et al. 2003). However, overexpression toxicity effects are not enriched among protein complex components (Sopko et al. 2006). Furthermore, gene overexpression effects fail to correlate with gene deletion effects. Thus, disruption of protein complex stoichiometry is unlikely to be a general explanation for gene overexpression toxicity.

Limitations of Yeast Recombinant Genetics

The reverse genetics paradigm serves as the fundament for yeast phenomics. However, it is generating increasing unease. One cause of this concern is the enormous genetic and phenotypic diversity observed in the species (Liti et al. 2009a, Schacherer et al. 2009, Warringer et al. 2011). The population structure is very pronounced and phenotypic variation tends to follow population boundaries as defined by the phylogenetic history of strains. This means that the genetic background is likely to have a profound impact on the contribution of individual genes to particular phenotypes. The focus on the artificial lab hybrid S288C, whose alleles have never jointly been exposed to natural selection and has poor sporulation (Cubillos et al. 2009, Deutschbauer and Davis 2005), poor respiration (Gaisne et al. 1999) and a transposon-infested genome (Liti et al. 2009a) is particularly troubling. S288C actually emerged as the phenotypically most atypical yeast in a large screen of strains in hundreds of environments (Warringer et al. 2011), suggesting that results obtained in this genetic background may be hard to translate into an understanding of the species as a whole. Compounding such concerns, yeast recombinant collections have been established using genetic backgrounds that lack several key genes involved in nitrogen metabolism (Brachmann et al. 1998). Removal of these genes not only confounds any studies on nitrogen metabolism and utilization, but also appears to have profound effects on the overall genotype-phenotype map. A second concern is the mutagenic nature of reverse genetics and the resulting secondary site mutations (Scherens and Goffeau 2004). An estimated 8% of gene deletions retain a copy of the deleted gene, presumably due to chromosomal or chromosomal segment amplifications (Hughes et al. 2000). Unwanted phenotypic influences may also arise from deletion cassettes (Goldstein and McCusker 1999, Wach et al. 1994) or from selection markers introduced together with tags and overexpressed genes (Wach et al. 1997). Marker genes, as well as their heterologous promoters, which are typically taken from the distantly related yeast *Ashbya gossypii* in order to avoid uncontrolled recombination with a locus already existing in the genome, are known to affect gene-phenotype associations (Babazadeh et al. 2011). These insertions may also disturb non-targeted genetic elements in the vicinity of the targeted locus. Seven to fifteen percent of deletion phenotypes are estimated to be due to such neighboring gene effects (Ben-Shitrit et al. 2012).

Phenomics on Natural Yeast Isolates

The emergence of next generation sequencing, potentiating nucleotide level dissection of the genotype-phenotype map of natural yeasts, has shifted the focus of yeast phenomics from reverse genetics in the direction of mapping natural variation. High resolution mapping of natural variation not only facilitates assignment of function to real-life polymorphisms, but also offers a potential to interpret the yeast genome in the face of what is known regarding its ecology and historical selective pressures (Dalziel et al. 2009). In addition, capitalization on natural variation may allow a circumvention of the consumer aversion against genetically modified organisms. This aversion currently impedes the use of reverse-engineered yeast in most industrial applications, such as food (baking) and beverage (wine, beer) production. Several studies aiming to identify the genetic basis of natural variation in industrial traits have also recently been published (Ambroset et al. 2011, Marullo et al. 2007, Salinas et al. 2012).

Population Phenomics in Yeast

The genotypic and phenotypic landscapes of both *S. cerevisiae* and its closest relative *Saccharomyces paradoxus* have recently been mapped to reasonably high resolution, providing a general view of patterns of yeast variation (Liti et al. 2009a, Schacherer et al. 2009). Five phylogenetically distinct gene pools, corresponding to Malaysian, Sake, West African, North American and European populations, were originally identified, with several deep-branching populations from mainland China recently joining that group (Wang et al. 2012). However, most industrial strains emerged as genomic mosaics and are apparent hybrids with blocks of sequence deriving from more than one of these gene pools. Haploids from diverged populations mate, but meiotic success is cross-dependent, with *S. cerevisiae* hybrids involving Malaysian isolates approaching sterility (Cubillos et al. 2011) and hybrids between *S. paradoxus* populations producing only unfit offspring. A mapping of mitotic properties over hundreds of environments revealed a surprisingly large trait variation in *S. cerevisiae* (Warringer et al. 2011), which largely followed gene pool boundaries and only to a marginal extent was influenced by human domestication for baking, lab or fermentation (Warringer et al. 2011). This disparity between traits and recent human domestication for industrial purposes suggests limited overall impact of recent human imposed selection. This leaves either genetic drift or selection preceding the association of yeast with human activities as the primary phenotypic determinants. Much of the phenotypic variation is accounted for by inferior performance of a particular gene pool in a certain environment (Warringer et al. 2011). This implies that genotype-phenotype associations are disproportionately influenced by population-specific mutations with a negative influence on performance in specific environments. Frequent stop codon and frameshift mutations partially explain this observation (Doniger et al. 2008). Also, taking amino acid mutations with less blatant detrimental effects, as estimated from the degree of evolutionary conservation (Ng and Henikoff 2003), yeast phenotypes can be predicted (Jelier et al. 2011). Consequently, it appears that loss-of-function mutations are key contributors to yeast phenotypic natural variation. Crosses of yeast haploids across gene pool boundaries tend to produce diploid hybrids whose phenotypes mimic those of the best performing parent. This is consistent with detrimental alleles causing the variability in traits, and with complete dominance in the hybrids masking these alleles (Zörgö et al. 2012).

Mapping the Genetic Basis of Yeast Natural Variation

Classical forward genetics, based on chemically induced mutagenesis and allele complementation using plasmid libraries, enjoyed early successes in yeast. However, approaches aimed at mapping the natural genetic variation that segregates in existing gene pools failed to attract much attention. Yeast was therefore a late arrival to the arena of quantitative genetics. Nevertheless, the experimental advantages of yeast, allowing straightforward establishment of causality, has recently seen yeast move to the forefront of the field. Linkage analysis using designed crosses and controlled pedigrees is now in widespread use in the community. A typical yeast linkage analysis with a marker density of 150 genetic markers distributed over the 12 Mbp genome allows mapping of reasonably strong QTLs to intervals of 10–50 genes, if a sizeable number of offspring are genotyped (Ehrenreich et al. 2009). Homing in on the culprit genes in intervals is often achieved using engineered reciprocal hemizygotes, i.e., the two parents are repeatedly crossed to produce diploid hybrids, in which either of the two parental alleles of a candidate gene has been deleted (Steinmetz et al. 2002) (Fig. 4). Since the two hybrids are isogenic, except at the locus of interest, any phenotypic difference between them can be assigned to genetic differences in the gene in question. In contrast to linkage analysis, genome-wide association studies (GWAS) are challenging in yeast. This is due to the confounding influence of the pronounced population structure, meaning that many thousands of alleles share a distribution pattern. Correcting for this population structure may be feasible (Diao and Chen 2012). However, so far, only a single published study has relied exclusively on association (Muller et al. 2011). Bulk segregant analysis, which relies on the breaking up of blocks of co-inherited alleles through the establishment of randomly mating populations from two or more parents, partially circumvents the issue of population structure (Ehrenreich et al. 2010). Initially introduced in *Drosophila* (Macdonald and Long 2007), the approach may be extraordinarily suitable for yeast genetics, given the organism's small genome size, high recombination rate and easily controllable life cycle (Parts et al. 2011). Both GWAS and linkage analysis suffer from being limited to consider a single, or at the most two interacting loci, at a time. In reality, many traits are likely to be determined by large sets of alleles, which interact in complex epistatic relationships. Multivariate analysis, for example in the form of partial least squares (PLS) analysis, which simultaneously considers the impact of many genetic variants on many traits, was recently introduced in yeast (Mehmood et al. 2011), and may help to resolve this issue.

Lessons Learned from Mapping of Natural Yeast Variation

Multivariate analyses suggest that a relatively small fraction of the yeast genome contains polymorphisms affecting mitotic properties in common environments (Mehmood et al. 2011). However, the complexity of individual traits varies enormously (Cubillos et al. 2011). Traits such as cupper resistance (*CUP1*, copy number variation) (Warringer et al. 2011), galactose non-utilization (*GAL3*, nonsense mutation) (Warringer et al. 2011), xylose utilization (*XDH1*, gene gain) (Wenger et al. 2010) and sensitivity to the DNA damaging agent 4-nitroquinolone (*RAD5*, missense mutation) (Demogines et al. 2008) are close to monogenic. Such large effect alleles tend to penetrate independently of the genetic context. Other traits such as heat tolerance (Parts et al. 2011, Steinmetz et al. 2002) and sporulation efficiency (Gerke et al. 2006) are exceedingly complex and involve substantial numbers of alleles, each contributing only a fraction of the phenotypic variation. The penetrance of small effect size alleles tends to be cross-dependent (Sinha et al. 2006), and thus exposed to epistatic effects, which in a few cases have been mapped (Gerke et al. 2009). Some natural alleles, e.g., missense mutations in *CYS4* (drug tolerance (Kim and Fay 2007)), *PHO84* (drug tolerance (Perlstein et al. 2007)) and *MKT1* (drug and heat tolerance, sporulation efficiency (Ehrenreich et al. 2010)) and gene amplifications of *ENA* (cation tolerance (Warringer et al. 2011)) appear to be highly pleiotropic. Pleiotropy in these cases appears to arise from environmental similarity, rather than from general roles of these genes in organism-environment control. Nevertheless, most QTLs are environment specific, suggesting exposure to selection only in certain habitats (Cubillos et al. 2011). Spatially, yeast QTLs are clearly non-randomly distributed (Ehrenreich et al. 2009). Distinct hot- and coldspots occur with distal chromosomal regions containing >30% of all QTLs (Cubillos et al. 2011). Part of the reason may lie in subtelomeric repeats, mediating epigenetic silencing of adjacent regions in a length-

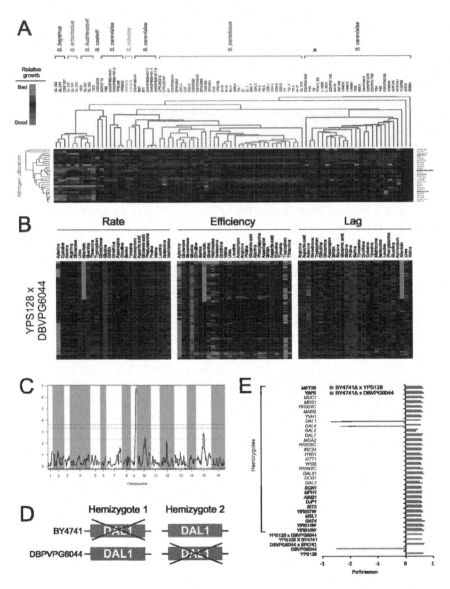

Figure 4. contd....

dependent manner (Pryde and Louis 1999). Subtelomeric lengths also vary substantially, primarily due to polymorphisms in the protein coding genes *yKU80* and *TLC1* controlling subtelomere structure (Gatbonton et al. 2006, Liti et al. 2009b). Part of the reason may also lie in enhanced recombination in regions adjacent to subtelomeres (Barton et al. 2008) leading to elevated rates of gene gain and loss (Brown et al. 2010). Given the QTL richness of subtelomeric regions, the difficulty of correctly assembling genome sequences close to chromosomal ends is unfortunate.

Yeast Evolutionary Phenomics

With the exclusion of a handful of metabolic traits inferred from gene family expansion and contraction (Dujon 2006), the phenotypic history of different yeast species and the degree to which genotype-phenotype links can be extrapolated between them remains poorly understood. The phenotypic diversity in *S. paradoxus* has been mapped to considerable detail, revealing much lower trait variation than in *S. cerevisiae* (Warringer et al. 2011).

Figure 4. Mapping the genetic basis for variation in yeast nitrogen source utilization using phenomics. (A) Variation in the ability to utilize low complexity nitrogen sources in *S. cerevisiae* and *S. paradoxus*, with other *sensu stricto* yeasts included for comparison. All strains were microcultivated individually in the presence of a single nitrogen source and mean population doubling times were extracted from high-density growth curves. Colors indicate the growth rate as compared to the S288C universal reference strain (black). Name color indicates species (Liti et al. 2009a). (B) Two natural strains with different phylogenetic origins, the North American YPS128 and the West African DBVPG6044, were repeatedly crossed, the resulting diploid hybrids were sporulated, and 96 recombined F1 haploids were obtained. The variation in ability to utilize different nitrogen sources among F1 haploids were quantified using microcultivation as in A. Mean population lag phases, doubling times (rate) and total changes in population density (efficiency), as compared to S288C (black), are shown. (C) The 96 F1 haploids from the crosses between YPS128 and DBVPG6044 were genotyped at 130 marker loci distributed over the yeast genome and the co-inheritance of each marker and the ability to utilize the nitrogen source allantoin (mean population doubling time with allantoin as sole nitrogen source) was tested using linkage analysis. The LOD score plot indicates a single strong significance peak corresponding to the most distal markers on the right arm of chromosome nine. (D) Principles for reciprocal hemizygosity. To test causality of individual genes in a chromosomal region associated with variation in a phenotype, parent strains with diverging traits are mated to create two hybrids in which the two alleles of each gene are individually deleted. A difference in phenotype between the two reciprocal hemizygotes for a gene is strong support for causality. (E) Hemizygotes of all individual genes in the distal end of the right arm of chromosome nine were constructed by mating YPS128 and DBVPG6044 respectively to the BY4741 deletion collection. Resulting hemizygotes were probed for the ability to utilize the nitrogen source allantoin (mean population doubling time). Hemizygotes containing only the DBVPG6044 allele of either the allantoinase *DAL1* or the allantoin importer *DAL4* were unable to utilize allantoin as sole nitrogen source, indicating that defects in both these genes underlie the allantoin defect of DBVPG6044.

Color image of this figure appears in the color plate section at the end of the book.

Extreme population structure (Liti et al. 2009a) and close to reproductive isolation of gene pools (Warringer et al. 2011) makes crosses across *S. paradoxus* population boundaries unsuitable for elucidating the genetic basis of phenotypic variation. However, reverse genetics of *S. paradoxus* is possible (Cubillos et al. 2009) and a *S. paradoxus* deletion collection is under way. Some efforts have also been made at understanding overall phenotypic variation within the *Saccharomyces sensu stricto* clade at large. *S. mikatae* resembles *S. paradoxus* and *S. cerevisiae* phenotypically, whereas *S. arboricolus*, *S. kudriavzevii* and *S. bayanus* form a distinct phenetic group where species are hard to distinguish phenotypically (Warringer et al. 2011). Mostly sterile hybrids between these species can form (Greig 2009) and the phenotypes of these hybrids are receiving increasing industrial attention. The lager beer yeast, *S. pastorianus*, is a tetraploid hybrid between *S. cerevisiae* and *S. bayanus*, with its combination of cold tolerance and ethanol production properties being of critical importance to lager brewing (Libkind et al. 2011). At the other end of the yeast phylogenetic spectrum, a study of trait variation within fission yeast, *Schizosaccharomyces pombe*, uncovered much less trait variation than within *S. cerevisiae* (Brown et al. 2011). However, the population structure of *S. pombe* is largely unknown, and deeper interpretation of the limited trait variation is currently not possible. Although reverse genetics is less facile in *S. pombe* than in *S. cerevisiae*, a nearly genome-wide deletion collection was recently established in the universal reference strain 972h-, revealing similar patterns of gene dispensability of non-duplicated *S. cerevisiae* and *S. pombe* orthologs (Kim et al. 2010). Outside the *S. pombe* and *S. cerevisiae* clades, few yeast species have been lab-domesticated and information regarding their genotype-phenotype maps is limited. A partial (11%) homozygotic deletion collection in diploid *Candida albicans* was recently reported (Noble et al. 2010) and transposon-mediated gene disruptions promise a potential way towards genome-wide phenomics in this pathogenic yeast (Smith et al. 2011). Despite drawbacks in the form of transposon mutagenesis coldspots, unequal impairment of gene function, and inability to disrupt both gene copies in diploids, such a collection can provide valuable information on drug tolerance and pathogenicity traits (Oh et al. 2010).

Future Developments and Trends in Yeast Phenomics

The Power of Experimental Evolution

The emergence of next generation sequencing techniques in combination with standardized analysis packages, allowing sequence assembly and annotation by scientists with only basic training in bioinformatics, is now opening an existing opportunity-window for linking phenotypic and genetic variation in yeast. This promises to shift the focus from reverse genetics to forward genetics and to techniques such as experimental evolution, for which the yeast with its ease of handling, small haploid genome and rapid mitotic life cycle is well suited. Artificial selection for enhancement of specific phenotypes in controlled laboratory conditions, coupled to sequencing of evolved populations, has the potential to reveal molecular relationships that go beyond what has been established by the use of gene deletions, which mimic loss-of-function mutations but say nothing about gain or extension of gene function. This may prove especially revealing when it comes to the intricacies of regulatory circuits, which still are poorly understood. It also has the potential to increase the resolution of our understanding of molecular relationships by pinpointing specific nucleotides/amino acids, thereby facilitating easier interpretations of structure–function relationships and of pleiotropy in multi-domain proteins. In contrast to reverse genetics, forward genetics techniques are not limited to distinct sets of species and strains and promise to expand our knowledge to encompass a much wider section of the yeast genotype and phenotype space. The genetic mapping of both natural variation, and of artificially generated variation, e.g., through experimental evolution, will not only increase our understanding of the species as a whole, beyond the universal reference strain, but also uncover genotype-phenotype relationships that may be exploited for biotechnological purposes. Because of public concerns regarding the use of genetically modified organisms, reverse genetics has been, and will for the foreseeable future be, of limited value for the beer, wine, and baking industries and may prove similarly prohibitive when it comes to the use of yeast in bioethanol production. Forward genetics also promises to expand our understanding of genotype-phenotype maps to unconventional yeasts that are not as amenable to reverse genetics techniques as the two model yeasts, *S. cerevisiae* and *S. pombe*. Biotechnologically interesting properties, such as ability to degrade compounds constituting environmental and human health hazards, and thus to bioremediate contaminated industrial sites, have been uncovered in many such yeasts.

Enhanced Technical Precision in Phenomics

We foresee that the rapid pace of technology development in genomics is likely to stimulate a shift in focus towards phenomics, and call for increases in both measurement precision and scale. The latter will not only expand the throughput but also, and perhaps more importantly, allow for high replication, thereby reducing not only the influence of noise but also that of bias, provided a proper experiment design with randomization over space and time. Spatial and temporal bias, which is rarely sufficiently controlled for in yeast phenomics due to the usually low level of replication and lack of randomization, currently leads to an unacceptably high rate of both false positives and negatives. Reducing especially the sizeable fraction of false positives should be a high priority in yeast phenomics. A similar high priority should be given to expanding our understanding of yeast fitness. The current approximation of yeast fitness to the rate of mitotic reproduction, or to a composite of unknown and sometimes contrasting reproductive parameters, clearly does not reflect the full extent of selection having acted, and acting, on the yeast genome. High-throughput techniques for precise quantification of phenomena such as mating, sporulation, germination, and viability, of both spores and vegetative cells, promise to vastly increase our understanding of gene-function relationships in yeast.

Phenomics in High-throughput Surrogate Genetics

Large-scale high-resolution phenotypic analysis in yeast also has a number of biomedical implications. Among other things, it opens up heterologous studies on genes and alleles from other organisms. In particular, this seems relevant for studies linked to human diseases where whole genome association studies and similar sequence-based analyses unravel large repertoires of genetic alterations that link to certain phenotypes. In almost all of these cases, the clinical studies only provide a first set of interesting hypotheses that await further functional characterization. There is a need for a methodology that could systematically test the causality for the indicated alleles in relation to the disease/phenotype.

Yeast appears to provide such a potent system and it has over the years been used quite extensively in various aspects of medical- and medicinal-relevant research. Expression of human cDNA in yeast has in the past been employed primarily for complementation of orthologous mutants (Mager and Winderickx 2005). In this case, the phenotype of a yeast mutant is rescued by expression of the human protein. This approach has a huge potential since roughly 30% of known genes involved in human disease have functional homologs in yeast. However, more-sophisticated assays have also been developed that aim at identifying genes that interfere with the functioning of the expressed human protein. These methodologies are usually based on sensitive growth assays, where growth defects from expression of a human protein in yeast can be readily scored. In addition, drug-induced inhibition of the human protein can also be analysed. In these drug-development studies, the problem with permeability or efflux/transport of drugs can be circumvented by utilizing yeast strains with the major efflux pumps (e.g., Pdr5 and Snq2) deleted or with enhanced membrane permeability (e.g., *ERG6* deletion).

In fact, it has turned out that yeast offers an attractive system to investigate even disease-related proteins that have no apparent homology to yeast genes. Several interesting cases are reported in the literature where this humanized yeast approach has been fruitful and provided valuable mechanistic information: (i) human steroid receptors expressed in yeast are responsive to their cognate ligands and can be functionally analysed (Lind et al. 1999), (ii) the human somatostatin receptor has been reconstituted. In this case, a functional system in yeast was established by co-expression of a heterologous human/yeast hybrid G_a-subunit of the trimeric G protein (Pausch 1997), (iii) the analysis of human ligand-binding domains via the use of the endogenous yeast dihydrofolate reductase (Dhfrp) where the hybrid-protein exhibits a temperature sensitivity that can be rescued by binding of a ligand (Tucker and Fields 2001), (iv) yeast has been used to identify genes that enhance the toxicity of a mutant huntingtin protein with an expanded polyQ domain (Zhang et al. 2005), (v) human secretase, that is linked to Alzheimer's disease, has been reconstituted in yeast and used in high-throughput selection of secretase inhibitors (Yonemura et al. 2011) and (vi) genetic as well as environmental parameters that trigger aggregation of the human protein tau have been elucidated (De Vos et al. 2011). It is to be expected that the repertoire of humanized yeast approaches for the analysis of human proteins will be drastically extended in the coming years.

In these times of genome-wide clinical association studies for many, if not most, diseases, there is still a lack of a clear molecular determinant for roughly 4,000 inherited traits of medical interest (Online Mendelian Inheritance in Man [OMIM] http://omim.org (Amberger et al. 2011)). Recently, a report was published with great future implications for how to enhance our understanding of human diseases and demonstrates the use of yeast as a test-bed for elucidating involved molecular mechanisms. Mayfield et al. (2012) used what they call "surrogate genetics" to assay the function of allelic variants linked to one particular human metabolic disease in yeast. This defect is based on a deficiency in the enzyme cystathionine-b-synthase (CBS), causing homocystinuria, thrombosis, mental retardation, and some other severe complications. However, a mechanistic understanding of the defect for these various alleles had previously not been obtained. The authors collected and analysed 84 missense mutations of the CBS protein known to be linked to this disease, including potential alterations in the heme-binding, vitamin B6, catalytic and AdoMet-binding regulatory domains. Using yeast growth, they quantified the relative function of CBS alleles and classified them according to growth rate and their ability to be rescued by the addition of vitamin B6 or heme (Fig. 5). Among other important findings, the authors concluded that heme deficiencies could complicate the diagnosis and treatment of homocystinuria. This report

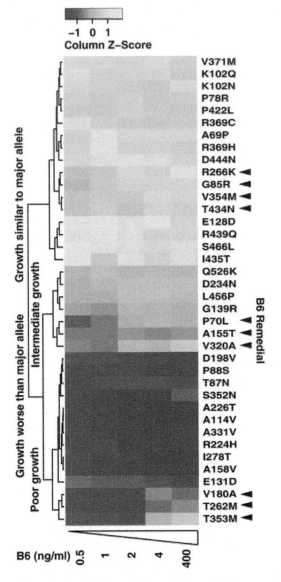

Figure 5. Humanized yeasts expressing different alleles of the enzyme cystathionine-b-synthase show different growth responses. Growth responses in relation to the concentration of vitamine B6, one of the co-factors of the enzyme (Mayfield et al. 2012). Heat maps of growth rates normalized to the growth of the major allele. The column Z-score indicates the mean growth rate (Z-score of 0) and standard deviation (Z-score of 61) of all alleles per column. Arrowheads indicate alleles that respond to cofactor titration more strongly than other alleles in their cluster. Published with permission from The Genetics Society of America.

Color image of this figure appears in the color plate section at the end of the book.

neatly exemplifies how quantitative phenotypic assays for human alleles in a surrogate organism like yeast should be broadly applicable to many genes, from humans or from other organisms. This will certainly open up new avenues for the use of high-throughput, high-resolution yeast phenotyping.

Acknowledgement

The financial support to AB from the Swedish National Research Council (VR) is acknowledged.

References Cited

Addinall, S.G., E.M. Holstein, C. Lawless, M. Yu, K. Chapman, A.P. Banks, H.P. Ngo, L. Maringele, M. Taschuk, A. Young, A. Ciesiolka, A.L. Lister, A. Wipat, D.J. Wilkinson and D. Lydall. 2011. Quantitative fitness analysis shows that NMD proteins and many other protein complexes suppress or enhance distinct telomere cap defects. PLoS Genet. 7: e1001362.

Amberger, J., C. Bocchini and A. Hamosh. 2011. A new face and new challenges for Online Mendelian Inheritance in Man (OMIM®). Human Mutation 32: 564–567.

Ambroset, C., M. Petit, C. Brion, I. Sanchez, P. Delobel, C. Guerin, H. Chiapello, P. Nicolas, F. Bigey, S. Dequin and B. Blondin. 2011. Deciphering the molecular basis of wine yeast fermentation traits using a combined genetic and genomic approach. G3 (Bethesda) 1: 263–281.

Babazadeh, R., S. Moghadas Jafari, M. Zackrisson, A. Blomberg, S. Hohmann, J. Warringer and M. Krantz. 2011. The Ashbya gossypiiEF-1alpha promoter of the ubiquitously used MX cassettes is toxic to Saccharomyces cerevisiae. FEBS Lett. 585: 3907–3913.

Banks, A., C. Lawless and D. Lydall. 2012a. A quantitative fitness analysis workflow. Journal of Visualized Experiments 66: e4018.

Barton, A.B., M.R. Pekosz, R.S. Kurvathi and D.B. Kaback. 2008. Meiotic recombination at the ends of chromosomes in Saccharomyces cerevisiae. Genetics 179: 1221–1235.

Baryshnikova, A., M. Costanzo, Y. Kim, H. Ding, J. Koh, K. Toufighi, J.-Y. Youn, J. Ou, B.-J. San Luis, S. Bandyopadhyay, M. Hibbs, D. Hess, A.-C. Gingras, G.D. Bader, O.G. Troyanskaya, G.W. Brown, B. Andrews, C. Boone and C.L. Myers. 2010. A genome-scale approach for quantitative analysis of fitness and genetic interactions in yeast. Nature Methods 7: 1017–1024.

Becher, P.G., M. Bengtsson, B.S. Hansson and P. Witzgall. 2010. Flying the fly: long-range flight behavior of Drosophila melanogaster to attractive odors. Journal of Chemical Ecology 36: 599–607.

Ben-Aroya, S., C. Coombes, T. Kwok, K.A. O'Donnell, J.D. Boeke and P. Hieter. 2008. Toward a comprehensive temperature-sensitive mutant repository of the essential genes of Saccharomyces cerevisiae. Mol. Cell 30: 248–258.

Ben-Aroya, S., X. Pan, J.D. Boeke and P. Hieter. 2010. Making temperature-sensitive mutants. Methods Enzymol. 470: 181–204.

Ben-Shitrit, T., N. Yosef, K. Shemesh, R. Sharan, E. Ruppin and M. Kupiec. 2012. Systematic identification of gene annotation errors in the widely used yeast mutation collections. Nat. Methods 9: 373–378.

Birrell, G.W., J.A. Brown, H.I. Wu, G. Giaever, A.M. Chu, R.W. Davis and J.M. Brown. 2002. Transcriptional response of Saccharomyces cerevisiae to DNA-damaging agents does not identify the genes that protect against these agents. Proc. Natl. Acad. Sci. USA 99: 8778–8783.

Blomberg, A. 2011. Measuring growth rate in high-throughput growth phenotyping. Curr. Opin. Biotechnol. 22: 94–102.

Brachmann, C.B., A. Davies, G.J. Cost, E. Caputo, J. Li, P. Hieter and J.D. Boeke. 1998. Designer deletion strains derived from Saccharomyces cerevisiae S288C: a useful set of strains and plasmids for PCR-mediated gene disruption and other applications. Yeast 14: 115–132.

Breslow, D.K., D.M. Cameron, S.R. Collins, M. Schuldiner, J. Stewart-Ornstein, H.W. Newman, S. Braun, H.D. Madhani, N.J. Krogan and J.S. Weissman. 2008. A comprehensive strategy enabling high-resolution functional analysis of the yeast genome. Nat. Methods 5: 711–718.

Briza, P., M. Breitenbach, A. Ellinger and J. Segall. 1990. Isolation of two developmentally regulated genes involved in spore wall maturation in *Saccharomyces cerevisiae*. Genes Dev. 4: 1775–1789.

Brown, C.A., A.W. Murray and K.J. Verstrepen. 2010. Rapid expansion and functional divergence of subtelomeric gene families in yeasts. Curr. Biol. 20: 895–903.

Brown, W.R.A., G. Liti, C. Rosa, S. James, I. Roberts, V. Robert, N. Jolly, W. Tang, P. Baumann, C. Green, K. Schlegel, J. Young, F. Hirchaud, S. Leek, G. Thomas, A. Blomberg and J. Warringer. 2011. A geographically diverse collection of schizosaccharomyces pombe isolates shows limited phenotypic variation but extensive karyotypic diversity. G3: Genes, Genomes, Genetics 1: 615–626.

Buchs, J. 2001. Introduction to advantages and problems of shaken cultures. Biochem Eng J 7: 91–98.

Cap, M., L. Stepanek, K. Harant, L. Vachova and Z. Palkova. 2012. Cell differentiation within a yeast colony: metabolic and regulatory parallels with a tumor-affected organism. Mol. Cell 46: 436–448.

Causton, H.C., B. Ren, S.S. Koh, C.T. Harbison, E. Kanin, E.G. Jennings, T.I. Lee, H.L. True, E.S. Lander and R.A. Young. 2001. Remodeling of yeast genome expression in response to environmental changes. Mol. Biol. Cell 12: 323–337.

Collins, S.R., M. Schuldiner, N.J. Krogan and J.S. Weissman. 2006. A strategy for extracting and analyzing large-scale quantitative epistatic interaction data. Genome Biol. 7: R63.

Coluccio, A. and A.M. Neiman. 2004. Interspore bridges: a new feature of the Saccharomyces cerevisiae spore wall. Microbiology 150: 3189–3196.

Coluccio, A.E., R.K. Rodriguez, M.J. Kernan and A.M. Neiman. 2008. The yeast spore wall enables spores to survive passage through the digestive tract of Drosophila. PLoS ONE 3: e2873.

Conant, G.C. and K.H. Wolfe. 2007. Increased glycolytic flux as an outcome of whole-genome duplication in yeast. Mol. Syst. Biol. 3: 129.

Cubillos, F.A., E. Billi, E. Zörgö, L. Parts, P. Fargier, S. Omholt, A. Blomberg, J. Warringer, E.J. Louis and G. Liti. 2011. Assessing the complex architecture of polygenic traits in diverged yeast populations. Mol. Ecol. 20: 1401–1413.

Cubillos, F.A., E.J. Louis and G. Liti. 2009. Generation of a large set of genetically tractable haploid and diploid Saccharomyces strains. FEMS Yeast Res 9: 1217–1225.

Dalziel, A.C., S.M. Rogers and P.M. Schulte. 2009. Linking genotypes to phenotypes and fitness: how mechanistic biology can inform molecular ecology. Mol. Ecol. 18: 4997–5017.

Dawes, I.W. and I.D. Hardie. 1974. Selective killing of vegetative cells in sporulated yeast cultures by exposure to diethyl ether. Mol. Gen. Genet 131: 281–289.

De Vos, A., J. Anandhakumar, J. Van den Brande, M. Verduyckt, V. Franssens, J. Winderickx and E. Swinnen. 2011. Yeast as a model system to study tau biology. International journal of Alzheimer's disease 2011: 428970.

Delneri, D., D.C. Hoyle, K. Gkargkas, E.J. Cross, B. Rash, L. Zeef, H.S. Leong, H.M. Davey, A. Hayes, D.B. Kell, G.W. Griffith and S.G. Oliver. 2008. Identification and characterization of high-flux-control genes of yeast through competition analyses in continuous cultures. Nat. Genet 40: 113–117.

DeLuna, A., K. Vetsigian, N. Shoresh, M. Hegreness, M. Colon-Gonzalez, S. Chao and R. Kishony. 2008. Exposing the fitness contribution of duplicated genes. Nat. Genet. 40: 676–681.

Demogines, A., E. Smith, L. Kruglyak and E. Alani. 2008. Identification and dissection of a complex DNA repair sensitivity phenotype in Baker's yeast. PLoS Genet 4: e1000123.

Deutschbauer, A.M. and R.W. Davis. 2005. Quantitative trait loci mapped to single-nucleotide resolution in yeast. Nat. Genet 37: 1333–1340.

Deutschbauer, A.M., D.F. Jaramillo, M. Proctor, J. Kumm, M.E. Hillenmeyer, R.W. Davis, C. Nislow and G. Giaever. 2005. Mechanisms of haploinsufficiency revealed by genome-wide profiling in yeast. Genetics 169: 1915–1925.

Di Talia, S., J.M. Skotheim, J.M. Bean, E.D. Siggia and F.R. Cross. 2007. The effects of molecular noise and size control on variability in the budding yeast cell cycle. Nature 440. 947 951.

Diao, L. and K.C. Chen. 2012. Local ancestry corrects for population structure in Saccharomyces cerevisiae genome-wide association studies. Genetics 192: 1503.

Dittmar, J.C., R.J. Reid and R. Rothstein. 2010. ScreenMill: a freely available software suite for growth measurement, analysis and visualization of high-throughput screen data. BMC Bioinformatics 11: 353.

Doniger, S.W., H.S. Kim, D. Swain, D. Corcuera, M. Williams, S.P. Yang and J.C. Fay. 2008. A catalog of neutral and deleterious polymorphism in yeast. PLoS Genet 4: e1000183.

Dudley, A.M., D.M. Janse, A. Tanay, R. Shamir and G.M. Church. 2005. A global view of pleiotropy and phenotypically derived gene function in yeast. Mol. Syst. Biol. 1: 2005 0001.

Dujon, B. 2006. Yeasts illustrate the molecular mechanisms of eukaryotic genome evolution. Trends Genet 22: 375–387.

Ehrenreich, I.M., J.P. Gerke and L. Kruglyak. 2009. Genetic dissection of complex traits in yeast: insights from studies of gene expression and other phenotypes in the BYxRM cross. Cold Spring Harb. Symp. Quant. Biol. 74: 145–153.

Ehrenreich, I.M., N. Torabi, Y. Jia, J. Kent, S. Martis, J.A. Shapiro, D. Gresham, A.A. Caudy and L. Kruglyak. 2010. Dissection of genetically complex traits with extremely large pools of yeast segregants. Nature 464: 1039–1042.

Ericson, E., I. Pylvanainen, L. Fernandez-Ricaud, O. Nerman, J. Warringer and A. Blomberg. 2006. Genetic pleiotropy in Saccharomyces cerevisiae quantified by high-resolution phenotypic profiling. Molecular Genetics and Genomics 275: 605–614.

Fay, J.C. and J.A. Benavides. 2005. Evidence for domesticated and wild populations of Saccharomyces cerevisiae. PLoS Genet 1: 66–71.

Fleet, G.H. 2003. Yeast interactions and wine flavour. International Journal of Food Microbiology 86: 11–22.

Fogleman, J.C., W.T. Starmer and W.B. Heed. 1981. Larval selectivity for yeast species by Drosophila mojavensis in natural substrates. Proc. Natl. Acad. Sci. USA 78: 4435–4439.

Gaisne, M., A.M. Becam, J. Verdiere and C.J. Herbert. 1999. A 'natural' mutation in Saccharomyces cerevisiae strains derived from S288c affects the complex regulatory gene HAP1 (CYP1). Curr. Genet. 36: 195–200.

Gasch, A.P., P.T. Spellman, C.M. Kao, O. Carmel-Harel, M.B. Eisen, G. Storz, D. Botstein and P.O. Brown. 2000. Genomic expression programs in the response of yeast cells to environmental changes. Mol. Biol. Cell 11: 4241–4257.

Gatbonton, T., M. Imbesi, M. Nelson, J.M. Akey, D.M. Ruderfer, L. Kruglyak, J.A. Simon and A. Bedalov. 2006. Telomere length as a quantitative trait: genome-wide survey and genetic mapping of telomere length-control genes in yeast. PLoS Genet 2: e35.

Gerke, J., K. Lorenz and B. Cohen. 2009. Genetic interactions between transcription factors cause natural variation in yeast. Science 323: 498–501.

Gerke, J.P., C.T. Chen and B.A. Cohen. 2006. Natural isolates of Saccharomyces cerevisiae display complex genetic variation in sporulation efficiency. Genetics 174: 985–997.

Gerstein, A.C. and S.P. Otto. 2009. Ploidy and the causes of genomic evolution. J. Hered 100: 571–581.

Ghaemmaghami, S., W.K. Huh, K. Bower, R.W. Howson, A. Belle, N. Dephoure, E.K. O'Shea and J.S. Weissman. 2003. Global analysis of protein expression in yeast. Nature 425: 737–741.

Giaever, G., A.M. Chu, L. Ni, C. Connelly, L. Riles, S. Veronneau, S. Dow, A. Lucau-Danila, K. Anderson, B. Andre, A.P. Arkin, A. Astromoff, M. El-Bakkoury, R. Bangham, R. Benito, S. Brachat, S. Campanaro, M. Curtiss, K. Davis, A. Deutschbauer, K.D. Entian, P. Flaherty, F. Foury, D.J. Garfinkel, M. Gerstein, D. Gotte, U. Guldener, J.H. Hegemann, S. Hempel, Z. Herman, D.F. Jaramillo, D.E. Kelly, S.L. Kelly, P. Kotter, D. LaBonte, D.C. Lamb, N. Lan, H. Liang, H. Liao, L. Liu, C. Luo, M. Lussier, R. Mao, P. Menard, S.L. Ooi, J.L. Revuelta, C.J. Roberts, M. Rose, P. Ross-Macdonald, B. Scherens, G. Schimmack, B. Shafer, D.D. Shoemaker, S. Sookhai-Mahadeo, R.K. Storms, J.N. Strathern, G. Valle, M. Voet, G. Volckaert, C.Y. Wang, T.R. Ward, J. Wilhelmy, E.A. Winzeler, Y. Yang, G. Yen, E. Youngman, K. Yu, H. Bussey, J.D. Boeke, M. Snyder, P. Philippsen, R.W. Davis and M. Johnston. 2002. Functional profiling of the Saccharomyces cerevisiae genome. Nature 418: 387–391.

Goffeau, A. 2000. Four years of post-genomic life with 6,000 yeast genes. FEBS Lett. 480: 37–41.

Goffeau, A., B.G. Barrell, H. Bussey, R.W. Davis, B. Dujon, H. Feldmann, F. Galibert, J.D. Hoheisel, C. Jacq, M. Johnston, E.J. Louis, H.W. Mewes, Y. Murakami, P. Philippsen, H. Tettelin and S.G. Oliver. 1996. Life with 6000 genes. Science 274: 546, 563–547.

Goldstein, A.L. and J.H. McCusker. 1999. Three new dominant drug resistance cassettes for gene disruption in Saccharomyces cerevisiae. Yeast 15: 1541–1553.

Greig, D. 2009. Reproductive isolation in Saccharomyces. Heredity (Edinb) 102: 39–44.

Gygi, S.P., Y. Rochon, B.R. Franza and R. Aebersold. 1999. Correlation between protein and mRNA abundance in yeast. Mol. Cell Biol. 19: 1720–1730.

Hartman, J.L.T. and N.P. Tippery. 2004. Systematic quantification of gene interactions by phenotypic array analysis. Genome Biol. 5: R49.

Hermann, R., M. Lehmann and J. Buchs. 2003. Characterization of gas-liquid mass transfer phenomena in microtiter plates. Biotechnol Bioeng 81: 178–186.

Hillenmeyer, M.E., E. Fung, J. Wildenhain, S.E. Pierce, S. Hoon, W. Lee, M. Proctor, R.P. St.Onge, M. Tyers, D. Koller, R.B. Altman, R.W. Davis, C. Nislow and G. Giaever. 2008. The chemical genomic portrait of yeast: uncovering a phenotype for all genes. Science 320: 362–365.

Hittinger, C.T., A. Rokas and S.B. Carroll. 2004. Parallel inactivation of multiple GAL pathway genes and ecological diversification in yeasts. Proc. Natl. Acad. Sci. USA 101: 14144–14149.

Ho, C.H., L. Magtanong, S.L. Barker, D. Gresham, S. Nishimura, P. Natarajan, J.L. Koh, J. Porter, C.A. Gray, R.J. Andersen, G. Giaever, C. Nislow, B. Andrews, D. Botstein, T.R. Graham, M. Yoshida and C. Boone. 2009. A molecular barcoded yeast ORF library enables mode-of-action analysis of bioactive compounds. Nat. Biotechnol. 27: 369–377.

Hughes, T.R., C.J. Roberts, H. Dai, A.R. Jones, M.R. Meyer, D. Slade, J. Burchard, S. Dow, T.R. Ward, M.J. Kidd, S.H. Friend and M.J. Marton. 2000. Widespread aneuploidy revealed by DNA microarray expression profiling. Nat. Genet. 25: 333–337.

Jelier, R., J.I. Semple, R. Garcia-Verdugo and B. Lehner. 2011. Predicting phenotypic variation in yeast from individual genome sequences. Nat. Genet. 43: 1270–1274.

Jin, Y.H., P.E. Dunlap, S.J. McBride, H. Al-Refai, P.R. Bushel and J.H. Freedman. 2008. Global transcriptome and deletome profiles of yeast exposed to transition metals. PLoS Genet. 4: e1000053.

Kaizu, K., H. Moriya and H. Kitano. 2010. Fragilities caused by dosage imbalance in regulation of the budding yeast cell cycle. PLoS Genet. 6: e1000919.

Kanemaki, M., A. Sanchez-Diaz, A. Gambus and K. Labib. 2003. Functional proteomic identification of DNA replication proteins by induced proteolysis *in vivo*. Nature 423: 720–724.

Kim, D.U., J. Hayles, D. Kim, V. Wood, H.O. Park, M. Won, H.S. Yoo, T. Duhig, M. Nam, G. Palmer, S. Han, L. Jeffery, S.T. Baek, H. Lee, Y.S. Shim, M. Lee, L. Kim, K.S. Heo, E.J. Noh, A.R. Lee, Y.J. Jang, K.S. Chung, S.J. Choi, J.Y. Park, Y. Park, H.M. Kim, S.K. Park, H.J. Park, E.J. Kang, H.B. Kim, H.S. Kang, H.M. Park, K. Kim, K. Song, K.B. Song, P. Nurse and K.L. Hoe. 2010. Analysis of a genome-wide set of gene deletions in the fission yeast Schizosaccharomyces pombe. Nat. Biotechnol. 28: 617–623.

Kim, H.S. and J.C. Fay. 2007. Genetic variation in the cysteine biosynthesis pathway causes sensitivity to pharmacological compounds. Proc. Natl. Acad. Sci. USA 104: 19387–19391.

Knop, M. 2006. Evolution of the hemiascomycete yeasts: on life styles and the importance of inbreeding. Bioessays 28: 696–708.

Krantz, M., D. Ahmadpour, L.G. Ottosson, J. Warringer, C. Waltermann, B. Nordlander, E. Klipp, A. Blomberg, S. Hohmann and H. Kitano. 2009. Robustness and fragility in the yeast high osmolarity glycerol (HOG) signal-transduction pathway. Mol. Syst. Biol. 5: 281.

Lamprecht, M.R., D.M. Sabatini and A.E. Carpenter. 2007. CellProfiler: free, versatile software for automated biological image analysis. Biotechniques 42: 71–75.

Langkjaer, R.B., P.F. Cliften, M. Johnston and J. Piskur. 2003. Yeast genome duplication was followed by asynchronous differentiation of duplicated genes. Nature 421: 848–852.

Lawless, C., D.J. Wilkinson, A. Young, S.G. Addinall and D.A. Lydall. 2010. Colonyzer: automated quantification of micro-organism growth characteristics on solid agar. BMC Bioinformatics 11: 287.

Levy, S.F., N. Ziv and M.L. Siegal. 2012. Bet hedging in yeast by heterogeneous, age-correlated expression of a stress protectant. PLoS Biol. 10: e1001325.

Li, Z., F.J. Vizeacoumar, S. Bahr, J. Li, J. Warringer, F.S. Vizeacoumar, R. Min, B. Vandersluis, J. Bellay, M. Devit, J.A. Fleming, A. Stephens, J. Haase, Z.Y. Lin, A. Baryshnikova, H. Lu, Z. Yan, K. Jin, S. Barker, A. Datti, G. Giaever, C. Nislow, C. Bulawa, C.L. Myers, M. Costanzo, A.C. Gingras, Z. Zhang, A. Blomberg, K. Bloom, B. Andrews and C. Boone. 2011. Systematic exploration of essential yeast gene function with temperature-sensitive mutants. Nat. Biotechnol. 29: 361.

Libkind, D., C.T. Hittinger, E. Valerio, C. Goncalves, J. Dover, M. Johnston, P. Goncalves and J.P. Sampaio. 2011. Microbe domestication and the identification of the wild genetic stock of lager-brewing yeast. Proc. Natl. Acad. Sci. USA 108: 14539–14544.

Lind, U., P. Greenidge, J.A. Gustafsson, A.P. Wright and J. Carlstedt-Duke. 1999. Valine 571 functions as a regional organizer in programming the glucocorticoid receptor for differential binding of glucocorticoids and mineralocorticoids. J. Biol. Chem. 274: 18515–18523.

Liti, G., D.M. Carter, A.M. Moses, J. Warringer, L. Parts, S.A. James, R.P. Davey, I.N. Roberts, A. Burt, V. Koufopanou, I.J. Tsai, C.M. Bergman, D. Bensasson, M.J. O'Kelly, A. van Oudenaarden, D.B. Barton, E. Bailes, A.N. Nguyen, M. Jones, M.A. Quail, I. Goodhead, S. Sims, F. Smith, A. Blomberg, R. Durbin and E.J. Louis. 2009a. Population genomics of domestic and wild yeasts. Nature 458: 337–341.

Liti, G., S. Haricharan, F.A. Cubillos, A.L. Tierney, S. Sharp, A.A. Bertuch, L. Parts, E. Bailes and E.J. Louis. 2009b. Segregating YKU80 and TLC1 alleles underlying natural variation in telomere properties in wild yeast. PLoS Genet 5: e1000659.

Lopez, A., A.B. Parsons, C. Nislow, G. Giaever and C. Boone. 2008. Chemical-genetic approaches for exploring the mode of action of natural products. Prog. Drug Res. 66: 237: 239–271.

Macdonald, S.J. and A.D. Long. 2007. Joint estimates of quantitative trait locus effect and frequency using synthetic recombinant populations of Drosophila melanogaster. Genetics 176: 1261–1281.

Mager, W.H. and J. Winderickx. 2005. Yeast as a model for medical and medicinal research. Trends in Pharmacological Sciences 26: 265–273.

Magtanong, L., C.H. Ho, S.L. Barker, W. Jiao, A. Baryshnikova, S. Bahr, A.M. Smith, L.E. Heisler, J.S. Choy, E. Kuzmin, K. Andrusiak, A. Kobylianski, Z. Li, M. Costanzo, M.A. Basrai, G. Giaever, C. Nislow, B. Andrews and C. Boone. 2011. Dosage suppression genetic interaction networks enhance functional wiring diagrams of the cell. Nat. Biotechnol. 29: 505–511.

Makanae, K., R. Kintaka, T. Makino, H. Kitano and H. Moriya. 2013. Identification of dosage-sensitive genes in Saccharomyces cerevisiae using the genetic tug-of-war method. Genome Res. 23: 300.

Marullo, P., M. Aigle, M. Bely, I. Masneuf-Pomarede, P. Durrens, D. Dubourdieu and G. Yvert. 2007. Single QTL mapping and nucleotide-level resolution of a physiologic trait in wine Saccharomyces cerevisiae strains. FEMS Yeast Res. 7: 941–952.

Mayfield, J.A., M.W. Davies, D. Dimster-Denk, N. Pleskac, S. McCarthy, E.A. Boydston, L. Fink, X.X. Lin, A.S. Narain, M. Meighan and J. Rine. 2012. Surrogate genetics and metabolic profiling for characterization of human disease alleles. Genetics 190: 1309–1323.

Mehmood, T., H. Martens, S. Saebo, J. Warringer and L. Snipen. 2011. Mining for genotype-phenotype relations in Saccharomyces using partial least squares. BMC Bioinformatics 12: 318.

Mnaimneh, S., A.P. Davierwala, J. Haynes, J. Moffat, W.T. Peng, W. Zhang, X. Yang, J. Pootoolal, G. Chua, A. Lopez, M. Trochesset, D. Morse, N.J. Krogan, S.L. Hiley, Z. Li, Q. Morris, J. Grigull, N. Mitsakakis, C.J. Roberts, J.F. Greenblatt, C. Boone, C.A. Kaiser, B.J. Andrews and T.R. Hughes. 2004. Exploration of essential gene functions via titratable promoter alleles. Cell 118: 31–44.

Moriya, H., Y. Shimizu-Yoshida and H. Kitano. 2006. In vivo robustness analysis of cell division cycle genes in Saccharomyces cerevisiae. PLoS Genet 2: e111.

Mortimer, R.K. and J.R. Johnston. 1986. Genealogy of principal strains of the yeast genetic stock center. Genetics 113: 35–43.

Muller, L.A., J.E. Lucas, D.R. Georgianna and J.H. McCusker. 2011. Genome-wide association analysis of clinical vs. nonclinical origin provides insights into Saccharomyces cerevisiae pathogenesis. Mol. Ecol. 20: 4085–4097.

Neiman, A.M. 2011. Sporulation in the budding yeast Saccharomyces cerevisiae. Genetics 189: 737–765.

Newman, J.R., S. Ghaemmaghami, J. Ihmels, D.K. Breslow, M. Noble, J.L. DeRisi and J.S. Weissman. 2006. Single-cell proteomic analysis of S. cerevisiae reveals the architecture of biological noise. Nature 441: 840–846.

Ng, P.C. and S. Henikoff. 2003. SIFT: Predicting amino acid changes that affect protein function. Nucleic Acids Res. 31: 3812–3814.

Noble, S.M., S. French, L.A. Kohn, V. Chen and A.D. Johnson. 2010. Systematic screens of a Candida albicans homozygous deletion library decouple morphogenetic switching and pathogenicity. Nat. Genet. 42: 590–598.

North, M. and C.D. Vulpe. 2010. Functional toxicogenomics: mechanism-centered toxicology. International Journal of Molecular Sciences 11: 4796–4813.

Oh, J., E. Fung, U. Schlecht, R.W. Davis, G. Giaever, R.P. St Onge, A. Deutschbauer and C. Nislow. 2010. Gene annotation and drug target discovery in Candida albicans with a tagged transposon mutant collection. PLoS Pathogens 6: e1001140.

Osterberg, M., H. Kim, J. Warringer, K. Melen, A. Blomberg and G. von Heijne. 2006. Phenotypic effects of membrane protein overexpression in Saccharomyces cerevisiae. Proc. Natl. Acad. Sci. USA 103: 11148–11153.

Papp, B., C. Pal and L.D. Hurst. 2003. Dosage sensitivity and the evolution of gene families in yeast. Nature 424: 194–197.

Parsons, A.B., R.L. Brost, H. Ding, Z. Li, C. Zhang, B. Sheikh, G.W. Brown, P.M. Kane, T.R. Hughes and C. Boone. 2004. Integration of chemical-genetic and genetic interaction data links bioactive compounds to cellular target pathways. Nat. Biotechnol. 22: 62–69.

Parsons, A.B., A. Lopez, I.E. Givoni, D.E. Williams, C.A. Gray, J. Porter, G. Chua, R. Sopko, R.L. Brost, C.H. Ho, J. Wang, T. Ketela, C. Brenner, J.A. Brill, G.E. Fernandez, T.C. Lorenz,

G.S. Payne, S. Ishihara, Y. Ohya, B. Andrews, T.R. Hughes, B.J. Frey, T.R. Graham, R.J. Andersen and C. Boone. 2006. Exploring the mode-of-action of bioactive compounds by chemical-genetic profiling in yeast. Cell 126: 611–625.

Parts, L., F.A. Cubillos, J. Warringer, K. Jain, F. Salinas, S.J. Bumpstead, M. Molin, A. Zia, J.T. Simpson, M.A. Quail, A. Moses, E.J. Louis, R. Durbin and G. Liti. 2011. Revealing the genetic structure of a trait by sequencing a population under selection. Genome Res. 21: 1131–1138.

Pausch, M.H. 1997. G-protein-coupled receptors in Saccharomyces cerevisiae: high-throughput screening assays for drug discovery. Trends in Biotechnology 15: 487–494.

Perlstein, E.O., D.M. Ruderfer, D.C. Roberts, S.L. Schreiber and L. Kruglyak. 2007. Genetic basis of individual differences in the response to small-molecule drugs in yeast. Nat. Genet. 39: 496–502.

Piskur, J. and R.B. Langkjaer. 2004. Yeast genome sequencing: the power of comparative genomics. Mol Microbiol 53: 381–389.

Pryde, F.E. and E.J. Louis. 1999. Limitations of silencing at native yeast telomeres. Embo. J. 18: 2538–2550.

Replansky, T., V. Koufopanou, D. Greig and G. Bell. 2008. Saccharomyces sensu stricto as a model system for evolution and ecology. Trends Ecol. Evol. 23: 494–501.

Roemer, T., J. Davies, G. Giaever and C. Nislow. 2012. Bugs, drugs and chemical genomics. Nat. Chem. Biol. 8: 46–56.

Ross-Macdonald, P., P.S. Coelho, T. Roemer, S. Agarwal, A. Kumar, R. Jansen, K.H. Cheung, A. Sheehan, D. Symoniatis, L. Umansky, M. Heidtman, F.K. Nelson, H. Iwasaki, K. Hager, M. Gerstein, P. Miller, G.S. Roeder and M. Snyder. 1999. Large-scale analysis of the yeast genome by transposon tagging and gene disruption. Nature 402: 413–418.

Ross-Macdonald, P., A. Sheehan, G.S. Roeder and M. Snyder. 1997. A multipurpose transposon system for analyzing protein production, localization, and function in Saccharomyces cerevisiae. Proc. Natl. Acad. Sci. USA 94: 190–195.

Ruderfer, D.M., S.C. Pratt, H.S. Seidel and L. Kruglyak. 2006. Population genomic analysis of outcrossing and recombination in yeast. Nat. Genet. 38: 1077–1081.

Salinas, F., F.A. Cubillos, D. Soto, V. Garcia, A. Bergstrom, J. Warringer, M.A. Ganga, E.J. Louis, G. Liti and C. Martinez. 2012. The genetic basis of natural variation in oenological traits in Saccharomyces cerevisiae. PLoS One 7: e49640.

Schacherer, J., J.A. Shapiro, D.M. Ruderfer and L. Kruglyak. 2009. Comprehensive polymorphism survey elucidates population structure of Saccharomyces cerevisiae. Nature 458: 342–345.

Scherens, B. and A. Goffeau. 2004. The uses of genome-wide yeast mutant collections. Genome Biol. 5: 229.

Schmitt, M.J. and F. Breinig. 2006. Yeast viral killer toxins: lethality and self-protection. Nat. Rev. Microbiol. 4: 212–221.

Shah, N.A., R.J. Laws, B. Wardman, L.P. Zhao and J.L.T. Hartman. 2007. Accurate, precise modeling of cell proliferation kinetics from time-lapse imaging and automated image analysis of agar yeast culture arrays. BMC Syst. Biol. 1: 3.

Sinha, H., B.P. Nicholson, L.M. Steinmetz and J.H. McCusker. 2006. Complex genetic interactions in a quantitative trait locus. PLoS Genet. 2: e13.

Slot, J.C. and A. Rokas. 2010. Multiple GAL pathway gene clusters evolved independently and by different mechanisms in fungi. Proc. Natl. Acad. Sci. USA 107: 10136–10141.

Smith, A.M., T. Durbic, J. Oh, M. Urbanus, M. Proctor, L.E. Heisler, G. Giaever and C. Nislow. 2011. Competitive genomic screens of barcoded yeast libraries. Journal of Visualized Experiments Aug 11; (54). doi:pii: 2864.

Sopko, R., D. Huang, N. Preston, G. Chua, B. Papp, K. Kafadar, M. Snyder, S.G. Oliver, M. Cyert, T.R. Hughes, C. Boone and B. Andrews. 2006. Mapping pathways and phenotypes by systematic gene overexpression. Mol. Cell 21: 319–330.

Springer, M., J.S. Weissman and M.W. Kirschner. 2010. A general lack of compensation for gene dosage in yeast. Mol. Syst. Biol. 6: 368.

Stamps, J.A., L.H. Yang, V.M. Morales and K.L. Boundy-Mills. 2012. Drosophila regulate yeast density and increase yeast community similarity in a natural substrate. PLoS One 7: e42238.

Steinmetz, L.M., H. Sinha, D.R. Richards, J.I. Spiegelman, P.J. Oefner, J.H. McCusker and R.W. Davis. 2002. Dissecting the architecture of a quantitative trait locus in yeast. Nature 416: 326–330.

Taxis, C., P. Keller, Z. Kavagiou, L.J. Jensen, J. Colombelli, P. Bork, E.H. Stelzer and M. Knop. 2005. Spore number control and breeding in Saccharomyces cerevisiae: a key role for a self-organizing system. J Cell Biol 171: 627–640.

Thatcher, J.W., J.M. Shaw and W.J. Dickinson. 1998. Marginal fitness contributions of nonessential genes in yeast. Proc. Natl. Acad. Sci. USA 95: 253–257.

Thorpe, G.W., C.S. Fong, N. Alic, V.J. Higgins and I.W. Dawes. 2004. Cells have distinct mechanisms to maintain protection against different reactive oxygen species: oxidative-stress-response genes. Proc. Natl. Acad. Sci. USA 101: 6564–6569.

Tsai, I.J., D. Bensasson, A. Burt and V. Koufopanou. 2008. Population genomics of the wild yeast Saccharomyces paradoxus: quantifying the life cycle. Proc. Natl. Acad Sci. USA 105: 4957–4962.

Tucker, C.L. and S. Fields. 2001. A yeast sensor of ligand binding. Nat Biotechnol 19: 1042–1046.

Vachova, L., M. Cap and Z. Palkova. 2012. Yeast colonies: a model for studies of aging, environmental adaptation, and longevity. Oxidative Medicine and Cellular Longevity 2012: 601836.

Wach, A., A. Brachat, C. Alberti-Segui, C. Rebischung and P. Philippsen. 1997. Heterologous HIS3 marker and GFP reporter modules for PCR-targeting in Saccharomyces cerevisiae. Yeast 13: 1065–1075.

Wach, A., A. Brachat, R. Pohlmann and P. Philippsen. 1994. New heterologous modules for classical or PCR-based gene disruptions in Saccharomyces cerevisiae. Yeast 10: 1793–1808.

Wang, Q.M., W.Q. Liu, G. Liti, S.A. Wang and F.Y. Bai. 2012. Surprisingly diverged populations of Saccharomyces cerevisiae in natural environments remote from human activity. Mol. Ecol. 21: 5404–5417.

Warringer, J., D. Anevski, B. Liu and A. Blomberg. 2008. Chemogenetic fingerprinting by analysis of cellular growth dynamics. BMC Chem. Biol. 8: 3.

Warringer, J. and A. Blomberg. 2003. Automated screening in environmental arrays allows analysis of quantitative phenotypic profiles in *Saccharomyces cerevisiae*. Yeast 20: 53–67.

Warringer, J., E. Ericson, L. Fernandez, O. Nerman and A. Blomberg. 2003. High-resolution yeast phenomics resolves different physiological features in the saline response. Proc. Natl. Acad. Sci. USA 100: 15724–15729.

Warringer, J., M. Hult, S. Regot, F. Posas and P. Sunnerhagen. 2010. The HOG pathway dictates the short-term translational response after hyperosmotic shock. Mol Biol Cell 21: 3080–3092.

Warringer, J., E. Zörgö, F.A. Cubillos, A. Zia, A. Gjuvsland, J.T. Simpson, A. Forsmark, R. Durbin, S.W. Omholt, E.J. Louis, G. Liti, A. Moses and A. Blomberg. 2011. Trait variation in yeast is defined by population history. PLoS Genet. 7: e1002111.

Wenger, J.W., K. Schwartz and G. Sherlock. 2010. Bulk segregant analysis by high-throughput sequencing reveals a novel xylose utilization gene from Saccharomyces cerevisiae. PLoS Genet. 6: e1000942.

Winston, F., C. Dollard and S.L. Ricupero-Hovasse. 1995. Construction of a set of convenient Saccharomyces cerevisiae strains that are isogenic to S288C. Yeast 11: 53–55.

Winzeler, E.A., B. Lee, J.H. McCusker and R.W. Davis. 1999a. Whole genome genetic-typing in yeast using high-density oligonucleotide arrays. Parasitology 118: S73–S80.

Winzeler, E.A., D.D. Shoemaker, A. Astromoff, H. Liang, K. Anderson, B. Andre, R. Bangham, R. Benito, J.D. Boeke, H. Bussey, A.M. Chu, C. Connelly, K. Davis, F. Dietrich, S.W. Dow, M. El Bakkoury, F. Foury, S.H. Friend, E. Gentalen, G. Giaever, J.H. Hegemann, T. Jones, M.

Laub, H. Liao, R.W. Davis and et al. 1999b. Functional characterization of the S. cerevisiae genome by gene deletion and parallel analysis. Science 285: 901–906.

Wolfe, K.H. and D.C. Shields. 1997. Molecular evidence for an ancient duplication of the entire yeast genome. Nature 387: 708–713.

Wong, S. and K.H. Wolfe. 2005. Birth of a metabolic gene cluster in yeast by adaptive gene relocation. Nat. Genet 37: 777–782.

Yan, Z., M. Costanzo, L.E. Heisler, J. Paw, F. Kaper, B.J. Andrews, C. Boone, G. Giaever and C. Nislow. 2008. Yeast Barcoders: a chemogenomic application of a universal donor-strain collection carrying bar-code identifiers. Nat. Methods 5: 719–725.

Yonemura, Y., E. Futai, S. Yagishita, S. Suo, T. Tomita, T. Iwatsubo and S. Ishiura. 2011. Comparison of presenilin 1 and presenilin 2 gamma-secretase activities using a yeast reconstitution system. J. Biol. Chem. 286: 44569–44575.

Zhang, X., D.L. Smith, A.B. Meriin, S. Engemann, D.E. Russel, M. Roark, S.L. Washington, M M Maxwell, J.L. Marsh, L.M. Thompson, E.E. Wanker, A.B. Young, D.E. Housman, G.P. Bates, M.Y. Sherman and A.G. Kazantsev. 2005. A potent small molecule inhibits polyglutamine aggregation in Huntington's disease neurons and suppresses neurodegeneration *in vivo*. Proc. Natl. Acad. Sci. USA 102: 892–897.

Zörgö, E., A. Gjuvsland, F.A. Cubillos, E.J. Louis, G. Liti, A. Blomberg, S.W. Omholt and J. Warringer. 2012. Life history shapes trait heredity by accumulation of loss-of-function alleles in yeast. Mol. Biol. Evol.

Zörgö, E., K. Chwialkowska, A.B. Gjuvsland, P. Sunnerhagen, G. Liti, A. Blomberg, S.W. Omholt and J. Warringer. 2013. Ancient evolutionary trade-offs between yeast ploidy states Plos Genetics 9:e1003388.

9

Phenomics in Bacteria

Robert J. Nichols[1,2] and Carol A. Gross[2,3,] *

Introduction

The genomic revolution of the last 20 years has launched the biosciences into a new frontier. For scientists working on many organisms, the availability of a genome sequence has dramatically accelerated research. The effects have been profound: Genetic engineering is now standard laboratory practice; evolutionary biologists can trace entire genomes; and genetic risk factors have been defined for many human conditions. However, our rapidly accelerating ability to collect and assemble DNA sequence information has greatly outpaced our ability to assign biological meaning to it. This imbalance creates a need for high-throughput methods aimed at developing leads to gene function: phenomic technologies. Phenomic technologies seek to associate DNA sequences with phenotypes in a high-throughput manner to gain a functional understanding of genetic code. In general, phenomic technologies require: (1) a method to create genomic variation; (2) an ability to adjust the experimental environment, and (3) a technique for observing and recording the phenotypes of interest (Fig. 1).

[1] Oral and Craniofacial Sciences Graduate Program, University of California at San Francisco, San Francisco, CA, 513 Parnassus Ave, CA-94143.
[2] Department of Microbiology and Immunology, University of California at San Francisco, San Francisco, CA, 600 16th Street, CA-94158.
[3] Department of Cell and Tissue Biology, University of California at San Francisco, San Francisco, CA, 513 Parnassus Ave, CA-94158.
* Corresponding author

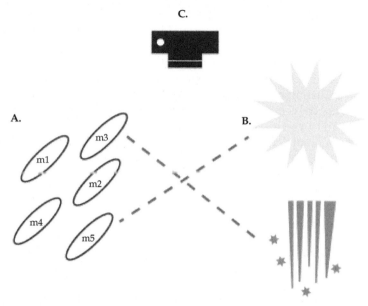

Figure 1. The cornerstones of phenomic analyses. All phenomic approaches require (a) a method to create a genomically variable strain collection; here, five isogenic mutants are depicted as m1–m5; (b) a means to vary the experimental environment; and (c) a technology to capture and record condition-gene interactions (phenotypes), depicted here as dotted lines.

Bacteria are obvious candidates for phenomics because there are easy ways to meet each phenomics requirement. Researchers have been manipulating bacterial genomes for more than 50 years using chemical and transposon mutagenesis, and, in some species, targeted chromosomal engineering. In addition to the relative ease of genetic engineering, bacteria do not present dosage or zygosity issues, as they are haploid organisms. Thus, the interpretation of genotype-phenotype relationships is relatively straightforward. Manipulating the environment of culturable bacteria is also quite simple. Chemical and drug stresses can be mixed directly into rich media, metabolic conditions can be tuned within defined media, and environmental factors like temperature, gas, and humidity can also be easily adjusted in incubators. *In vivo* analyses benefit from tools like germ-free and disease model mice. Techniques for capturing phenotypes resulting from genome-environment interactions have developed quickly over the last 10 years. Bacterial phenomic studies have now been published using a variety of readouts: colony size, growth rate, cellular respiration, strain abundance, survival, host colonization, and host clearance are at the beginning of a growing list. In all, a variety of options exist to manipulate bacterial genomes, control growth environments, and read out phenotypes in a high-throughput manner. The suite of available tools, and the unique biological qualities of bacteria, positions them as ideal subjects for developing new phenomic technologies.

This chapter will begin with a review of experimental technologies utilized in each of the three cornerstones of phenomics: creating genomic variation, adjusting growth environment, and capturing phenotypes. We briefly discuss the computational strategies used to analyze phenomic data, and then turn to a series of case studies that represent both the breadth and power of phenomic analyses in bacteria. We conclude with a look at the future and the coming advances in bacterial phenomics.

Experimental Technologies

Creating Genomic Variation

Bacterial phenomic approaches generally begin with the assembly of a genetically variable strain library, which is then screened or selected under a variety of growth conditions to search for specific condition-gene interactions, or phenotypes. Key to these strategies is that the genetic variability between strains is known or easily quantifiable, and as limited as possible. Thus, engineered libraries based on the techniques reviewed below should be constructed in a consistent strain background. Approaches based on natural genetic variation should put a premium on describing as much of the variation between strains as possible. This way, when phenotypes are detected, they can be traced to the known genetic variant(s) with high confidence. If significant genetic variability exists between strains in the same library (i.e., if mutant strains were constructed in varying strain backgrounds), it can be nearly impossible to detect meaningful condition-gene interactions.

There are several ways to establish genetic diversity for the screen, but all engineered genetic modifications can be classified as either loss-of-function (LOF) or gain-of-function (GOF). Simply put, LOF genetic approaches reduce the expression or activity of genomic DNA sequences, while GOF approaches either increase the function of native genes through overexpression or previously characterized GOF alleles, or add function by introducing non-native (heterologous) sequences into the cell. LOF mutant libraries are the most common means to establish genetic diversity for phenomic screens in bacteria, as LOF genetic analyses have a long history of revealing phenotypes that ultimately point toward gene function. GOF approaches based on overexpression have been widely used to search for drug-target relationships, while those based on heterologous expression have been employed in functional metagenomic applications. In this section, we first review each of the seven methodologies currently used to establish genomic variation in phenomics and then explore whether the resulting mutants are archived or pooled, whether they are used for screens or selections, and the relative merits of each strategy.

1. Transposon Mutagenesis

Transposon mutagenesis is widely employed to generate mutant libraries. Here, a transposon sequence containing a selectable marker is randomly inserted into the genome by a transposase enzyme. For competent bacteria, the selectable marker can be transposed into the host genome *in vitro*, followed by a transformation and selection procedure. For less competent bacteria, the transposon and transposase are encoded on a plasmid transformed into host cells, where translation of the transposase enzyme, and subsequent insertion of the transposon DNA into the host genome occur *in vivo*. Selection is then carried out to purify cells that have received a transposon insertion. Transposon mutagenesis technologies are reviewed in Mazurkiewicz et al. (2006).

Most random transposon insertions occur in a coding region because of the high density of open reading frames (ORFs) in bacterial genomes. Occasionally, transposon insertions occur in intergenic and promoter regions. Insertions of this type may have interesting genetic effects beyond the complete LOF phenotype offered by engineered deletion mutants and may also reveal previously unannotated functional elements. However, the randomness of transposon mutagenesis also presents challenges: the genomic insertion sites must be identified; follow-up experiments are necessary to determine whether insertions yield LOF or GOF proteins; and the "average" phenotype of independent insertions within a single gene must be determined. The issue of polarity is discussed in the section on targeted chromosomal engineering, and the strategies for mapping insertion sites are discussed in the pooled screening technology section.

2. Targeted Chromosomal Engineering

Genetic engineering of the chromosome via natural competence or induced recombineering is possible in some bacterial species. For these, many options exist for building LOF libraries. The most straightforward is to build a single-gene knockout library (Baba et al. 2006, de Berardinis et al. 2008, Santiviago et al. 2009). In these libraries, a single gene is deleted in each strain, covering all nonessential genes of the genome. The gene is replaced with an antibiotic resistance cassette for selection, and this cassette can also contain a DNA barcode for use in pooled approaches (discussed below). While comprehensive single-gene knockout libraries have proven to be extremely powerful tools in many organisms (Nichols et al. 2011, Roguev et al. 2008, Santiviago et al. 2009, Schuldiner et al. 2006), they are inherently limited by the annotation of the target genome. More specifically, functional DNA sequences yet to be annotated, like those coding for small proteins or small RNAs, will not be interrogated by most targeted engineering approaches. Herein lies a central tradeoff in targeted vs. random mutagenesis: targeted knockouts are clean (complete or controlled LOF), while random mutagenesis can reveal previously undetected functional sequences. Both random mutagenesis strategies and targeted knockouts may have unanticipated transcriptional and/or translational polar effects due to operon structure, potentially clouding interpretation of phenotypes. Follow-up studies are essential to verify the suspected genotype-phenotype relationship.

Chromosomal engineering is also used to modulate gene expression. In trackable multiplex recombineering (TRMR) (Warner et al. 2010), a synthetic promoter and ribosome-binding site (RBS) is recombined into the chromosome to replace the native promoter of a gene. These constructs may encode strong promoter/RBS sequences conferring GOF, or weak ones resulting in LOF. In addition, it is possible to knock in tunable promoter sequences, allowing for analysis of mutant strains under various promoter activities on a genome-wide scale.

It is also possible to establish genetic variability in the strain library through collection and manipulation of previously constructed alleles: point mutations, small deletions, insertions, truncations, etc. Such alleles can be either LOF or GOF, and need not be previously functionally characterized. Regardless, it is essential that the alleles be transferred into a common strain background by either phage transduction or targeted chromosomal engineering to maintain an isogenic library (Nichols et al. 2011).

3. Antisense RNA

In recent years, antisense RNA (asRNA) has been an effective tool to knock down gene expression in some Gram-positive bacteria. Very recently, the first asRNA-based screen in Gram-negative bacteria has been reported (Meng et al. 2012). Thus far, all approaches are plasmid based, with expression of the targeted asRNA induced by an exogenous small molecule. Presumably, the asRNA transcribed from the plasmid base pairs with its complementary message, and either causes transcript degradation or interferes directly with ribosome binding. Because a variety of asRNA mechanisms exist (and many asRNAs are ineffective), asRNAs must be screened and validated before being used in a phenomic screen or selection. Because asRNA mechanisms and effectiveness are poorly understood in bacteria, there is an "off target" caveat to all phenotypes. The effects of a single asRNA may reach beyond reduction of its target mRNA levels and impact the transcription/translation of other genes in the cellular network. Therefore, phenotypes detected using asRNA approaches must be verified through careful follow-up studies. In addition, all asRNA strains should be analyzed for response to the inducer molecule. Differential sensitivity across strains can complicate strain fitness measurements, especially in pooled approaches. One effective strategy to normalize for this effect is to pool only those strains sharing the same response kinetics to the inducer molecule (Donald et al. 2009).

4. Chemical Mutagenesis

Many bacterial species are culturable yet lack genetic tools. For these, chemical mutagenesis can be the only laboratory means to generate genomic diversity. Chemical mutagenesis has not previously been useful for phenomics, because of the lack of a high-throughput methodology to identify base changes. However, the increased efficiency and cost effectiveness of whole-genome sequencing (WGS) methodologies coupled with the small genome size of bacteria now allow chemically induced mutations to be identified through WGS. The first report of this approach is a study of the virulence determinants of the obligate intracellular bacterium Chlamydia trachomatis (Nguyen and Valdivia 2012).

5. Overexpression

Overexpression of genes has been used in phenomic approaches known as high-copy suppressor screens. Such screens have generally been used to search for genes whose overexpression increases resistance to a drug or stress of interest. These phenotypes can suggest drug-target relationships, and have proven quite powerful in drug discovery efforts. Overexpression is usually achieved through plasmid-based systems, but chromosomal alleles that increase the expression and/or stability of the mRNA and/or protein, or increase protein activity could also be used.

For plasmid-based systems, the plasmid bearing the ORF of interest is transformed into host cells and maintained with the appropriate selective agent. Overexpression of the ORF can be either constitutive or inducible. In the latter case, a small molecule added to the culture activates promoter expression. Overexpression screens must be optimized to find a level of overexpression that is effective, but not toxic to the cell. High levels of overexpression provide a potent selection for spontaneous mutations within the expression plasmid or elsewhere in the genome (suppressors) that can compromise the controlled genetic variability necessary to execute an effective phenomic screen. Though less common than LOF libraries, a

GOF library based on overexpression has been constructed in *Escherichia coli* (Kitagawa et al. 2005), and has been utilized for phenomic studies (Couce et al. 2012, Pathania et al. 2009).

6. Heterologous Expression

Metagenomic sequencing projects have been a major component of the bacterial sequencing boom in recent years. Using environmental samples with many species, these projects sequence all DNA to generate vast collections of ORF sequences, many of which have no known function or genomic source. However, many researchers have begun to recognize that such sequences may code for medically (antibiotic-resistance) or commercially (metabolic, remediation) valuable genes. Therefore, phenomic approaches aimed at deciphering the function of such genes are being developed. Functional metagenomic approaches rely on heterologous expression of the sequenced genes of interest in a controlled system (like *E. coli*). In these approaches, the genome of the host strain is held constant, and expression of the unknown ORF creates genetic diversity in the strain library (McGarvey et al. 2012, Sommer et al. 2010).

7. Natural Variation

Natural variation between strains can also create genetic diversity of the strain library. In some settings, it may be desirable to understand the genetic basis of a trait possessed by only certain strains of a given organism (i.e., host colonization or metal reduction). In such cases, harnessing the natural variation between strains can be a powerful route to functional discovery. Here, isogenicity between strains becomes impossible, but screening closely related strains can allow meaningful conclusions to be drawn.

Evolved strains can also be utilized for studying the functional impact of natural variation. Bacterial strains can be evolved in the lab (Tenaillon et al. 2012) or in a host (Fabich et al. 2011). Evolved strain approaches can be particularly powerful, as a high degree of isogenicity can be maintained between the evolved and parent strains, keeping phenotypic analyses relatively controlled. However, such strategies demand either WGS or polymorphic analyses (i.e., SNP) of all strains.

Adjusting the Experimental Environment

Phenotypes are classically defined as the output of interaction between genome and environment. By systematically varying the genome and the environment, phenotypes detected in high-throughput phenomic approaches represent functional interactions between genes and environments. A variety of approaches for modulating the genome are described above. Here, we review techniques for varying the environment.

The probability of phenotypic discovery scales with the degree of both genetic and environmental variability screened. Thus, the more unique environments under which the genetically diverse strain library is evaluated, the greater the likelihood of detecting an interaction between a given gene and environment. Therefore, it is desirable to screen as many environments as possible. However, careful selection of those environments may also increase the odds for discovery. The establishment of genetic diversity in the strain library gives the researcher power to query the function of nearly all parts of the genome. The selection of environments allows the researcher to test what each of those parts may have evolved to do.

Phenomic studies can apply various types of stresses to the strain library. Chemical stresses including antibiotics and other drugs, mutagens, detergents, etc. can be applied (Couce et al. 2012), as can existing natural product libraries. On the other hand, environmental stresses like temperature, pH, UV light exposure, and metabolic conditions can be effective in revealing phenotypes (Deutschbauer et al. 2011, Ishii et al. 2007). The screening environment may also be perturbed by the addition of another organism, such as the infection of bacteria with phage (Maynard et al. 2010), or the infection of an animal host (Bianconi et al. 2011, Gawronski et al. 2009, Goodman et al. 2009, Hensel et al. 1995, Hisert et al. 2004, Lawley et al. 2006, Potvin et al. 2003, Santiviago et al. 2009), or cell line (Camacho, Ensergueix et al. 1999, Nguyen and Valdivia 2012) with bacteria.

Importantly, not all stresses can be applied in all experimental systems. For example, solubility and/or volatility properties may dictate whether a chemical stress can be assayed in solid- and/or liquid-based setups. Chemicals and certain environmental conditions assayed *in vitro* may not be tolerated by an animal host, and may thus be incompatible with *in vivo* analyses. Therefore, selection of stresses for a given phenomic study must take into account the capacity and potential limitations of the experimental system. For all types of stress, the dose or concentration used in the assay can have major effects on the results. High stress load can increase the likelihood of spontaneous suppressor mutations, complicating the interpretation of phenotypes, while low stress load may fail to separate less or more fit mutants from the population. Therefore, pre-screens aimed at quantifying and optimizing the stress load for each unique stress can be extremely beneficial for downstream phenotypic discovery.

Finding the Phenotype: Screens/Selections, Readouts, and Scoring

A multitude of readouts can be used to monitor diverse processes and states for phenomic analysis. High-throughput microscopy can be used to monitor morphological phenotypes or protein localization. Fluorescent or other colorimetric reporters can be used to monitor diverse processes including transcriptional activity, metabolic capacity, developmental processes, and metal reduction. Additional phenotypes related to developmental cycles can be monitored by specialized readouts (i.e., sporulation by resistance to toxic chemicals). However, the most inclusive, and by far most-utilized, readout to this point has been fitness, which itself has been quantified in a variety of ways.

The typical objective of phenomic analyses is to identify specific condition-gene interactions. By normalizing out the mutation-specific effect, it is possible to identify environments where a distinct mutation causes a more (or less) severe phenotype than is normally observed for the same mutation under baseline conditions. Currently, there are two main ways to screen for these events. The first, called an arrayed approach, evaluates each strain in isolation under a battery of growth environments. The second, called a pooled approach, evaluates the ability of each strain in the library to compete with all other strains under the set of growth environments. In addition, it is possible to identify condition-gene interactions through a variety of selection techniques.

Just as a number of options exist for establishing genomic and environmental diversity, many experimental setups have been used successfully in bacterial phenomics, all of which can be divided roughly into forward and reverse genetic approaches. Forward genetic approaches begin with a phenotype of interest and search for genomic regions contributing to that phenotype, using technological advances that simplify identification of the genomic region(s) of interest. In general, these analyses use deletion or transposon libraries that are pooled and then subject to screens or selections. Reverse genetic approaches begin with a discrete set of genetically variant strains, and seek to associate phenotypes with each one. These analyses typically utilize arrayed screening approaches.

Arrayed Screens

In an arrayed approach, individual strains are grown in isolation, and evaluated relative to their baseline behavior. Phenotypes are defined as deviations from that baseline. The power of arrayed approaches, which can be conducted in liquid culture or on solid surfaces, comes in the control of the screen and potential robotic integration for high-throughput screening. Because the strains are grown and evaluated against themselves, the deviation of a given strain from its baseline behavior is clear: the interaction of a specific mutation and a specific environment produces the observed phenotype. While some strain competition does exist in solid surface arrayed approaches, it is tightly controlled relative to pooled approaches, as each strain competes only mildly with the strains that surround it. The surrounding strains are normally held constant throughout the entire screen, allowing low-level "neighborhood" competition to be normalized away. Strain competition is completely absent from liquid-arrayed approaches. Liquid-handling (for liquid culture setups) or colony-pinning (for solid surfaces) robots automate the screening process and provide technical accuracy and reproducibility (Fig. 2).

In liquid culture applications, growth is usually read by calculating the rate of exponential growth (Nakahigashi et al. 2009), endpoint culture density (Kohanski et al. 2008), or cellular respiration (Bochner 2009, Fabich et al. 2011). For solid surface approaches, fitness is normally read out as colony size. This non-traditional indicator of fitness is quite powerful as it reports on duration of lag phase, the rate of log phase growth, and the time of onset of stationary phase (Nichols et al. 2011).

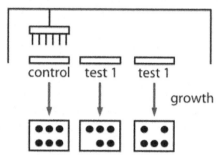

Figure 2. Arrayed approaches generally require colony-pinning or liquid-handling robotics. These platforms deliver high-throughput scaling and technical accuracy through highly reproducible aliquoting of cells to control and test conditions.

Arrayed Screen Scoring

Arrayed approaches are based on distributing a collection of strains onto a solid surface or into microtiter plates for liquid culture. In both cases, the media is infused with a condition/stress of interest, and growth is allowed to proceed for a pre-determined incubation period. For arrayed approaches utilizing a quantitative readout (colony size, culture density, growth rate), a global scaling normalization must be employed to account for differences in stress load and absolute growth. This normalization centers the distribution of all readouts observed for each condition on a common value, making cross-condition comparisons feasible.

It is important to accurately estimate the control, or baseline behavior of each strain in order to detect specific condition-gene interactions. Two techniques exist for estimating control behavior in the context of arrayed approaches: median-centering and control conditions. A central assumption of median-centering is that condition-gene interactions (phenotypes) are rare. Therefore, if enough unique conditions are screened, it is possible to estimate the control behavior of a strain by simply identifying the median of all quantitative readouts for that strain. This approach has been effective in large-scale phenomic and genetic interaction studies of bacteria and yeast, demonstrating that the median of a large number of experiments is a robust indicator of control behavior (Collins et al. 2007, Nichols et al. 2011, Roguev et al. 2008, Schuldiner et al. 2006, Typas et al. 2008).

When a smaller number of conditions are screened, the median of all readings is not a reliable estimate of control. In these cases, an appropriate control condition must be identified and screened in parallel with the stress conditions of interest. Here, the behavior of each strain under control conditions establishes its control behavior. One disadvantage of this method compared to median centering is that the control estimate is based on a comparably low number of readings (low "n"). Therefore, it is essential that the control condition is screened a sufficient number of times for each strain to establish a statistically robust number of readings (high "n"), as estimates of control behavior should always be accompanied by corresponding measures of variance.

After control behavior and variance have been estimated for each strain, fitness scores may be calculated for each condition screened relative to the control measures. Recently, this has been done using a modified t-statistic known as the "S score" (Nichols et al. 2011) or other statistical metrics based on the normal distribution (Kohanski et al. 2008). The S score procedure calculates fitness scores for a given strain by comparing the average growth of that strain across a replicate series of a given condition to the control growth of that strain. Variance estimates associated with the replicate series and control growth are used to modify the confidence associated with the estimated score (high variance functions as a penalty). Replicate experiments come in two forms: strain replicates and screen replicates. Strain replicates require construction and isolation of independent clones of the same mutant strain. These replicates help control for the possibility that additional mutations may arise either during strain construction or during the screening process as the cell evolves to compensate for the primary mutation. The more strain replicates screened, the lower the likelihood that the same secondary mutation(s) will arise and cloud the results. Screen replicates are technical replicates of the same screen to control for day-to-day variation of potentially variable components of the screening process (e.g., stress concentration, colony pinning accuracy, media quality, imaging parameters, and human error).

Pooled Screens

Pooled screens evaluate the ability of one strain to compete with all other strains in the library in batch culture under a given growth environment. Pooled approaches work very well with forward genetic approaches like transposon mutagenesis, and thus are available for use in a wide range of bacterial organisms. However, arrayed libraries can also be used for pooled analyses, and offer the dual benefits of increased control over starting cell inoculum for individual mutants and availability of individual mutants for follow-up study after the screen.

Pooled approaches are best positioned to evaluate strain fitness, or growth, but have also been used to evaluate cellular processes like motility (Girgis et al. 2007) and biofilm formation (Amini et al. 2009). Fitness evaluation is based on reading out the relative abundance of every mutant strain in the batch culture at the start and endpoint of each competition (Fig. 3). Just as described for arrayed approaches, phenotypes are defined as deviations from the baseline behavior of each strain. In pooled approaches based on growth, baseline behavior is estimated from the typical frequency of a given strain in the pool.

Pooled approaches are powerful for several reasons. First, they scale incredibly well without the need for fancy robotics. Since each condition screened requires a single batch culture, a single person can evaluate hundreds of conditions rather easily. Second, pooled approaches can evaluate thousands of strains simultaneously in the same culture. Therefore, a tremendous amount of genomic diversity can be assayed under many conditions. Pooled approaches therefore represent the most efficient way to screen the largest amount of condition-gene interactions for potential phenotypes.

The major challenge of such approaches is teasing apart the confounding factor of strain competition on phenotypic evaluation. Because pooled approaches often compete thousands of strains against each other, the ability to quantitatively estimate changes in a given strain's fitness is influenced by the potential fitness changes of every other strain in the pool. For dramatic fitness changes, this is unlikely to prevent detection. But for more moderate effects, especially for fitness loss, competition within the pool has been reported as a potential confounding factor (Girgis et al. 2009). An additional caveat to pooled analyses is that certain mutations may exhibit phenotypes only in the context of the complex community. For example, an LOF mutation that disables production of an essential metabolite may be tolerated, and even beneficial, if that mutant can scavenge the metabolite from other members of the community. While this scenario can complicate interpretation of pooled results, it also represents an advantage of pooled approaches. They report a physiologically relevant analysis of bacteria within a genetically diverse community.

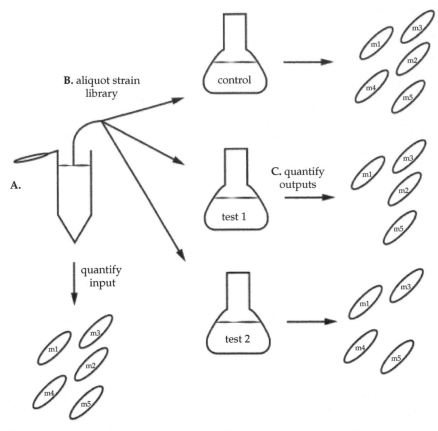

Figure 3. Workflow of pooled growth approaches. (A) The starting strain library is sampled and quantified via one of the techniques discussed in the pooled screen section. (B) Aliquots of the strain library are distributed to test and control condition cultures. (C) Competitive cultures are grown for a set time, and then sampled and quantified according to the procedure established for the input library.

All pooled approaches based on transposon mutant libraries require a readout technology capable of quantifying the abundance of each unique strain in the pool. Different transposon systems have been used for phenomic studies in bacteria, and have typically been paired with complementary readout technologies. Initial transposon studies were based on Signature-Tagged Mutagenesis (STM) systems. In these systems, each insertion mutant carries a unique DNA barcode, so that a given barcode sequence can be assigned to a specific genomic insertion site. Therefore, when quantifying strains after a screen or selection, one needs only to quantify the abundance of each barcode. STM approaches were the first to break through in phenomic applications, because these barcodes allowed a high-throughput readout based on DNA blotting (Hensel et al. 1995), multiplex PCR (Potvin et al. 2003), or microarray hybridizations (Deutschbauer et al. 2011). However, the need for a large amount of unique DNA barcodes requires a great deal of transposon template synthesis, making library creation a laborious task.

Second-generation transposon-based strategies focus on eliminating the need for a unique transposon sequence for each mutant. These technologies need only one transposon sequence, and rely on the genomic DNA flanking the insertion site to serve as a barcode. Three methods exist for the detection of transposon-flanking genomic DNA signal in insertion sequencing-based approaches: (1) type-IIS restriction enzyme (mariner Tn), (2) selective PCR using a Tn-specific primer, and (3) the circle method (described in the following paragraphs). For all three, the resulting transposon-flanking genomic DNA fragments can be analyzed via hybridization to a full-genome tiling array or short-read sequencing with transposon and/or adapter-specific primers.

Some engineered Mariner transposons have MmeI restriction sites in their inverted repeat sequences of the transposon ends (van Opijnen et al. 2009). MmeI is a type-IIS restriction enzyme, and cuts 20 base pairs downstream of its recognition site in the genomic DNA just outside the transposon-genome junction. Adapters can then be ligated to the cut ends, and the genomic DNA amplified with one primer specific to the transposon and the other specific to the adapter. This approach has been widely used.

Selective PCR can also be run without specific cutting by MmeI. Genomic DNA can be sheared, adapters ligated onto blunt ends, and selective PCR used to enrich transposon-genome junction fragments as described above for Mariner approaches. However, the absence of the MmeI digestion means that far more DNA fragments are ligated to adapters, and non-specific PCR can be problematic. Therefore, some studies have added an additional purification step by biotinylation of the transposon-specific primer and affinity purification following PCR (Gawronski et al. 2009).

More recently, the Tn-seq circle method has been developed (Gallagher et al. 2010). In this method, genomic DNA is sheared, and a single adapter is ligated to all free ends. After adapter ligation, fragments are digested with a restriction enzyme that cuts inside the transposon sequence (near one end), and circularized via templated ligation. Then, all circularized fragments contain part of the transposon sequence ligated to the generic adapter, and surrounded by the genomic DNA of the transposon-genome junction. PCR is then used to amplify those fragments.

Pooled Screen Scoring

In pooled competitions, the baseline behavior of a given strain reflects the fitness associated with that mutation as indicated by the ability of the mutant to compete with all other mutants in the pool. To estimate these controls, one may simply examine the frequency of each mutant in the pool before selection, as these values reflect the relative ability of each mutant to compete during selection and outgrowth of the library (Deutschbauer et al. 2011). A second strategy is to base control estimates on the endpoint strain abundance measures in competitive cultures without stress (Langridge et al. 2009). These "control cultures" can be run in parallel to competitions under stress. It is also possible to estimate control behavior based on both time-zero and control competition results (Girgis et al. 2009).

Pooled competitions normally utilize strain libraries generated via transposon mutagenesis, though such libraries can also be arrayed (Cameron et al. 2008, Deutschbauer et al. 2011, Gallagher et al. 2007, Liberati et al. 2006). As is the case with arrayed approaches, strain and screen replicates are necessary to accurately estimate phenotypes. Transposon libraries have an advantage over arrayed engineered libraries in that saturating mutagenesis can allow for hundreds of mutants per gene to be constructed and analyzed. Therefore, each experiment has many "biological replicates", providing a very high "n" for strain replicate readings in such approaches, and thus increased confidence in the effects of mutating a given gene. However, it is still important to perform "technical" replicate screens to control for day-to-day variation, innoculum effects, and human error.

Conditional strain fitness in pooled competition experiments is usually represented with a calculated Z score. Z scores have been calculated using different metrics, but all attempt to quantify the deviation of the conditional behavior of a strain from its control behavior by calculating the mean and standard deviation of all conditional and control frequency measures for that strain (Deutschbauer et al. 2011, Girgis et al. 2009). Thus, the Z score, S score (reviewed in the section on arrayed approaches), and t-statistic are extremely similar.

Selections—*In vivo* Transposons and Functional Metagenomics

In vivo phenomic approaches based on transposon mutant libraries often use positive or negative selections to identify insertion mutants of interest. The techniques used to identify the selected insertion mutants are discussed above in the pooled screen section. Negative selections can be used to identify genes necessary to establish infection or maintain host colonization, as the host organism can select against LOF insertion mutants in such genes (i.e., following infection/colonization particular transposon mutants are depleted in the population). Positive selections can identify genes detrimental to such processes and suggest host-adaptive strategies used by the bacteria.

Positive selections are ideally suited for functional metagenomic approaches. Metgenomic ORFs of interest may be cloned into expression vectors, and then transformed into a host strain unable to grow under the selective condition. Then, the transformed strains can be pooled and plated under the given condition to select for clones that have survived. The expression vectors in the surviving clones can then be sequenced to identify the gene conferring survival.

Data Analysis—Benchmarking and Exploration

Quantitative phenomic datasets can be large and complex, and require computational strategies to demonstrate data quality through benchmarking analyses, and then search for functional insights in the data via exploratory techniques.

Benchmarking

Upon completion of a phenomic screen and assembly of a compendium of phenotypic scores, benchmarking analyses are performed to assess data quality and potential functional predictive power. One commonly used metric to evaluate data quality is the correlation between replicate measurements. This can be calculated either as the correlation of all strain measurements across any pair of screen replicates (Deutschbauer et al. 2011, Nichols et al. 2011), or as the correlation of any pair of strain replicates across all screens. Such correlations can demonstrate both the reproducibility of the data and the strength of phenotypic signal (high correlations cannot generally be achieved without signal).

Receiver operating characteristic (ROC) plots are extremely powerful for determining the functional predictive power of the dataset. These plots show the relationship between the true positive rate and the false positive rate at regularly spaced intervals along a continuous scale (in this case, the correlation coefficient scale). Therefore, the calculation of such plots requires a "gold standard" index classifying pairs of genes as true positive (functionally related) or true negative (unrelated). For bacteria, a variety of classifiers may be used to define the gold standard index: shared operon membership can suggest a functional link between pairs of genes; protein-protein interaction datasets report on pairs of genes known to code for physically interacting proteins; transcriptomics datasets can indicate pairs of co-expressed genes; functional annotation classifiers (Gene Ontology, TIGR, COG) can be used to identify genes in a common class. For all of these classifiers, the precision and accuracy of prediction of true positive pairs by measures of similarity in phenomic datasets is solid evidence for potential of functional discovery.

Data Exploration

Complete phenomic datasets are large and require a strategy to systematically explore and extract testable hypotheses of gene function. Fortunately, the data itself shares many similarities with microarray data (many genes, many conditions, a null expectation of zero change) and therefore many tools have already been developed. We present a brief overview of basic techniques; see Gentleman (2005) for a comprehensive review.

(a) Specific Condition-Gene Interactions

The baseline level of phenomic data analysis attempts to tie genes to cellular processes manually by extrapolating from specific phenotypes of the mutant. For example, the study by Donald et al. (2009) discussed below in the *in vitro* competitive assay section generated a hypothesis that an essential gene of unknown function is a peptidoglycan flippase, based on the hypersensitivity of its knockdown strain to a set of cell wall-targeting drugs. The association of the gene with a broad process (cell wall biosynthesis) was the starting point for further exploration.

(b) Unsupervised Methods

The second tier of phenomic exploration approaches uses unsupervised computational methods to group mutant and condition profiles by similarity. "Unsupervised" refers to the fact that no prior knowledge about the functional relatedness of genes in the dataset is fed into the algorithm. Therefore, all relationships predicted by the clustering algorithm are based only on the phenomic data. The most commonly used unsupervised method is two-dimensional (2D) hierarchical clustering and heat map visualization. In 2D hierarchical clustering, rows and columns of an input data matrix are iteratively joined based on the similarity of their profiles. Relationships are represented by the spatial grouping and bootstrapping of rows and columns in the output matrix.

(c) Machine Learning

A higher level of phenomic data exploration can be achieved through machine learning, or supervised methods. Contrary to unsupervised methods, supervised methods predict relationships in the data based on prior knowledge. For example, functional annotation of genomes (Gene Ontology, TIGR, COG) can be used to "train" algorithms on functionally related genes. Then, patterns in the data that connect these known groups can be used to predict novel members.

Phenomic Applications in Bacteria

Phenomic technologies have been implemented to understand gene function in several different bacterial applications, ranging from cellular modeling of *E. coli* to identifying novel antibiotic resistance genes in complex environmental samples. Researchers have utilized forward and reverse genetics, selections and screens, and *in vitro* and *in vivo* approaches. Five major areas have been developed to date, and they are reviewed below.

In vitro Competitive Assays

In vitro competitive assays are generally based on growth, but can also be used to select for differences in traits of interest (i.e., motility). Most published studies are based on transposon mutant libraries, but other genetic techniques like asRNA have also been integrated. Regardless of genetic technique or readout, competition is used to reveal conditional phenotypic differences amongst a collection of strains. These condition-specific phenotypes have presented powerful leads for understanding gene function in model, pathogenic, and applied species.

Girgis et al. (2007) conducted a series of forward genetic positive selections on an *E. coli* transposon mutant library to identify genes involved in motility. By plating a pre-selected pool of mutants in the center of a soft agar plate and allowing motile mutants to chemotax away from the resource-depleted center, the authors enriched for non-motile mutants. Serial passages and selections of this non-motile pool led to further enrichment, and eventually the identification of a suite of genes implicated in motility via a microarray-based readout of the insertion-flanking genomic DNAs. Importantly, this suite contained more than 95% of all previously known motility genes, as well as more than 30 genes not previously associated with the process. By searching for suppressors of the motility-impaired phenotype of the novel candidates, the authors were able to make important biological insights regarding the involvement of the second-messenger c-di-GMP system in chemotaxis signaling. In all, this study represents an elegant fusion of classic mutagenesis and selection techniques with a high-throughput readout, and demonstrates that phenomic approaches can reveal new insights into even the most-studied processes in the best-studied organisms.

Gallagher et al. (2010) executed a transposon-based forward genetic screen of *P. aeruginosa* to identify genes involved in the intrinsic resistance of this opportunistic pathogen to the aminoglycoside antibiotic tobramycin. The authors competed nearly 100,000 mutants in batch culture both with and without a low concentration of tobramycin, and quantified the abundance of all mutants in the pool with a high-throughput sequencing-based readout. Because *P. aeruginosa* is normally resistant to the drug, most mutants were expected to be unaffected. Those strains harboring mutations in genes necessary for resistance were expected to decrease in abundance, or disappear from the population under tobramycin stress. This strategy proved effective, as many previously described tobramycin resistance genes, as well as a group of novel candidates, were captured. These new candidate genes collectively suggested important roles for both cell envelope and intracellular potassium homeostasis in regulating tobramycin resistance. This study illustrates the power of pooled growth assays and high-throughput sequencing to reveal important biology in clinically relevant species.

Deutschbauer et al. (2011) conducted an STM-based screen evaluating mutant fitness in a large battery of growth conditions. Almost 25,000 transposon insertion mutants of *S. oneidensis*, covering ~3,500 genes, were assayed for fitness under 121 growth conditions. Quantifying strain abundance with a microarray-based readout, the authors built a matrix of fitness scores for each gene under each condition. Using this matrix to identify specific mutant phenotypes and genes whose mutants behaved similarly throughout the study, the authors generated new evidence-based predictions of function for 40 genes, including many involved in metabolism and three involved in motility. Further, characterization and archiving of the transposon mutant strains created a reverse genetic resource, allowing specific mutant phenotypes to be pursued via targeted experiments at a later time. This allowed several of the 40 predictions to be validated experimentally. As *S. oneidensis* is of potential use in the bioremediation of heavy metals and energy generation, this study demonstrates the potential of transposon-based phenomic analyses to generate functional insights into important processes of applied organisms.

To this point, asRNA screens in bacteria have focused on defining drug mode-of-action (MOA), and therefore have not yet been exploited for phenomic screens. However, Donald et al. (2009) conducted a drug MOA study by screening a library of asRNA essential gene knockdown strains in *S. aureus* for sensitivity to a variety of known and unknown compounds. While most findings emphasized using the asRNA strain response patterns to predict drug MOA, a knockdown strain of an essential gene of unknown function (SAV1754) was found to be highly sensitive to a variety of cell-wall-targeting compounds. After additional experimentation, the authors suggest that SAV1754 may function as a peptidoglycan flippase, demonstrating the potential of asRNA-based screens to inform discovery of gene function.

In vitro Arrayed Approaches

Bacterial species for which individual mutants can be isolated are amenable to arrayed phenomic analyses. To this point, the limited availability of arrayed libraries and advanced robotics has made arrayed approaches much less common than pooled ones discussed above. However, exciting biology has been revealed by arrayed approaches using deletion libraries, overexpression libraries, and evolved strains. These tools have been used for large-scale screens for gene function, smaller-scale, targeted screens, and drug discovery efforts.

Nichols et al. (2011) executed a large-scale reverse genetic screen of the *E. coli* Keio Collection, a comprehensive single-gene knockout library. Using the endpoint colony size as a readout, almost 4,000 mutants were screened under 324 growth conditions. The large number of conditions screened and the sensitivity of the reverse genetic approach allowed many uncharacterized genes to be associated with well-studied functional modules through pairwise correlation analysis of knockout strains. In all, high-confidence functional predictions were generated for more than 300 previously uncharacterized genes. As a follow-up case, one uncharacterized gene that correlated very highly to a gene involved in peptidoglycan synthesis was characterized to be essential for peptidoglycan synthase activity *in vivo* using a series of genetic and biochemical experiments (Typas et al. 2010). In addition, the phenotypes identified in this study were used to generate insights into drug action and the evolution and genomic organization of *E. coli*.

Arrayed approaches have also been used for targeted questions. A study by Kohanski et al. (2008) used a liquid culture array of the Keio Collection. The authors used endpoint optical density as a readout of the growth of each mutant under gentamicin stress relative to control to identify gene mutants that sensitize the cell to the drug (Kohanski et al. 2008). Integration of the phenotypic data with gene expression data, collected under the same conditions, enabled the authors to make new insights into the mechanism by which aminoglycoside antibiotics trigger cell death.

Arrayed approaches can also be useful for analyzing a small number of genetically distinct strains. Biolog technology involves arraying of a single strain into microplate format, with subsequent analysis of the ability of that strain to grow under a variety of metabolic conditions. Using a colorimetric readout of cellular respiration, the Biolog system enables rapid analysis of a small number of strains under a large number of growth conditions. A recent study by Fabich et al. (2011) examined the relative ability of a host-adapted strain of *E. coli* to catabolize a variety of carbon sources relative to its parent strain. By screening the adapted strain, which contained an LOF mutation in the flhDC locus, a key master regulator of motility in *E. coli*, the authors were able to correlate this mutation with increased capacity to proliferate under a variety of carbon-source conditions. After constructing an flhDC mutant in an isogenic background to the wild-type parent strain, the authors concluded that loss of flhDC actually results in increased cellular metabolic capacity, and therefore inhibition of motility may represent an adaptive response for *E. coli* host colonization. These findings demonstrate the utility of Biolog technology for rapidly evaluating cellular fitness under varying metabolic conditions, and further show the ease with which the technology can be applied to naturally occurring or evolved bacterial strains.

Overexpression libraries can also be arrayed for phenomic analyses. Pathania and colleagues assembled a set of *E. coli* strains (Kitagawa et al. 2005), each overexpressing one of the ~300 essential genes of *E. coli* K12 (Pathania et al. 2009). The strains were arrayed into microtiter plates, grown in liquid rich media, and challenged with 49 compounds that had been selected from a library of more than 8,000 small molecules for growth inhibitory effects against the wild-type parent strain. Suppressors of the drug-induced growth inhibition were identified for 33 of the 49 compounds, pointing immediately to cellular targets of those compounds (growth-inhibitory compounds generally target essential gene products). In one case, a small molecule was identified that targeted LolA, a key protein in lipoprotein trafficking in Gram-negative bacteria. Importantly, no other inhibitors of this protein or process were previously known, and therefore the chemical-genetic interaction between the compound and LolA represented a new phenotype for the *lolA* gene. Although this screen was designed for drug discovery, such chemical-genetic interactions can generate new functional information about the gene, even when its function is already known. Overexpression or high-copy suppressor screens are quite powerful for drug discovery, but can also identify chemical-genetic relationships that open new avenues of experimentation regarding both drug action and gene function.

Natural Strain Approaches

Phenomic analyses have also been conducted on naturally occurring bacterial isolates. While it is more difficult to associate observed phenotypes with specific genetic changes using such isolates, the increasing efficiency of genome sequencing has made the identification of genetic differences between isolates much more routine. As long as these differences are known, phenomic analyses can be extremely powerful.

A recent study by Franz et al. (2011) examined the relative ability of 18 strains of *E. coli* O157 to proliferate in manure-infused soil. The authors were interested in the genetic basis of this phenotype, as survival in that environment is thought to be critical for strains ultimately found to contaminate agricultural crops. Amongst the 18 isolates, the authors observed a large variance in soil proliferation capacity. In parallel to the soil screen, Biolog technology was utilized to assess the metabolic capacity of the strains, and genotyping of known virulence genes was carried out. No correlation was found between the presence/absence of the virulence genes and the ability to proliferate in the soil. However, the authors observed that strains able to grow on several different acid substrates were likely to survive longer in the manure-infused soil than strains unable to grow in the acids. Therefore, it is possible that genes involved in metabolism underlie the soil survival phenotype, and sequencing of candidate loci (or of the entire genomes) may demonstrate this conclusively.

Another recent study by Kadali et al. (2012) utilized a selection followed by Biolog analysis to examine the ability of species able to grow on crude oil to remediate specific hydrocarbons. First, a complex environmental sample from a former oil refinery site was harvested, and a selection was carried out on this sample using crude oil as a complex carbon source. A collection of strains were selected based on their ability to survive the selection, and then Biolog analysis was used to characterize the ability of each selected strain to proliferate on specific hydrocarbons. The results represent significant progress in characterizing strains useful for the bioremediation of hydrocarbons, and mark the first step in drawing phenotypic connections between genes within the selected strains and the survival phenotype.

In vivo Approaches

In vivo phenomic approaches based on transposon mutagenesis have been extremely powerful in studies of infection biology and the human microbiome. Both negative and positive selections have been utilized to identify genes involved in infection, virulence, and colonization. In an STM study of *S. typhimurium*, Hensel et al. (1995) used a negative selection to identify a pathogenicity island coding for a novel type three secretion system (T3SS) essential for intracellular replication. To identify virulence genes, the authors created a library of transposon mutants, with each mutant carrying a unique DNA barcode in the transposon. Mutants were arranged into 96-well plates, and a single pool was created from each plate. Each pool was sampled at the start point and then used to infect two BALB/c mice. After three days of infection, the mice were sacrificed, and spleens were homogenized and plated onto selective media to isolate the transposon mutants. Colonies from this plating were then scraped into a pool, which represented the "selected" population. A DNA-blotting method was used to identify the presence or absence of each DNA barcode in both the starting and selected pools. Mutants lost from the selected pool represented candidate virulence genes. Genomic DNA flanking the insertion sites in these mutants was cloned into a pUC vector and sequenced to identify the mutated genes. This study represented a major advance in *in vivo* genetic selections, and has inspired an entire field of study over the last 15 years (as indicated by nearly 1,000 citations in PubMed). Over time, technological advances have made such approaches even more powerful by allowing quantitative measurement of individual mutants within a pool. Rather than simply examining loss of a mutant from the pool, these more sensitive methods can assess decreased mutant abundance in the pool.

Conversely, it is also possible to employ positive selections to identify mutants that expand in frequency in the host, which could point to adaptive strategies used by the bacteria to establish and maintain infection/ colonization. A recent study by Bianconi et al. (2011) utilized this approach to identify "patho-adaptive" mutations that promote chronic infection by *P. aeruginosa* in a mouse model of cystic fibrosis. The authors used a previously characterized *P. aeruginosa* STM mutant library, and pooled individual mutants to infect the mice. A total of 24 DNA barcode sequences from three unique vector backbones ($24 \times 3 = 72$ unique transposon sequences) were present in each pool. Following infection, mice were observed and scored for establishment of a chronic infection. Those animals in which a chronic infection was established were classified as positives, sacrificed, and lungs were homogenized and plated on selective media to count the CFU and screen for STM mutants. In this study, a previously developed multiplex PCR strategy (Potvin et al. 2003) was used to track the abundance of each unique mutant within the pools. Mutants that expanded in the population were flagged as candidates, and re-screened in subsequent rounds. Eventually, 16 mutants were identified that significantly increased chronic infection relative to wild-type. These mutants impacted a variety of processes, suggesting that "patho-adaptive" mutations promoting chronic infection may decrease motility, alter biofilm formation, or decrease secretion of virulence factors. To extend their analysis of patho-adaptive mutations, the authors sequenced ORFs of the 16 candidate genes in a cohort of early and late CF patient samples. Interestingly, an independent clinical study found mutations in 7/16 of the candidate genes. The convergence of STM and clinical sequence results are strong evidence that the authors have indeed identified "patho-adaptive" strategies used by *P. aeruginosa* to establish and maintain chronic infection in CF airways.

In vivo screens of transposon mutant populations can also focus on understanding the molecular determinants of establishing the symbiotic relationships between the host and its microbiome. A study by Goodman et al. (2009) utilized a library of *B. thetaiotaomicron* transposon mutant strains and germ-free mice to search for candidate genes involved in colonization of the mouse gut. Importantly, the authors utilized short-read sequencing technology and an engineered Mariner transposon to establish an insertion sequencing approach. Such approaches are based on a single transposon without STM. Instead, the genomic DNA flanking the insertion site is used to map and quantify the relative abundance of transposon mutants. This approach allowed the authors to carry out a forward genetic screen by assigning a quantitative measure of abundance to each unique insertion strain in the pool before and after various selections. To search for loci specifically required for *in vivo* colonization, the authors compared all *in vivo* strain fitness measures with a set of *in vitro* fitness measurements calculated from the results of a competition experiment conducted in rich media. More clearly, they used an *in vitro* competition to identify loci that affect growth generally, and therefore were able to filter for strains exhibiting fitness defects specifically *in vivo*. After conducting a series of *in vivo* screens in varying bacterial community contexts, the authors discovered a genomic locus containing several genes involved in vitamin B12 utilization that, when mutated, led to a significant fitness defect only when vitamin B12 was presumed to be of limited availability in the mouse gut. As vitamin B12 is essential for the mouse and is synthesized exclusively by members of the microbiome, these genes are likely to be necessary for normal colonization *in vivo*. This finding demonstrates the enormous potential of *in vivo* phenomic approaches to understand symbiotic relationships between bacteria and their hosts—a topic of broad appeal that has been relatively inaccessible to investigation until now.

Functional Metagenomics

A critical and emerging area of microbiology is the study of complex microbial communities via metagenomic sequencing. Initial studies in this area have focused on surveying the species composition of varying environmental communities. To date, such projects have deposited enormous amounts of genome sequences into databases, many of which code for ORFs of both unknown function and unknown species origin. Such sequences are of extremely limited utility, and therefore approaches aimed at assigning function to them are in high demand.

A recent study by McGarvey et al. (2012) utilized a heterologous expression system to select for metagenomic sequences conferring antibiotic resistance to *E. coli*. Briefly, the authors cloned a metagenomic DNA library into a pUC18 vector, yielding 1.4 million clones covering 2.8×10^9 bases of DNA. By transforming the plasmid library into *E. coli* and simply plating on a variety of antibiotics, the authors were able to search for metagenomic sequences that conferred antibiotic resistance phenotypes on the *E. coli* host. In all, the authors identified 39 antibiotic-resistance genes, of which all but one coded for members of previously defined drug-resistance protein families. However, interestingly, many of the newly discovered members shared little nucleotide sequence homology to the previously described family members. For example, the authors discovered one type II DHFR (conferring resistance to trimethoprim) having a maximum of 24% amino acid sequence identity to other known type II DHFRs. Evolutionary analysis of normally conserved amino acid residues indicates that the newly discovered DHFR is evolutionarily distant. These types of discoveries are incredibly valuable for further defining the bacterial resistome, allowing for a better understanding of drug-protein structure-function relationships, and potentially informing future drug design efforts.

An earlier study by Sommer et al. (2010) utilized a slightly different strategy to screen four different soil microbiomes for genetic elements that conferred resistance to growth-inhibitory compounds in *E. coli*. Similar to the McGarvey study, heterologous expression of the metagenomic DNA was carried out in *E. coli*, and selective pressure consisted of high doses of industrially relevant chemicals (organic acids, alcohols, and aldehydes). This study cloned metagenomic DNA into fosmid libraries, containing 40–50 kb of DNA rather than the 1–3 kb inserts in the pUC18-based approach of McGarvey. Advantages of this approach are that fewer clones need to be screened, and phenotypes based on more than 1 neighboring gene can be identified. Of course, in this approach, follow-up is necessary to determine which gene(s) are responsible for a phenotype of interest. In the Sommer study, identification of candidate genes was obtained by a second transposon mutagenesis on the resistant fosmid clones, followed by re-screening for loss of the phenotype. This approach resulted in discovery of novel genes conferring tolerance to biomass-conversion-related growth inhibitory compounds. Importantly, it offers another example of a phenomic strategy to identify meaningful functional genetic elements from complex (and perhaps unculturable) microbial sources.

Future Perspectives

The tremendous impact of the genomic revolution has inspired the burgeoning field of phenomics. Following the traditional model, the gateway technologies for high-throughput discovery of gene function were developed primarily in the model organisms, and recently have expanded into many applied systems. Going forward, the field will advance on three major fronts: screening readout technology, monoculture species application, and mixed community phenomics.

As more approaches are adapted to sequencing platforms, researchers will have the ability to survey increasing swaths of phenotypic space efficiently. Especially for arrayed approaches, the addition of diverse phenotypic readouts will push the field beyond its current "fitness-centric" state. Transcriptional reporters, indicator dyes for developmental processes, and metabolomic integration represent coming advances, but are likely just the tip of the iceberg.

Sequencing and genetic technologies will open up a growing list of bacterial genes and species for phenomic analyses. The recent demonstrations of phenomic analyses based on antisense RNA should allow LOF studies of essential genes in Gram-positive and Gram-negative species. Further engineering of the Mariner transposons will continue to expand their utility and allow for cutting-edge phenomic studies of many applied and environmental species.

Last, and perhaps most excitingly, phenomics will expand into mixed community applications. Exciting work has already been done examining the ability of genetically variant bacteria to colonize the mouse gut in controlled mixed communities (Goodman et al. 2009). *In vitro* coculture models also present exciting opportunities for phenomics aimed at deciphering complex interspecies and interkingdom relationships.

Overall, the future is extremely bright for phenomics. As genomics technologies continue to inspire and drive the development of functional genomics technologies, phenomic science will take center stage, as rapid and efficient techniques to assign meaning to sequence will be in greater demand than ever before.

References Cited

Amini, S., H. Goodarzi and S. Tavazoie. 2009. Genetic dissection of an exogenously induced biofilm in laboratory and clinical isolates of *E. coli*. PLoS Pathogens 5: e1000432.

Baba, T., T. Ara, M. Hasegawa, Y. Takai, Y. Okumura, M. Baba, K.A. Datsenko, M. Tomita, B.L. Wanner and H. Mori. 2006. Construction of *Escherichia coli* K-12 in-frame, single-gene knockout mutants: the Keio collection. Mol. Syst. Biol. 2: 2006 0008.

Bianconi, I., A. Milani, C. Cigana, M. Paroni, R.C. Levesque, G. Bertoni and A. Bragonzi. 2011. Positive signature-tagged mutagenesis in pseudomonas aeruginosa: tracking patho-adaptive mutations promoting airways chronic infection. PLoS Pathogens 7: e1001270.

Bochner, B.R. 2009. Global phenotypic characterization of bacteria. FEMS Microbiology Reviews 33: 191–205.

Cameron, D.E., J.M. Urbach and J.J. Mekalanos. 2008. A defined transposon mutant library and its use in identifying motility genes in Vibrio cholerae. Proceedings of the National Academy of Sciences 105: 8736.

Collins, S.R., K.M. Miller, N.L. Maas, A. Roguev, J. Fillingham, C.S. Chu, M. Schuldiner, M. Gebbia, J. Recht, M. Shales, H. Ding, H. Xu, J. Han, K. Ingvarsdottir, B. Cheng, B. Andrews, C. Boone, S.L. Berger, P. Hieter, Z. Zhang, G.W. Brown, C.J. Ingles, A. Emili, C.D. Allis, D.P. Toczyski, J.S. Weissman, J.F. Greenblatt and N.J. Krogan. 2007. Functional dissection of protein complexes involved in yeast chromosome biology using a genetic interaction map. Nature 446: 806–810.

Camacho, L.R., D. Ensergueix, E. Perez, B. Gicquel and C. Guilhot. 1999. Identification of a virulence gene cluster of Mycobacterium tuberculosis by signature-tagged transposon mutagenesis. Molecular Microbiology 34(2): 257–267.

Couce, A., A. Briales, A. Rodriguez-Rojas, C. Costas, A. Pascual and J. Blazquez. 2012. Genome-wide overexpression screen for fosfomycin resistance in Escherichia coli: MurA confers clinical resistance at low fitness cost. Antimicrob. Agents Chemother 56(5): 2767–9.

de Berardinis, V., D. Vallenet, V. Castelli, M. Besnard, A. Pinet, C. Cruaud, S. Samair, C. Lechaplais, G. Gyapay, C. Richez, M. Durot, A. Kreimeyer, F. Le Fevre, V. Schachter, V. Pezo, V. Doring, C. Scarpelli, C. Medigue, G.N. Cohen, P. Marliere, M. Salanoubat and J. Weissenbach. 2008. A complete collection of single-gene deletion mutants of Acinetobacter baylyi ADP1. Mol. Syst. Biol. 4: 174.

Deutschbauer, A., M.N. Price, K.M. Wetmore, W. Shao, J.K. Baumohl, Z. Xu, M. Nguyen, R. Tamse, R.W. Davis and A.P. Arkin. 2011. Evidence-based annotation of gene function in Shewanella oneidensis MR-1 using genome-wide fitness profiling across 121 conditions. PLoS Genetics 7: e1002385.

Donald, R.G.K., S. Skwish, R.A. Forsyth, J.W. Anderson, T. Zhong, C. Burns, S. Lee, X. Meng, L. LoCastro, L.W. Jarantow, J. Martín, S.H. Lee, I. Taylor, D. Robbins, C. Malone, L. Wang, C.S. Zamudio, P.J. Youngman and J.W. Phillips. 2009. A Staphylococcus aureus fitness test platform for mechanism-based profiling of antibacterial compounds. Chemistry & Biology 16: 826–836.

Fabich, A.J., M.P. Leatham, J.E. Grissom, G. Wiley, H. Lai, F. Najar, B.A. Roe, P.S. Cohen and T. Conway. 2011. Genotype and phenotypes of an intestine-adapted Escherichia coli K-12 mutant selected by animal passage for superior colonization. Infection and Immunity 79: 2430–2439.

Franz, E., A.H.A.M. van Hoek, E. Bouw and H.J.M. Aarts. 2011. Variability of *Escherichia coli* O157 strain survival in manure-amended soil in relation to strain origin, virulence profile, and carbon nutrition profile. Applied and Environmental Microbiology 77: 8088–8096.

Gallagher, L.A., E. Ramage, M.A. Jacobs, R. Kaul, M. Brittnacher and C. Manoil. 2007. A comprehensive transposon mutant library of Francisella novicida, a bioweapon surrogate. Proc. Natl. Acad. Sci. USA 104: 1009–1014.

Gallagher, L.A., J. Shendure and C. Manoil. 2010. Genome-scale identification of resistance functions in Pseudomonas aeruginosa using Tn-seq. mBio 2: e00315-00310-e00315-00310.

Gawronski, J.D., S.M.S. Wong, G. Giannoukos, D.V. Ward and B.J. Akerley. 2009. Tracking insertion mutants within libraries by deep sequencing and a genome-wide screen for Haemophilus genes required in the lung. Proc. Natl. Acad. Sci. USA 106: 16422–16427.

Gentleman, R. 2005. Bioinformatics and Computational Biology Solutions Using R and Bioconductor. Springer Science+Business Media, New York.

Girgis, H.S., A.K. Hottes and S. Tavazoie. 2009. Genetic architecture of intrinsic antibiotic susceptibility. PLoS One 4: e5629.

Girgis, H.S., Y. Liu, W.S. Ryu and S. Tavazoie. 2007. A comprehensive genetic characterization of bacterial motility. PLoS Genetics 3: e154.

Goodman, A.L., N.P. McNulty, Y. Zhao, D. Leip, R.D. Mitra, C.A. Lozupone, R. Knight and J.I. Gordon. 2009. Identifying genetic determinants needed to establish a human gut symbiont in its habitat. Cell Host and Microbe 6: 279–289.

Hensel, M., J.E. Shea, C. Gleeson, M.D. Jones, E. Dalton and D.W. Holden. 1995. Simultaneous identification of bacterial virulence genes by negative selection. Science 269: 400–403.

Hisert, K.B., M.A. Kirksey, J.E. Gomez, A.O. Sousa, J.S. Cox, W.R. Jacobs, C.F. Nathan and J.D. McKinney. 2004. Identification of Mycobacterium tuberculosis Counterimmune (cim) mutants in immunodeficient mice by differential screening. Infection and Immunity 72: 5315–5321.

Ishii, N., K. Nakahigashi, T. Baba, M. Robert, T. Soga, A. Kanai, T. Hirasawa, M. Naba, K. Hirai, A. Hoque, P.Y. Ho, Y. Kakazu, K. Sugawara, S. Igarashi, S. Harada, T. Masuda, N. Sugiyama, T. Togashi, M. Hasegawa, Y. Takai, K. Yugi, K. Arakawa, N. Iwata, Y.

Toya, Y. Nakayama, T. Nishioka, K. Shimizu, H. Mori and M. Tomita. 2007. Multiple high-throughput analyses monitor the response of E. coli to perturbations. Science 316: 593–597.

Kadali, K.K., K.L. Simons, P.P. Skuza, R.B. Moore and A.S. Ball. 2012. A complementary approach to identifying and assessing the remediation potential of hydrocarbonoclastic bacteria. Journal of Microbiological Methods 88(3): 348–55.

Kitagawa, M., T. Ara, M. Arifuzzaman, T. Ioka-Nakamichi, E. Inamoto, H. Toyonaga and H. Mori. 2005. Complete set of ORF clones of *Escherichia coli* ASKA library (a complete set of E. coli K-12 ORF archive): unique resources for biological research. DNA Research: an international journal for rapid publication of reports on genes and genomes 12: 291–299.

Kohanski, M.A., D.J. Dwyer, J. Wierzbowski, G. Cottarel and J.J. Collins. 2008. Mistranslation of membrane proteins and two-component system activation trigger antibiotic-mediated cell death. Cell 135: 679–690.

Langridge, G.C., M.D. Phan, D.J. Turner, T.T. Perkins, L. Parts, J. Haase, I. Charles, D.J. Maskell, S.E. Peters, G. Dougan, J. Wain, J. Parkhill and A.K. Turner. 2009. Simultaneous assay of every Salmonella Typhi gene using one million transposon mutants. Genome Research 19: 2308–2316.

Lawley, T.D., K. Chan, L.J. Thompson, C.C. Kim, G.R. Govoni and D.M. Monack. 2006. Genome-wide screen for Salmonella genes required for long-term systemic infection of the mouse. PLoS Pathogens 2: e11.

Liberati, N.T., J.M. Urbach, S. Miyata, D.G. Lee, E. Drenkard, G. Wu, J. Villanueva, T. Wei and F.M. Ausubel. 2006. An ordered, nonredundant library of Pseudomonas aeruginosa strain PA14 transposon insertion mutants. Proc. Natl. Acad. Sci. USA 103: 2833–2838.

Maynard, N.D., E.W. Birch, J.C. Sanghvi, L. Chen, M.V. Gutschow and M.W. Covert. 2010. A forward-genetic screen and dynamic analysis of lambda phage host-dependencies reveals an extensive interaction network and a new anti-viral strategy. PLoS Genet. 6: e1001017.

Mazurkiewicz, P., C.M. Tang, C. Boone and D.W. Holden. 2006. Signature-tagged mutagenesis: barcoding mutants for genome-wide screens. Nature Reviews Genetics 7: 929–939.

McGarvey, K.M., K. Queitsch and S. Fields. 2012. Wide variation in antibiotic resistance proteins identified by functional metagenomic screening of a soil DNA library. Applied and Environmental Microbiology 78: 1708–1714.

Meng, J., G. Kanzaki, D. Meas, C.K. Lam, H. Crummer, J. Tain and H.H. Xu. 2012. A genome-wide inducible phenotypic screen identifies antisense RNA constructs silencing *Escherichia coli* essential genes. FEMS Microbiology Letters n/a-n/a 329(1): 45–53.

Nakahigashi, K., Y. Toya, N. Ishii, T. Soga, M. Hasegawa, H. Watanabe, Y. Takai, M. Honma, H. Mori and M. Tomita. 2009. Systematic phenome analysis of *Escherichia coli* multiple-knockout mutants reveals hidden reactions in central carbon metabolism. Mol. Syst. Biol. 5: 306.

Nguyen, B. and R. Valdivia. 2012. Virulence determinants in the obligate intracellular pathogen Chlamydia trachomatis revealed by forward genetic approaches. Proceedings of the National Academy of Sciences 109(4): 1263–8.

Nichols, R.J., S. Sen, Y.J. Choo, P. Beltrao, M. Zietek, R. Chaba, S. Lee, K.M. Kazmierczak, K.J. Lee, A. Wong, M. Shales, S. Lovett, M.E. Winkler, N.J. Krogan, A. Typas and C.A. Gross. 2011. Phenotypic landscape of a bacterial cell. Cell 144: 143–156.

Pathania, R., S. Zlitni, C. Barker, R. Das, D.A. Gerritsma, J. Lebert, E. Awuah, G. Melacini, F.A. Capretta and E.D. Brown. 2009. Chemical genomics in *Escherichia coli* identifies an inhibitor of bacterial lipoprotein targeting. Nature Chemical Biology 5: 849–856.

Potvin, E., D.E. Lehoux, I. Kukavica-Ibrulj, K.L. Richard, F. Sanschagrin, G.W. Lau and R.C. Levesque. 2003a. *In vivo* functional genomics of Pseudomonas aeruginosa for high-throughput screening of new virulence factors and antibacterial targets. Environ. Microbiol. 5: 1294–1308.

Roguev, A., S. Bandyopadhyay, M. Zofall, K. Zhang, T. Fischer, S.R. Collins, H. Qu, M. Shales, H.O. Park, J. Hayles, K.L. Hoe, D.U. Kim, T. Ideker, S.I. Grewal, J.S. Weissman and N.J. Krogan. 2008. Conservation and rewiring of functional modules revealed by an epistasis map in fission yeast. Science 322: 405–410.

Santiviago, C.A., M.M. Reynolds, S. Porwollik, S.H. Choi, F. Long, H.L. Andrews-Polymenis and M. McClelland. 2009b. Analysis of pools of targeted Salmonella deletion mutants identifies novel genes affecting fitness during competitive infection in mice. PLoS Pathog. 5: e1000477.

Schuldiner, M., S.R. Collins, J.S. Weissman and N.J. Krogan. 2006. Quantitative genetic analysis in Saccharomyces cerevisiae using epistatic miniarray profiles (E-MAPs) and its application to chromatin functions. Methods 40: 344–352.

Sommer, M.O., G.M. Church and G. Dantas. 2010. A functional metagenomic approach for expanding the synthetic biology toolbox for biomass conversion. Molecular Systems Biology 6: 1–7.

Tenaillon, O., A. Rodriguez-Verdugo, R.L. Gaut, P. McDonald, A.F. Bennett, A.D. Long and B.S. Gaut. 2012. The molecular diversity of adaptive convergence. Science 335: 457–461.

Typas, A., M. Banzhaf, B. van den Berg van Saparoea, J. Verheul, J. Biboy, R.J. Nichols, M. Zietek, K. Beilharz, K. Kannenberg, M. von Rechenberg, E. Breukink, T. den Blaauwen, C.A. Gross and W. Vollmer. 2010. Regulation of peptidoglycan synthesis by outer-membrane proteins. Cell 143: 1097–1109.

Typas, A., R.J. Nichols, D.A. Siegele, M. Shales, S.R. Collins, B. Lim, H. Braberg, N. Yamamoto, R. Takeuchi, B.L. Wanner, H. Mori, J.S. Weissman, N.J. Krogan and C.A. Gross. 2008. High-throughput, quantitative analyses of genetic interactions in E. coli. Nature methods 5: 781–787.

van Opijnen, T., K.L. Bodi and A. Camilli. 2009. Tn-seq: high-throughput parallel sequencing for fitness and genetic interaction studies in microorganisms. Nat. Methods 6: 767–772.

Warner, J.R., P.J. Reeder, A. Karimpour-Fard, L.B.A. Woodruff and R.T. Gill. 2010. Rapid profiling of a microbial genome using mixtures of barcoded oligonucleotides. Nature Biotechnology 28: 856–862.

10

Phenotype Databases

Philip Groth[1] and *Bertram Weiss[2],***

Introduction

Historical Perspective

Phenotypes have been a subject of research ever since ancient Greek and Roman physicians such as Hippocrates (460–370 BC), Celsus (25 BC–AD 50), and Galen (AD 130–201) took an interest in meticulously describing and studying the human body and associated illnesses with physical causes (Delvey and Barbara 2005). It wasn't until the late 19th century, however, before pioneers such as Charles Darwin (1809–1882) and Gregor Mendel (1822–1884) started to systematically examine evolutionary selection and trait inheritance, setting the scene for modern genetics and phenotype research.

After the rediscovery of Mendel's systematic analyses of phenotypes in 1900 (DeVries 1900, Tschermak 1900), and Walter Sutton's findings that chromosomes form "the physical basis of the Mendelian law of heredity" in 1902 (Sutton 1902), the most fundamental impact on our understanding of phenotypes and their genetic association began in 1972 with the advent of genetic engineering. From the creation of the first recombinant DNA molecules by Paul Berg in 1972 (Nobel Prize awarded in 1980) to its contemporary peak of the discovery of RNA strands that can selectively

---

[1] Therapeutic Research Group Oncology, Bayer Pharma AG, 13353 Berlin, Germany.
[2] Target Discovery Technologies, Bayer Pharma AG, 13353 Berlin, Germany.
* Corresponding author: bertram.weiss@bayer.com

silence genes by Andrew Fire and Craig Mello in 1998 (Nobel Prize awarded jointly in 2006), an era of generating and studying phenotypes through altering the genome and observing the outcome has begun.

But it took 60 years from Sutton's insights plus the discovery of the double helix (Watson and Crick 1953) before answers to questions like, "How much and what genetic information is carried by the X chromosome of man?", were first collected systematically and in larger scale. In 1960, Victor A. McKusick, often referred to as "the father of medical genetics", started publishing findings on mendelian disorders in a collection of annual annotated reviews on Medical Genetics (McKusick 1960). Upon that, a "catalog of X-linked traits" (1962) and the autosomal recessive catalogue (1964) were initiated, culminating in the first edition of the "Mendelian Inheritance in Man—Catalogs of Autosomal Dominant, Autosomal Recessive and X-Linked Phenotypes" in 1966 (Cousin et al. 1998). This book contained in its first edition 1,487 entries divided into the three categories named in the title, autosomal dominant, autosomal recessive, and X-linked. In 1992, an anniversary issue (10th edition) was published in two volumes, comprising 5,710 disease entries on 2,320 pages (McKusick 1992). Over the years, catalogues for Y-linked and mitochondrial phenotypes and genes were added, and by the 12th and to date final edition of 1998, the subtitle had been changed to "A Catalog of Human Genes and Genetic Disorders" (Amberger et al. 2009). Its electronic counterpart, the "Online Mendelian Inheritance in Man" (OMIM) was started in 1987 and became directly linked to the Genome Database (GDB), a prototype repository for gene-mapping data in 1989 (Brandt 1993). It was the first electronic phenotype data repository in history and is today the largest resource of human monogenic diseases.

Just shortly after the release of OMIM, in-kind resources for model organisms followed suit. In the 1980s, scientists at The Jackson Laboratory published the first resource for mouse genetic information, the "Genomic Database of the Mouse" (GBASE). This was developed further and integrated with other in-house resources, culminating in the release of the "Encyclopedia of the Mouse Genome" in 1989, a suite of software tools for browsing mouse genetic data and displaying genetic and cytogenetic maps. Today, the Mouse Genome Database (MGD) is a renowned resource of mouse genotypes and phenotypes.

Durbin and Mieg (1991) presented "ACEDB, a C. elegans database" at a conference and made it publicly available. Their prototype resulted in the creation of WormBase with its first traceable data release (WB5) in 1998.

In 1993, Michael Ashburner first reported on "Flybase, a genome database." The earliest still active release of November 2004 (Dmel 3.2.2), already contained 93,777 entries connecting 28,206 genotypes to 3,074 phenotypes.

In 1994, a group of zebrafish researchers established an on-line database of zebrafish information. Their efforts have resulted in the Zebrafish Information Network (ZFIN). In its first release in 1998, 220 genes were included, mapping to 1,851 phenotypes.

In 2001, another important mouse phenotype resource has entered the scene, the Mouse Phenome Database (MPD), a collaborative effort of inbred mouse strain characteristics. The project originated from a Mouse Strain Characterization Workshop at The Jackson Laboratory in May 1999 and was initiated with data from body weight measurements for 24 inbred strains of mice from the JAX Mice Catalog of 2000.

It is not the scope of this section to list all phenotype repositories and when they have first entered the scene (see Table 1 in another section of this chapter or also (Groth and Weiss 2006) for a more comprehensive list, also of other resources and species). However, the developments outlined above have resulted in a grand total of approximately 500,000 phenotype entries in these resources alone, showing that phenotype data have been collected with great efforts in the past and today form a considerable and highly valuable resource in the field of biomedical science.

Phenotype Data Types

Naturally, 50 years of systematic phenotype studies have driven the creation of phenotype data repositories for many more different kinds of data types, species and purposes than listed above (Groth and Weiss 2006). Along with the phenotypes, we have collected a growing knowledge about the complexity of genetic interactions and the inter-individual genetic variance, also reflected by our daily experience of the high degree of phenotypic individuality in humans. Therefore, to define the mere concept of what comprises a phenotype is difficult. We have ultimately become accustomed to defining a phenotype simply as a *change* in appearance, i.e., as a deviation from the so-called wild-type regarded as an average, hence hypothetical, individual from that species. In consequence, phenotype data can comprise any observable characteristic of an organism, the description of a disease, the characterization of a natural mutation, a result of a gene knockout or knockdown experiment, or an artificially induced mutation at any clinical, cellular, or molecular level. In some cases, even the results of microarray studies are labelled "phenotype". Even if this view is arguable, in case of the newly released Electronic Health Records (I), it is not. Here, yet another complex phenotype has entered the field.

These different levels of granularity of what is being regarded a phenotype by different communities have added to the complexity and number of phenotype data resources, each with their specific scope. Adding to that, a phenotype's genetic origin is of interest as well. In its

Table 1. Overview over major phenotype resources of metazoans.

Organism	Name	URL	Reference
cross-species	**PhenomicDB**—Multi-species Phenotype-Genotype Database	http://www.phenomicdb.de	(Groth et al. 2007)
	OMIA—Online Mendelian Inheritance in Animals	http://omia.angis.org.au	(Lenffer et al. 2006)
	GenomeRNAi—a database for RNAi phenotypes and reagents	http://genomernai.de	(Gilsdorf et al. 2010)
Homo sapiens	**OMIM**—Online Mendelian Inheritance in Man	http://www.omim.org	(McKusick 2007)
	HGMD—Human Gene Mutation Database	http://www.hgmd.cf.ac.uk	(Cooper et al. 2006)
	PharmGKB—Pharmakogenetics and Pharmakogenomics Knowledge Base	http://www.pharmgkb.org	(Altman 2007)
	GenAtlas – Gene and Phenotype database	http://www.genatlas.org	(Frezal 1998)
Rattus norvegicus	**RGD**—Rat Genome Database	http://rgd.mcw.edu	(de la Cruz et al. 2005)
Mus musculus	**MGD**—The Mouse Genome Database	http://www.informatics.jax.org	(Bult et al. 2008)
	MPD—The Mouse Phenome Database	http://www.jax.org/phenome	(Bogue et al. 2007)
Caenorhabditis	**WormBase**—Biology and Genome of *Caenorhabditis elegans*	http://www.wormbase.org	(Rogers et al. 2008)
	RNAiDB—*Caenorhabditis elegans* RNAi Database	http://www.rnai.org	(Gunsalus et al. 2004)
	PhenoBank—*Caenorhabditis elegans* RNAi Screens	http://www.worm.mpi-cbg.de/phenobank/cgi-bin/ProjectInfoPage.py	(Sonnichsen et al. 2005)
Drosophila	**FlyBase**—Database of the *Drosophila* Genome	http://www.flybase.org	(Wilson et al. 2008)
	FlyMine—Database for *Drosophila* Genomics	http://www.flymine.org	(Lyne et al. 2007)
	FlyRNAi—*Drosophila* RNAi Screening Center (DRSC)	http://www.flyRNAi.org	(Flockhart et al. 2006)
Saccharomyces	**PROPHECY**—Profiling of Phenotypic Characteristics in Yeast	http://prophecy.lundberg.gu.se	(Fernandez-Ricaud et al. 2007)
	SGD—*Saccharomyces* Genome Database	http://www.yeastgenome.org	(Hong et al. 2008)
	CYGD—Comprehensive Yeast Genome Database	http://mips.helmholtz-muenchen.de/genre/proj/yeast	(Guldener et al. 2005)
	PomBase—A fission yeast database *Schizosaccharomyces pombe*	http://www.pombase.org	(Wood et al. 2012)
Danio rerio	**ZFIN**—The Zebrafish Model Organism Database	http://www.zfin.org	(Sprague et al. 2008)

"simplest" form, it can be monogenic (i.e., "mendelian"). Alternatively, the disease-causing signal may spread over several loci. For many such multifactorial, "complex" diseases like diabetes, Alzheimer's disease, stroke, psychiatric disorders, or obesity, the complete picture of the genotype-phenotype relationships remains largely unsolved. Here, the contributions of a gene to the disease are usually detected through studies of larger populations (e.g., in genome-wide associations studies, GWAS) and are rather termed "association" or "susceptibility", underlining the lack of a true understanding of the contribution. Further phenotypes may originate from descriptions of the outcome of RNA interference (RNAi) knockdown experiments in model organisms, knockout studies in mice, or the descriptions of genetic diseases of humans in OMIM.

Currently, the phenotype community places particular emphasis on phenotype data from RNAi screens. The concept of introducing one RNA sequence into a cell to knock down one single gene and thus the abundance of its protein derivative at a time is groundbreaking (Tuschl and Borkhardt 2002). Its high specificity (a single base mismatch prevents the silencing effect (Brummelkamp et al. 2002)) makes its application cost-effective and thus gives ground to the possibility of knocking down the expression of mutated alleles, e.g., in cancer and neurodegenerative diseases (Shi 2003). The development of RNAi technologies for mammals (Shi 2003) has enabled generating large amounts of mammalian *in vitro* data (Downward 2004, Kolfschoten et al. 2005, Stephens et al. 2005, Westbrook et al. 2005). As RNAi is applicable at the cellular level, it overcomes the limiting generation time of genetically modified higher mammals (e.g., knockout mice) and has proven to be useful in filling gaps in our understanding of genotype-phenotype relationships (Shi 2003, Tuschl 2003, Tuschl and Borkhardt 2002). The "range of biological readouts that can be used to infer function" (Wheeler et al. 2005) is actually not limited and is one of the most important aspects of large RNAi screens. Due to the availability of whole genome sequences for many model organisms as well as for humans, the number of projects relating phenotypes with genotypes using RNAi is rising steadily (see the review by Friedman and Perrimon (2004)). From the number of RNAi-based phenotypes elucidated in the last few years, a large amount of RNAi-based phenotype data can be extrapolated: for instance, 25 RNAi phenotypic assays in a genome-wide screen (~ 20,000 genes) in eight of the most important model organisms would lead to the accumulation of four million cellular phenotypes.

These low-level cellular "phenotypes" are a blessing and a curse in our understanding of phenotype data. Their abundance in almost all sequenced model organisms, each adding their part to a mosaic, will eventually give a phenotypic picture in unparalleled high resolution. On the other hand, they are very simple, often quantitative data points in such vast amounts but low granularity that makes a comparison to "classical" phenotypes, e.g., from knockout mice or human genetic diseases, notoriously difficult. Yet, they are too valuable to be ignored. The co-existence of qualitative (i.e., "classical" phenotype data) and quantitative phenotype data leads to a heterogeneity that can explain why there have been few attempts to integrate phenotype data, leaving us with a large number of data resources. Given the plethora of species-specific and locus-specific databases with phenotype content, genome-wide and cross-species databases are the next logical step. Large-scale projects are now taking off, like the Mouse Phenome Project (Bogue 2003, Bogue and Grubb 2004) and Eumorphia (Brown et al. 2005). They have started coordinating global efforts to generate standardized phenotype data. In 2009, the European Union granted €12 million in order to support GEN2PHEN, a collaborative genotype-phenotype effort involving 19 institutions across Europe (Brookes 2008), an early effort expected to deliver eventually "holisitic views into genotype-phenotype data". Together with the establishment of a cross-species phenotype ontology (Mungall et al. 2010), the field has clearly started joining efforts and moving into a more collaborative mode to address those issues across species.

Phenotype descriptions as used here comprise all of the concepts named above. In summary, the scope of what can be considered "phenotype data" is poorly defined, if at all, and can range from data at the molecular level to Electronic Health Records at organism and even population level.

Technical Setups of Phenotype Databases

Different technical solutions have evolved for phenotype data, not all of them being actually specifically developed for that purpose but are rather based on adaptations of specific needs for this data type.

One of the first published approaches on how to create a phenotype database was accomplished by ACEDB (http://www.acedb.org). Even though originally developed for WormBase (http://www.sanger.ac.uk/Projects/C_elegans) the database scheme and tools have been generalized to be utilizable for any life science database application from bacteria to man (Durbin and Mieg 1991). Its universal applicability is probably best reflected by its adaptation for ESTHER, a database of sequences, biological annotations and experimental biochemical results related to enzymes of the alpha/beta hydrolase family (Cousin et al. 1998). Its current stable version 4.9.39 is mainly distributed under the GNU General Public License, showing that it is actively developed by a dedicated community with several noteworthy extensions and features, such as its own query language AQL (a new Acedb query language) and interfaces for Java, Perl,

C and CORBA. This activity is also reflected in other developments in the WormBase community, e.g., the Web2.0-like interface of WormBase 2—beta (http://beta.wormbase.org), showing that new developments, like user interactions, workspaces, Blogs and Forums, have long since reached the phenotype communities.

Another user-friendly development that has recently found its way into most phenotype databases is the advent of BioMarts (http://www.biomart.org). In principle, BioMart is a unique open source data federation technology that provides unified access to distributed databases storing a wide range of data (Landsman 2011). Notably, some resources do not directly provide a BioMart of their data, i.e., OMIM, FlyBase, PhenomicDB, and the ZebraFish Information Network (ZFIN). Some of these data are, however, available as DAS tracks in ENSEMBL and therefore available through EnsMart (e.g., ZFIN SNP and indel analyses).

Almost all resources allow for flat-file or XML download of their data (and sometimes also of their database schemes) to enable in-depth data analyses that may not be covered within the offered technical setup.

Phenotype Data Repositories

Overview

As explained above, the list of "phenotype" resources is highly dependent on the view of what should be considered a phenotype. We therefore focus on the following overview to databases with entries adhering to the following minimal concept: Only an observable *change in appearance*, i.e., a visible deviation from the so-called wild-type, will be considered a phenotype. Furthermore, such a change of a wild-type needs to be associated with a traceable change in a genotype, either by mutation, disease or genetic interference (e.g., induced mutation, silencing knockdown, etc.). Still even in this narrow view, the number and scope of phenotype resources is considerable (see Table 1).

We restrict ourselves further to the largest and most mature databases of human phenotypes and the most frequently studied model organisms, i.e., yeast, zebrafish, worm, fruit fly, mouse and rat.

In these resources, the publicly available genotype-phenotype data has increased steadily, especially after large amounts of RNAi data came into the public domain between 2004 and today (see Fig. 1 for a direct comparison of current data availability for different species). The next section gives a more detailed overview over these resources, especially with regard to their data content and typical use cases.

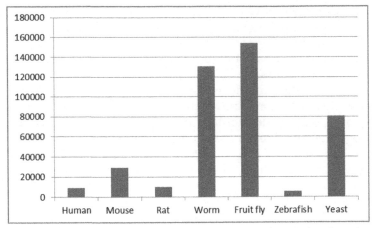

Figure 1. Number of phenotype entries for different species in 2012 (taken from the sources detailed in Table 1).

Single Species

For mice, the current release of the MGI (MGI_4.42 of February 29, 2012) associates 11,678 genes with 710,444 mutant alleles to 29,914 phenotypes and disease models (Bogue and Grubb 2004, Bult et al. 2008, Grubb et al. 2004, MGI 2012, Smith et al. 2005). Furthermore, 2,816 phenotypic measurements in 777 strains are featured in the most recent release (March 2, 2012) of MPD. Most of these data are derived from genetically engineered KO mice or naturally occurring mutants. Notably, a consortium exists to integrate mouse phenotypes from various resources (Hancock et al. 2007).

The Rat Genome Database (RGD) aims at integrating its genomics data with phenome data, covering ~10,000 phenotype or disease annotations for 23,599 rat genes, 1,771 rat and 1,911 human quantitative trait loci (QTLs) and 2,549 rat strains (de la Cruz et al. 2005, Laulederkind et al. 2011, Twigger et al. 2005, Twigger et al. 2008). It also offers a special disease portal presenting integrated entry points for researchers interested in Diabetes, Cardiovascular, Neurological, or Respiratory disorders, as well as Obesity/ Metabolic Syndromes and Cancer.

In its most recent release of WormBase (WB220) (Rogers et al. 2008, Yook et al. 2012), 20,758 entries of 1,203 phenotypes are linked to 7,369 genotypes. Interestingly, there are an additional 81,311 RNAi entries (data counted using WormMart (Schwarz et al. 2006) in March 2012), gathered from public screens (Fraser et al. 2000, Kamath and Ahringer 2003, Kamath et al. 2003, Piano et al. 2002, Rual et al. 2004, Simmer et al. 2003, Walhout et al. 2002) and data resources like PhenoBank (24,671 RNAi phenotypes) (Gonczy et al. 2000, Gonczy et al. 1999, Sonnichsen et al. 2005), making WormBase a rich resource of this novel phenotype. Also an integrative resource for RNAi screens for *Caenorhabditis elegans* is RNAiDB (Gunsalus et al. 2004). In their version 5, this resource contained 59,991 RNAi phenotypes from several screens (Fernandez et al. 2005, Piano et al. 2000, Piano et al. 2002, Sonnichsen et al. 2005). Further updates seem not have taken place since 2007.

Large mutant screens based on different methodologies (reviewed by Carroll et al. (2003)) have led to a rich database for *Drosophila melanogaster*. Flybase has grown significantly in the past few years, now containing 309,148 entries of 71,262 genotypes associated with 93,914 phenotypes (version FB2012_02 of March 2, 2012) (Tweedie et al. 2009, Wilson et al. 2008). Drysdale and Crosby (2005) have given a detailed guide on how to access phenotype data in FlyBase. In contrast to WormBase, RNAi data from genome-wide screens in *Drosophila melanogaster* are being kept separately in FlyRNAi (Flockhart et al. 2006) which is run by the Drosophila RNAi Screening Center (DRSC) at Harvard Medical School where in its version DRSC 2.0, 2,768,614 raw data points have resulted in 29,886 hits from 67 RNAi knockdown screens since 2003 (DRSC 2012).

Data on *Danio rerio* (zebrafish), a helpful model organism, e.g., for angiogenesis, can be found at the Zebrafish Information Network database ZFIN. On March 7, 2012, 52,378 entries link 6,427 genotypes to 500 unique phenotype descriptors (Sprague et al. 2008, ZFIN 2012).

For *Homo sapiens*, the phenotype data resources are more diverse (purely clinical data as a phenotype resource are being omitted here). For a long while, the first address has been OMIM, the Online Mendelian Inheritance in Man database (McKusick 2007), a rich hand-curated free-text catalogue linking human genes to genetic disorders, containing 19,829 entries in its current release (March 12, 2012). OMIM is good at giving excellent textual reviews of the literature but its shallow structure makes the text corpus difficult for automatic parsing into categorized facts. In contrast to OMIM, however, GenAtlas (22,655 genes and 4,869 phenotypes in the February 2012 release) (Frezal 1998, Roux-Rouquie et al. 1999) is rich in gene details and the phenotype data are listed in the "Variant and Pathology" section, which is further subdivided into several well-structured fields. This substructure is also reflected in the query interface allowing narrow and specific filtering. More recently, GWAS studies have been integrated into PheGenI, the Phenotype-Genotype Integrator of the NCBI (http://www.ncbi.nlm.nih.gov/gap/PheGenI). GWAS data from the NHGRI GWAS catalog is integrated with Entrez Gene, dbGaP, OMIM, GTEx and dbSNP. Of the 7,614 GWAS association records, 78% can be mapped to one of only eight disease categories (i.e., Chemicals and Drugs, Digestive System Diseases, Eye Diseases, Immune System Diseases, Mental Disorders, Neoplasms, Nervous System Diseases, Skin and Connective Tissue Diseases) (see: http://www.ichg2011.org/cgi-bin/showdetail.pl?absno=21633). In another large collaborative effort, GEN2PHEN (http://www.gen2phen. org) (Brookes 2008) has since 2009 collected phenotype data on 4,108 genes from a plethora of locus-specific databases.

Also mutation databases often contain interesting phenotype data as long as the link between the mutation and the phenotype has been clearly established. This, however, is true only for a minor part of the entries. Still, these resources provide a valuable data source in the research of phenotypes, especially with regard to complex diseases, such as cancer. Here, the Human Genome Mutation Database (HGMD) (Cooper et al. 2006) has been collecting and manually curating disease-causing mutations for over 20 years and covers 85,840 mutations of 3,253 genes in the public version by now. The extended "HGMD professional release" 2011.4 (license required) covers 120,004 mutations of 4,411 genes (HGMD 2012). Since 2003, somatic mutations have been collected by COSMIC (Catalog of Somatic Mutations in Cancer) of the Wellcome Trust Sanger Institute. COSMIC in its version 57 of January 2012 covers 75,109 unique somatic mutations, 7,428 fusions, and 2,752 structural variants for 217,031 mutated cancer samples (including cell lines). Whereas dbSNP (Wheeler et al. 2008) is the repository for human SNPs, the Human Genome Variation Database (HGVbase) group (Estivill et al. 2008) has announced they will go one step further in the future, annotating these SNPs with their phenotypic consequences (Patrinos and Brookes 2005). HGVbase and dbSNP exchange their SNP content bidirectionally. SNPeffect (Reumers et al. 2008, Reumers et al. 2006), in contrast, tries to predict the effects of SNPs on functional or physicochemical properties of the corresponding proteins. The Pharmacogenetics and Pharmacogenomics Knowledge Base (PharmGKB) compiles data on how genetic variation contributes to variation in drug response (Altman 2007, Hodge et al. 2007). Currently, PharmGKB holds 10,483 publicly accessible entries connecting 838 diseases to 3,806 genes (PharmGKB 2012).

There are additional databases on diseases, genetic variation and phenotypes but they are either too specialized to be mentioned here, or do not fit the constraints defined above, or have restricted access. For further or special interest on such databases, we refer to the extensive review on 1,188 locus-specific databases by Mitropoulou et al. (2010). Others are reviewed elsewhere (Ayme 2000, Claustres et al. 2002, Horaitis and Cotton 2004, Mitropoulou et al. 2010).

Cross Species

Besides OMIA (Lenffer et al. 2006, Nicholas 2003), an animal equivalent of OMIM on a smaller scale, the first cross-species phenotype database on a larger scope was PhenomicDB (Groth et al. 2007, Kahraman et al. 2005). Since its creation in 2004, only very few other explicitly cross-species databases have emerged, e.g., the phenotype tracks in Ensembl (Flicek et al. 2008), and GenomeRNAi (Gilsdorf et al. 2010, Horn et al. 2007). GenomeRNAi is an integrative effort, collecting readouts from RNA interference screens from the public domain across species. Currently (version 8.0, build 1039 of June 5, 2012), 291 screens in *Drosophila melanogaster* and *Homo sapiens* are included. For *Drosophila melanogaster*, the database contains 1,359 unique phenotypic readouts connected to 9,716 unique genes, forming 34,713 unique genotype-phenotype pairs. For *Homo sapiens*, 238 readouts are reported for 10,712 genes, yielding 21,060 unique genotype-phenotype pairs. However, the full potential of cross-species phenotype analysis is leveraged by the knowledge transfer between species through exploiting the concept of orthology, which was implemented for the first time in a systematic manner in PhenomicDB.

The first version of PhenomicDB was created by Kahraman et al. (2005) "in order to remedy" that "there is no integrative system of genotypes and phenotypes across species and screening methods". They gathered data from the different public resources and mapped them semantically into a single data model. The first productive version was released in 2004 and is being regularly updated since then. The publicly available genotype-phenotype data have increased steadily since then, especially by large amounts of RNAi data. To accommodate these large-scale data, PhenomicDB was extended and remodelled moving towards version 2 in 2006 (Groth et al. 2007). This large cross-species phenotype data set was then used to develop a method to gain novel insights from systematic analyses by for example identifying and comparing similar phenotypes. One of the methods is "phenoclustering" introduced in 2008 (Groth et al. 2008), which results in large phenotype networks. This was subsequently implemented as an interactive and graphical web-application within PhenomicDB's frontend in 2010 allowing to explore the phenotype-genotype network intuitively (PhenomicDB version 3) (Groth et al. 2010). As of February 2012, PhenomicDB v. 3.5 contains 443,185 phenotypes, only 2% of which (8,944) are from humans (see Fig. 2).

Use Cases for Phenotype Data

Overview

For a long time, phenotypes have been regarded solely as indicators for changes in genotypes or diseases. The ability to interfere with the genetic component in a systematic manner, e.g., by gene knockout or RNA interference (Hannon 2002, Shi 2003), has raised the importance of phenotypes as a tool to understand biological processes on the molecular level. Even though whole-genome RNAi screens have created large

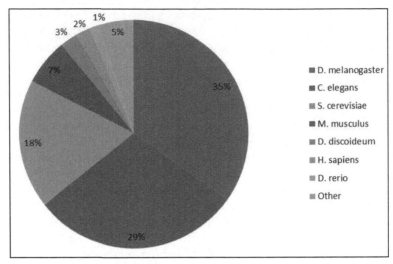

Figure 2. Distribution of phenotype entries in PhenomicDB v. 3.5 by species.

Color image of this figure appears in the color plate section at the end of the book.

amounts of publicly available phenotype data, very few attempts have been reported to systematically analyze these data beyond single gene effects. It is noteworthy that even eight years after the availability of RNAi screens for mammals and many calls for standardization of data types and analysis methods in phenomics (Freimer and Sabatti 2003, Scriver 2004), such data are still poorly organized and difficult to access. Only the Eumorphia project has released standard operating procedures for phenotype screening in mouse and has created PhenoStat, a tool for visualization and systematic statistical analysis of standardized phenotype data (Brown et al. 2005, Stephens et al. 2005).

One of the reasons for this lack of standardization lies in the data heterogeneity. The term "phenotype" in itself is used for a broad variety of concepts which makes an integrated analysis of phenotypes from different experiments and laboratories particularly challenging (see Groth and Weiss (2006)). Another issue was that until very recently, no comprehensive set of phenotypes with associated genes was available. Despite these challenges, several phenotype studies go beyond the analysis of single genotype-phenotype relations. As those use cases not only are of scientific interest but should be the key drivers of phenotype repository development, we have collected some enlightening examples.

Comparative Phenomics

Some of the groundwork of comparative phenomics studies has been laid by Piano et al. (2002) who used manually curated data sets from one RNAi screen to describe a phenotype as the sum of 45 phenotypic features, each represented by either absence or presence calls. They coined the term "phenoclusters" to describe groups of such vectorized phenotypes that "correlate well with sequence-based functional predictions and thus may be useful in predicting functions of uncharacterized genes" (Piano et al. 2002). By this method termed "PhenoBLAST" (Gunsalus et al. 2004), phenotypes can be compared within this data set according to the sum of absence or presence of features in the vector. In its version 5.0 from January 2007, PhenoBLAST supports 191 phenotypic features. Furthermore, the well-structured manually curated *Caenorhabditis elegans* data set of Sonnichsen et al. (2005) was used to create a "disease map", a graphical display of 45 disease categories—like "meiotic arrest"—with values characterizing each category—like "passage through meiosis". Such categories are ideally taken from phenotype ontologies or other adequate vocabularies/ontologies (e.g., functional classes in GO). This profiling system allows using for example bi-clustering to group genes based on their common phenotypic feature patterns as distance measure, associating genes of unknown function directly with specific disease categories. Other clustering methods based on feature vectors have found broad application in the analysis of post-genomic data and are reviewed elsewhere (Handl et al. 2005).

In a study by van Driel et al. (2006) human phenotype descriptions from OMIM were compared. They found that grouping such descriptions reflects biological modules of interacting, functionally related genes. Lage et al. (2007) have developed a phenotype similarity score based on text-mining. They show that 90% of their similar phenotypes are equally found to be similar by human curators. They build a "human phenome-interactome network", integrating interactions of human proteins with this phenotype score for identifying protein complexes ranked as candidates for disease models (Lage et al. 2007).

Gaulton et al. (2007) have developed a computational system to suggest new genes contributing towards a "complex trait" (i.e., a phenotype). They use ontologies and entity recognition to extract genes and proteins from phenotype descriptions and rank them to corresponding biological data from online resources. Butte and Kohane (2006) clustered keywords from the Unified Medical Language System (UMLS) annotated to gene expression data and interpret the resulting connection between these terms and the associated genes (termed "phenome-genome network").

Using phenotype data for more than annotation prediction, Eggert et al. (2004) compared phenotypes from RNAi as well as chemical genetic screens to find genes responsible for the same cellular phenotype. Thus, they could identify new members of known pathways as well as small molecules with an effect on the very same pathway nicely showing the potential of comparative phenomics.

Prank et al. (2005) have compared methods to determine phenotype-genotype relationships in order to predict genetic alterations that lead to adrenal hyperplasia from complex biochemical data. Using serum level profiles of steroid intermediates from 54 patients with heterozygous 21-hydroxylase (CYP21B) mutations versus healthy controls, they compared traditional clinical methods, traditional linear discriminant analysis, support vector machines (SVMs) and nonlinear methods, i.e., artificial neural networks, and k-nearest neighbour (kNN) classifiers. They showed that the nonlinear statistical analyses performed with an accuracy of up to 83%, in contrast to prediction accuracy by clinical methods of 39% and of 64% by classical linear analysis.

Generally, in order to classify phenotype data based on vectorization of their phenotypic profile as illustrated in the above examples, various supervised machine learning methods are available. The kNN classification maintains a set of training cases in predefined classes (clusters) where each data point is nearest to the mean feature vector of that class. For a test case, the k nearest data points are computed and this new point is allocated to a class, depending on the prior classification of these k points by majority vote (Chaudhuri et al. 1993). Artificial neural networks (ANN) are an extension of the standard k-means clustering procedure and take into account a "neighbourhood ranking" of the nearest vectors. The dynamic neighbourhood ranking takes place during an input-driven adaptation procedure of the reference vectors (Martinetz et al. 1993). SVMs realize pattern recognition between two classes by finding a decision function (hyperplane) determined by selected points from the training data, termed support vectors. In general, this hyperplane corresponds to a linear decision boundary in the input space. While traditional techniques for pattern recognition are based on minimizing the empirical risk (i.e., on the attempt to optimize the performance of the training set), SVMs minimize the structural risk (i.e., the probability of yet-to-be-seen patterns to be classified correctly for an unknown probability distribution of the data) (Vapnik and Chapelle 2000). These machine-learning approaches can grasp well the typically nonlinear nature of the underlying complex genetic interactions by learning from a training set. For example, Rodin et al. (2005) have applied Bayesian belief networks to phenotype data consisting of plasma apolipoprotein E (apoE) levels from 702 African-Americans and 854 non-Hispanic whites. From 72 individuals, 20 variable sites in the apoE gene were included in the belief networks. Three SNPs could be singled out as most likely responsible for plasma apoE levels. This method can be used to reduce the number of candidates in an association study for a phenotype of interest, provided that reliable phenotype data are at hand. A belief network's topology shows a graphical relationship among variables (here SNPs and genes), or nodes, thus showing which variables are dependent on other variables or conditionally independent of them. Edges connecting nodes are therefore undirected and indicate dependence. The edge strength indicates the relative magnitude of the dependency between two variables, given the other interrelationships within the network. It therefore reflects a joint probability distribution among the nodes. Conveniently, an edge between two SNPs also indicates linkage disequilibrium. By employing this approach, Rodin et al. (2005) could "simultaneously take into account linkage disequilibrium while performing genotype–phenotype association analyses".

Clare and King (2002) have applied supervised machine learning methods to the problem of predicting the functional classes of genes in *Saccharomyces cerevisiae* from phenotypic growth data. The data are combined from three different sources (TRIPLES (Kumar et al. 2002), EUROFAN (Dujon 1998) and MIPS (Mewes et al. 2008)) and represented as a vectorization of attributes (growth medium) and values (observed sensitivity or resistance of the mutant compared to the wild type). The classes were assigned from the MIPS functional catalogue. The accuracy of the learned rules was then estimated using phenotype data from deletion mutants of genes with known function. Eventually, Clare and King (2002) could predict the function of 83 genes of hitherto unknown function with an estimated precision of at least 80%.

Troyanskaya et al. (2003) developed MAGIC (Multisource Association of Genes by Integration of Clusters) as a general and flexible probabilistic framework to combine heterogeneous data sets for integrated analysis based on Bayesian networks. To illustrate its utility, clusters of *Saccharomyces cerevisiae* genes were formed using data about genetic and physical interactions, microarray, and transcription factor binding sites with methods like k-means clustering, self-organizing maps, and hierarchical clustering. For these clusters, MAGIC created a posterior belief for whether a gene pair has a functional relationship, identifying a cluster of genes involved in ubiquitin-dependent protein catabolism, which provides "potential functional annotation for an ORF present in that cluster (YGL004C), and confirms the recently added annotation for YNL311C". Furthermore, they discovered a gene group involved in protein biosynthesis. In this cluster, 49 genes already annotated as involved in protein biosynthesis were found as well as 10 unknown genes.

In order to better understand *Caenorhabditis elegans* embryogenesis at a systems level, a large-scale integrative approach has been employed by Gunsalus et al. (2005). Data from protein interactions, gene expression clusters, and phenotypic RNAi profile similarities were incorporated to model one large gene/protein network said to have "a high predictive value for novel gene functions". To integrate three different types of functional relationships, graphs were built representing 661 embryogenesis genes as nodes connected by edges suggested by any evidence from the three data sets. Integration was accomplished by finding correlation among pairs of the same nodes in the different graphs. This last high-profile multi-source approach gives a first taste of the emerging power of comparative phenomics.

Almost all approaches described above have in common that they work either with smaller sets of phenotype data (usually only one data set from one screen) or with a large but very unspecific set of "phenotypes" (such as a selected subset of Medline abstracts or ontology terms). To go beyond these limitations was the next logical step in comparative phenomics. We have therefore undertaken a study where we use a large set of phenotype data from PhenomicDB to predict gene function (Groth et al. 2008). To this end, we have used text clustering of phenotype descriptions ("phenoclusters") to group genes. We could show these clusters correlate well with several indicators for biological coherence in gene groups, such as functional annotations from the Gene Ontology (GO) and protein-protein interactions. Predicting gene function then has been done by transferring annotations from well-annotated genes to other, less-characterized genes in the same cluster. For a subset of groups selected by applying objective criteria, we were able to predict GO-term annotations from the biological process sub-ontology with up to 72.6% precision and 16.7% recall. We have then made available this system as an integrative part of PhenomicDB (Groth et al. 2010), providing a tool to mine its wealth of information.

Our phenoclusters and other comparative phenomics approaches have in common the employment of similarity metrics across phenotypes. To this end, we have utilized term-frequency counts. Other frequently published similarity measures utilize ontologies, taking advantage of the hierarchy of terms within an ontology. Recently, Espinosa and Hancock (2011) have shown that genotype-phenotype associations can be represented in phenotype networks using the Mammalian Phenotype ontology. Further similar efforts have been reviewed elsewhere (Schofield et al. 2010).

Pathway Reconstruction

Another interesting field in phenotype analysis is pathway reconstruction. From only microarray expression profiles ("global transcriptional phenotypes"), groups have successfully used epistasis analysis to reconstruct topologies of pathways in organisms such as *Dictyostelium discoideum* (Van

Driessche et al. 2005) or *Saccharomyces cerevisiae* (van de Peppel et al. 2005). To that end, double mutants are generated and the distance of all their expression profiles to each of the profiles of the corresponding single mutants is determined. The single mutant closer to the double mutant is topologically downstream from the other single mutant. However, full reconstruction of pathways with components not transcriptionally regulated is only feasible if additionally external interventions such as RNAi or gene knockouts are applied and used as "single-gene phenotypes" as shown by Markowetz et al. (2005).

Prodiotion of Phenotypes

The first to predict phenotypes were Famili et al. (2003), conducting an analysis of a genome-scale metabolic network from *Saccharomyces cerevisiae*. Computations of functions within such a network were consistent with observed phenotypic functions for 70–80% of the considered conditions.

In a more comprehensive study, Lee et al. (2008) have built networks of genes essential for viability in *Caenorhabditis elegans* for which edges indicate the probability of involvement in the same biological process. To calculate this functional relationship probability, they integrate gene expression profiles, physical or genetic interactions, literature-mined associations, functional associations and co-inherited or operon-related homologs, building an integrated network model from all of these relationships. They then examined 43 different loss-of-function phenotypes from genome-wide RNAi screens and their responsible genes and ranked connected genes within their network by connectivity as candidates for the same phenotype. They have shown that 29 of the 43 phenotypes can be predicted with high accuracy, another 10 with accuracy better than random. The network and data are freely accessible in WormNet (currently version 2), covering 15,139 genes with 999,367 linkages (75% of all protein-coding loci of WB170) (Lee et al. 2008, Lee et al. 2010).

Summary and Outlook

The development and status of phenomics—defined here as the field of collecting and comparing systematically larger sets of phenotypes typically crossing the species boundary—is progressing slowly but steadily. Only in the last decade, phenomics has gained momentum mainly with the availability of complete genomes for a given species in combination with higher throughput methods to knock-in, -down, or -out a gene and observe the resulting phenotype. Given the appearance of more and more phenotypes per species, researchers have started not only dealing with a few phenotypes at a time but also investigating systematically how larger sets of phenotypes can be compared to each other within a species but also across species and how knowledge can be gained from such analyses (Groth et al. 2010, Groth et al. 2011, Groth et al. 2008, Lage et al. 2007, Schulz et al. 2011).

Beyond accumulating ever more data in the databases, the advent of phenotype ontologies has started formalizing the process of describing phenotypes in a more systematic manner easing large-scale analysis. Here the mouse (and also the Drosophila) community with the mammalian phenotype ontology (MPO) (Smith et al. 2005) and the phenotypic quality ontology (PATO) (Gkoutos et al. 2005) have been at the forefront. In consequence, systematic phenotype analyses are found to be most abundant for mouse whereas the other (mostly species-focused) communities are still struggling with the lack of systematic annotation.

Surprisingly, human phenotype repositories and their analysis, despite their importance to man, are less advanced. It appears to the authors that this is mainly due to the fact that the communities dealing with human phenotypes (i.e., the medical sector) have been traditionally separated from the human genomics community whereas within the much smaller model organism communities, researchers have collaborated more closely from the beginning. Electronic Health Records albeit being a huge natural data collection of human phenotypes have been and still continue to be difficult to access and to analyze systematically (Roque et al. 2011). Reasons are multiple, ranging from logistics to collect the data from so many different places, ethical issues, different languages, lack of controlled vocabulary and in many countries a strong resistance to providing access to such sensitive personal data. In light of those obstacles, it is obvious that despite some smaller successes, a solution cannot be expected near-term. Many of these issues do not exist in the "model organism world" where English is the main scientific language, access to the data is readily available and ethical issues or data sensitivity is of little relevance. The "human phenotype community", however, currently deals with a lack of resources, with OMIM (a hardly structured and rather free-text database for human diseases with a genetic background) being the only human genotype-phenotype repository we know of. This lack of adequate systematic resources for human phenotypes does not meet the current needs and it is actually surprising that the mouse resources are more advanced than the human repositories. Hence, inferring data from mouse to man is the best pursued option (Chen et al. 2012).

Mapping of disease, anatomy, or phenotype (ontology) terms between species is a further hurdle currently being tackled by groups involved in phenomics. Uberon and the cross-species phenotype ontology are good examples of such efforts (Mungall et al. 2010, Mungall et al. 2012). The development of a human phenotype ontology (Robinson and Mundlos 2010) has started in 2008 and is still in its infancy but this effort has joined forces with members of the mouse community not only to profit from each other but also to make resources more interoperable. Improving the human phenotype resources requires as a first step the development of ontologies and controlled vocabularies but later also a better exploitation of the "written output" from the medical sector (EHRs) (Roque et al. 2011) and rigorous data collection in an integrative manner. This will pay off by developments improving for example medical expert systems aiding diagnostics (Kohler et al. 2009).

In general, the fuzzy concept of phenotype as outlined in the introduction hinders the formation of a clear "phenotype community" taking responsibility to further advance the field. Communities restricted to a model organism can typically benefit from genetic engineering and shorter life cycle times, allowing more efficient elaboration of a genotype-phenotype relationship. As a consequence they take the lead in developing systematic approaches in phenomics. A big deal of these efforts is to understand how organisms do work, with the goal to understand how the human organism works. To eventually help patients, at least an equally strong human phenotype community should join forces from both the medical and the genomics side to tackle the fuzziness of what is called a phenotype.

References Cited

Altman, R.B. 2007. PharmGKB: a logical home for knowledge relating genotype to drug response phenotype. Nat. Genet. 39: 426.

Amberger, J., C.A. Bocchini, A.F. Scott and A. Hamosh. 2009. McKusick's Online Mendelian Inheritance in Man (OMIM). Nucleic Acids Res. 37: D793–D796.

Ayme, S. 2000. Bridging the gap between molecular genetics and metabolic medicine: access to genetic information. Eur J Pediatr 159 Suppl 3: S183–S185.

Bogue, M. 2003. Mouse phenome project: understanding human biology through mouse genetics and genomics. J Appl Physiol 95: 1335–1337.

Bogue, M.A. and S.C. Grubb. 2004. The mouse phenome project. Genetica 122: 71–74.

Bogue, M.A., S.C. Grubb, T.P. Maddatu and C.J. Bult. 2007. Mouse Phenome Database (MPD). Nucleic Acids Res. 35: D643–D649.

Brandt, K.A. 1993. The GDB Human Genome Data Base: a source of integrated genetic mapping and disease data. Bull Med Libr Assoc 81: 285–292.

Brookes, A.J. 2008. Major European Project to create new Knowledgebase of Gene-Disease Relationships. http://www.gen2phen.org/docs/PressRelease_FINAL.pdf. Accessed: March 2012.

Brown, S.D., P. Chambon and M.H. de Angelis. 2005. EMPReSS: standardized phenotype screens for functional annotation of the mouse genome. Nat. Genet. 37: 1155.

Brummelkamp, T.R., R. Bernards and R. Agami. 2002. A system for stable expression of short interfering RNAs in mammalian cells. Science 296: 550–553.

Bult, C.J., J.T. Eppig, J.A. Kadin, J.E. Richardson and J.A. Blake. 2008. The Mouse Genome Database (MGD): mouse biology and model systems. Nucleic Acids Res. 36: D724–D728.

Butte, A.J. and I.S. Kohane. 2006. Creation and implications of a phenome-genome network. Nat. Biotechnol 24: 55–62.

Carroll, P.M., B. Dougherty, P. Ross-Macdonald, K. Browman and K. FitzGerald. 2003. Model systems in drug discovery: chemical genetics meets genomics. Pharmacol Ther 99: 183–220.

Chaudhuri, B.B., N. Sarkar and P. Kundu. 1993. Improved fractal geometry based texture segmentation technique. IEE Proc. E 140: 233–241.

Chen, C.K., C.J. Mungall, G.V. Gkoutos, S.C. Doelken, S. Kohler, B.J. Ruef, C. Smith, M. Westerfield, P.N. Robinson, S.E. Lewis, P.N. Schofield and D. Smedley. 2012. MouseFinder: candidate disease genes from mouse phenotype data. Hum Mutat. 33(5): 858–66.

Clare, A. and R.D. King. 2002. Machine learning of functional class from phenotype data. Bioinformatics 18: 160–166.

Claustres, M., O. Horaitis, M. Vanevski and R.G. Cotton. 2002. Time for a unified system of mutation description and reporting: a review of locus-specific mutation databases. Genome Res. 12: 680–688.

Cooper, D.N., P.D. Stenson and N.A. Chuzhanova. 2006. The Human Gene Mutation Database (HGMD) and its exploitation in the study of mutational mechanisms. Curr. Protoc. Bioinform. 39:1.13.1–1.13.20.

Cousin, X., T. Hotelier, K. Giles, J.P. Toutant and A. Chatonnet. 1998. aCHEdb: the database system for ESTHER, the alpha/beta fold family of proteins and the Cholinesterase gene server. Nucleic Acids Res. 26: 226–228.

de la Cruz, N., S. Bromberg, D. Pasko, M. Shimoyama, S. Twigger, J. Chen, C.F. Chen, C. Fan, C. Foote, G.R. Gopinath, G. Harris, A. Hughes, Y. Ji, W. Jin, D. Li, J. Mathis, N. Nenasheva, J. Nie, R. Nigam, V. Petri, D. Reilly, W. Wang, W. Wu, A. Zuniga-Meyer, L. Zhao, A. Kwitek, P. Tonellato and H. Jacob. 2005. The Rat Genome Database (RGD): developments towards a phenome database. Nucleic Acids Res. 33: D485–D491.

Delvey, J. and R.H. Barbara. 2005. Hippocrates. Virtual Museum San Jose State University. http://www.biologie.uni-hamburg.de/b-online/library/history/hippoc.html. Accessed: April 2012.

DeVries, H. 1900. Sur la loi de disjonction des Hybrides. Comptes Rendus de l'Académie des Sciences 130: 845–847.

Downward, J. 2004. Use of RNA interference libraries to investigate oncogenic signalling in mammalian cells. Oncogene 23: 8376–8383.

DRSC. 2012. DRSC: Completed Screens by Topic. http://flyrnai.org/DRSC-PTO.html. Accessed: March 2012.

Drysdale, R.A. and M.A. Crosby. 2005. FlyBase: genes and gene models. Nucleic Acids Res. 33: D390–D395.

Dujon, B. 1998. European Functional Analysis Network (EUROFAN) and the functional analysis of the Saccharomyces cerevisiae genome. Electrophoresis 19: 617–624.

Durbin, R. and J.T. Mieg. 1991. ACEDB—A C. elegans Database. Documentation, code and data available from anonymous FTP servers. lirmm.lirmm.fr, cele.mrc-lmb.cam.ac.uk and ncbi.nlm.nih.gov.

Eggert, U.S., A.A. Kiger, C. Richter, Z.E. Perlman, N. Perrimon, T.J. Mitchison and C.M. Field. 2004. Parallel chemical genetic and genome-wide RNAi screens identify cytokinesis inhibitors and targets. PLoS Biol. 2: e379.

Espinosa, O. and J.M. Hancock. 2011. A gene-phenotype network for the laboratory mouse and its implications for systematic phenotyping. PLoS One 6: e19693.

Estivill, X., N.J. Cox, S.J. Chanock, P.Y. Kwok, S.W. Scherer and A.J. Brookes. 2008. SNPs meet CNVs in genome-wide association studies: HGV2007 meeting report. PLoS Genet. 4: e1000068.

Famili, I., J. Forster, J. Nielsen and B.O. Palsson. 2003. Saccharomyces cerevisiae phenotypes can be predicted by using constraint-based analysis of a genome-scale reconstructed metabolic network. Proc. Natl. Acad. Sci. USA 100: 13134–13139.

Fernandez, A.G., K.C. Gunsalus, J. Huang, L.S. Chuang, N. Ying, H.L. Liang, C. Tang, A.J. Schetter, C. Zegar, J.F. Rual, D.E. Hill, V. Reinke, M. Vidal and F. Piano. 2005. New genes with roles in the C. elegans embryo revealed using RNAi of ovary-enriched ORFeome clones. Genome Res. 15: 250–259.

Fernandez-Ricaud, L., J. Warringer, E. Ericson, K. Glaab, P. Davidsson, F. Nilsson, G.J. Kemp, O. Nerman and A. Blomberg. 2007. PROPHECY—a yeast phenome database, update 2006. Nucleic Acids Res. 35: D463–D467.

Flicek, P., B.L. Aken, K. Beal, B. Ballester, M. Caccamo, Y. Chen, L. Clarke, G. Coates, F. Cunningham, T. Cutts, T. Down, S.C. Dyer, T. Eyre, S. Fitzgerald, J. Fernandez-Banet, S. Graf, S. Haider, M. Hammond, R. Holland, K.L. Howe, K. Howe, N. Johnson, A. Jenkinson, A. Kahari, D. Keefe, F. Kokocinski, E. Kulesha, D. Lawson, I. Longden, K. Megy, P. Meidl, B. Overduin, A. Parker, B. Pritchard, A. Prlic, S. Rice, D. Rios, M. Schuster, I. Sealy, G. Slater, D. Smedley, G. Spudich, S. Trevanion, A.J. Vilella, J. Vogel, S. White, M. Wood, E. Birney, T. Cox, V. Curwen, R. Durbin, X.M. Fernandez-Suarez, J. Herrero, T.J. Hubbard, A. Kasprzyk, G. Proctor, J. Smith, A. Ureta-Vidal and S. Searle. 2008. Ensembl 2008. Nucleic Acids Research 36: D707–D714.

Flockhart, I., M. Booker, A. Kiger, M. Boutros, S. Armknecht, N. Ramadan, K. Richardson, A. Xu, N. Perrimon and B. Mathey-Prevot. 2006. FlyRNAi: the Drosophila RNAi screening center database. Nucleic Acids Res. 34: D489–D494.

Fraser, A.G., R.S. Kamath, P. Zipperlen, M. Martinez-Campos, M. Sohrmann and J. Ahringer. 2000. Functional genomic analysis of C. elegans chromosome I by systematic RNA interference. Nature 408: 325–330.

Freimer, N. and C. Sabatti. 2003. The human phenome project. Nat. Genet. 34: 15–21.

Frezal, J. 1998. Genatlas database, genes and development defects. C R Acad. Sci. III 321: 805–817.

Friedman, A. and N. Perrimon. 2004. Genome-wide high-throughput screens in functional genomics. Curr. Opin Genet. Dev 14: 470–476.

Gaulton, K.J., K.L. Mohlke and T.J. Vision. 2007. A computational system to select candidate genes for complex human traits. Bioinformatics 23: 1132–1140.

Gilsdorf, M., T. Horn, Z. Arziman, O. Pelz, E. Kiner and M. Boutros. 2010. GenomeRNAi: a database for cell-based RNAi phenotypes. 2009 update. Nucleic Acids Res. 38: D448–D452.

Gkoutos, G.V., E.C. Green, A.M. Mallon, J.M. Hancock and D. Davidson. 2005. Using ontologies to describe mouse phenotypes. Genome Biol. 6: R8.

Gonczy, P., C. Echeverri, K. Oegema, A. Coulson, S.J. Jones, R.R. Copley, J. Duperon, J. Oegema, M. Brehm, E. Cassin, E. Hannak, M. Kirkham, S. Pichler, K. Flohrs, A. Goessen, S. Leidel, A.M. Alleaume, C. Martin, N. Ozlu, P. Bork and A.A. Hyman. 2000. Functional genomic analysis of cell division in C. elegans using RNAi of genes on chromosome III. Nature 408: 331–336.

Gonczy, P., H. Schnabel, T. Kaletta, A.D. Amores, T. Hyman and R. Schnabel. 1999. Dissection of cell division processes in the one cell stage Caenorhabditis elegans embryo by mutational analysis. J Cell Biol. 144: 927–946.

Groth, P., I. Kalev, I. Kirov, B. Traikov, U. Leser and B. Weiss. 2010. Phenoclustering: online mining of cross-species phenotypes. Bioinformatics 26: 1924–1925.

Groth, P., U. Leser and B. Weiss. 2011. Phenotype mining for functional genomics and gene discovery. Methods Mol. Biol. 760: 159–173.

Groth, P., N. Pavlova, I. Kalev, S. Tonov, G. Georgiev, H.D. Pohlenz and B. Weiss. 2007. PhenomicDB: a new cross-species genotype/phenotype resource. Nucleic Acids Res. 35: D696–D699.

Groth, P. and B. Weiss. 2006. Phenotype data: a neglected resource in biomedical research? Current Bioinformatics 1: 347–358.

Groth, P., B. Weiss, H.D. Pohlenz and U. Leser. 2008. Mining phenotypes for gene function prediction. BMC Bioinformatics 9: 136.

Grubb, S.C., G.A. Churchill and M.A. Bogue. 2004. A collaborative database of inbred mouse strain characteristics. Bioinformatics 20: 2857–2859.

Guldener, U., M. Munsterkotter, G. Kastenmuller, N. Strack, J. van Helden, C. Lemer, J. Richelles, S.J. Wodak, J. Garcia-Martinez, J.E. Perez-Ortin, H. Michael, A. Kaps, E. Talla, B. Dujon, B. Andre, J.L. Souciet, J. De Montigny, E. Bon, C. Gaillardin and H.W. Mewes. 2005. CYGD: the Comprehensive Yeast Genome Database. Nucleic Acids Res. 33: D364–D368.

Gunsalus, K.C., H. Ge, A.J. Schetter, D.S. Goldberg, J.D. Han, T. Hao, G.F. Berriz, N. Bertin, J. Huang, L.S. Chuang, N. Li, R. Mani, A.A. Hyman, B. Sonnichsen, C.J. Echeverri, F.P. Roth, M. Vidal and F. Piano. 2005. Predictive models of molecular machines involved in Caenorhabditis elegans early embryogenesis. Nature 436: 861–865.

Gunsalus, K.C., W.C. Yueh, P. MacMenamin and F. Piano. 2004. RNAiDB and PhenoBlast: web tools for genome-wide phenotypic mapping projects. Nucleic Acids Res. 32: D406–D410.

Hancock, J.M., N.C. Adams, V. Aidinis, A. Blake, M. Bogue, S.D. Brown, E.J. Chesler, D. Davidson, C. Duran, J.T. Eppig, V. Gailus-Durner, H. Gates, G.V. Gkoutos, S. Greenaway, M. Hrabe de Angelis, G. Kollias, S. Leblanc, K. Lee, C. Lengger, H. Maier, A.M. Mallon, H. Masuya, D.G. Melvin, W. Muller, H. Parkinson, G. Proctor, E. Reuveni, P. Schofield, A. Shukla, C. Smith, T. Toyoda, L. Vasseur, S. Wakana, A. Walling, J. White, J. Wood and M. Zouberakis. 2007. Mouse Phenotype Database Integration Consortium: integration [corrected] of mouse phenome data resources. Mamm Genome 18: 157–163.

Handl, J., J. Knowles and D.B. Kell. 2005. Computational cluster validation in post-genomic data analysis. Bioinformatics 21: 3201–3212.

Hannon, G.J. 2002. RNA interference. Nature 418: 244–251.

HGMD. 2012. Number of entries in HGMD by type. http://www.hgmd.cf.ac.uk/ac/index.php. Accessed: March 2012.

Hodge, A.E., R.B. Altman and T.E. Klein. 2007. The PharmGKB: integration, aggregation, and annotation of pharmacogenomic data and knowledge. Clin Pharmacol Ther 81: 21–24.

Hong, E.L., R. Balakrishnan, Q. Dong, K.R. Christie, J. Park, G. Binkley, M.C. Costanzo, S.S. Dwight, S.R. Engel, D.G. Fisk, J.E. Hirschman, B.C. Hitz, C.J. Krieger, M.S. Livstone, S.R. Miyasato, R.S. Nash, R. Oughtred, M.S. Skrzypek, S. Weng, E.D. Wong, K.K. Zhu, K. Dolinski, D. Botstein and J.M. Cherry. 2008. Gene Ontology annotations at SGD: new data sources and annotation methods. Nucleic Acids Res. 36: D577–D581.

Horaitis, O. and R.G. Cotton. 2004. The challenge of documenting mutation across the genome: the human genome variation society approach. Hum Mutat 23: 447–452.

Horn, T., Z. Arziman, J. Berger and M. Boutros. 2007. GenomeRNAi: a database for cell-based RNAi phenotypes. Nucleic Acids Res. 35: D492–D497.

Kahraman, A., A. Avramov, L.G. Nashev, D. Popov, R. Ternes, H.D. Pohlenz and B. Weiss. 2005. PhenomicDB: a multi-species genotype/phenotype database for comparative phenomics. Bioinformatics 21: 418–420.

Kamath, R.S. and J. Ahringer. 2003. Genome-wide RNAi screening in Caenorhabditis elegans. Methods 30: 313–321.

Kamath, R.S., A.G. Fraser, Y. Dong, G. Poulin, R. Durbin, M. Gotta, A. Kanapin, N. Le Bot, S. Moreno, M. Sohrmann, D.P. Welchman, P. Zipperlen and J. Ahringer. 2003. Systematic functional analysis of the Caenorhabditis elegans genome using RNAi. Nature 421: 231–237.

Kohler, S., M.H. Schulz, P. Krawitz, S. Bauer, S. Dolken, C.E. Ott, C. Mundlos, D. Horn, S. Mundlos and P.N. Robinson. 2009. Clinical diagnostics in human genetics with semantic similarity searches in ontologies. Am J Hum Genet. 85: 457–464.

Kolfschoten, I.G., B. van Leeuwen, K. Berns, J. Mullenders, R.L. Beijersbergen, R. Bernards, P.M. Voorhoeve and R. Agami. 2005. A genetic screen identifies PITX1 as a suppressor of RAS activity and tumorigenicity. Cell 121: 849–858.

Kumar, A., K.H. Cheung, N. Tosches, P. Masiar, Y. Liu, P. Miller and M. Snyder. 2002. The TRIPLES database: a community resource for yeast molecular biology. Nucleic Acids Res. 30: 73–75.

Lage, K., E.O. Karlberg, Z.M. Storling, P.I. Olason, A.G. Pedersen, O. Rigina, A.M. Hinsby, Z. Tumer, F. Pociot, N. Tommerup, Y. Moreau and S. Brunak. 2007. A human phenome-interactome network of protein complexes implicated in genetic disorders. Nature Biotechnology 25: 309–316.

Landsman, D. 2011. Database—BioMart Virtual Issue. Journal of Biological Databases and Curation. http://www.oxfordjournals.org/our_journals/databa/biomart_virtual_issue. html. Accessed: March 2012.

Laulederkind, S.J., M. Shimoyama, G.T. Hayman, T.F. Lowry, R. Nigam, V. Petri, J.R. Smith, S.J. Wang, J. de Pons, G. Kowalski, W. Liu, W. Rood, D.H. Munzenmaier, M.R. Dwinell, S.N. Twigger and H.J. Jacob. 2011. The Rat Genome Database curation tool suite: a set of optimized software tools enabling efficient acquisition, organization, and presentation of biological data. Database (Oxford) 2011: bar002.

Lee, I., B. Lehner, C. Crombie, W. Wong, A.G. Fraser and E.M. Marcotte. 2008. A single gene network accurately predicts phenotypic effects of gene perturbation in Caenorhabditis elegans. Nat. Genet. 40: 181–188.

Lee, I., B. Lehner, T. Vavouri, J. Shin, A.G. Fraser and E.M. Marcotte. 2010. Predicting genetic modifier loci using functional gene networks. Genome Res. 20: 1143–1153.

Lenffer, J., F.W. Nicholas, K. Castle, A. Rao, S. Gregory, M. Poidinger, M.D. Mailman and S. Ranganathan. 2006. OMIA (Online Mendelian Inheritance in Animals): an enhanced platform and integration into the Entrez search interface at NCBI. Nucleic Acids Res. 34: D599–D601.

Lyne, R., R. Smith, K. Rutherford, M. Wakeling, A. Varley, F. Guillier, H. Janssens, W. Ji, P. McLaren, P. North, D. Rana, T. Riley, J. Sullivan, X. Watkins, M. Woodbridge, K. Lilley, S. Russell, M. Ashburner, K. Mizuguchi and G. Micklem. 2007. FlyMine: an integrated database for Drosophila and Anopheles genomics. Genome Biol. 8: R129.

Markowetz, F., J. Bloch and R. Spang. 2005. Non-transcriptional pathway features reconstructed from secondary effects of RNA interference. Bioinformatics 21: 4026–4032.

Martinetz, T.M., S.G. Berkovich and K.J. Schulten. 1993. 'Neural-gas' network for vector quantization and its application to timeseries prediction. IEEE Transactions on Neural Networks 4: 558–569.

McKusick, V.A. 1960. Medical Genetics 1958–1960: An Annotated Review. Mosbay, St. Louis.

McKusick, V.A. 1992. Mendelian Inheritance in Man: Catalogs of Autosomal Dominant, Autosomal Recessive, and X-Linked Phenotypes. Johns Hopkins University Press, Baltimore.

McKusick, V.A. 2007. Mendelian inheritance in man and its online version, OMIM. Am J Hum Genet. 80: 588–604.

Mewes, H.W., S. Dietmann, D. Frishman, R. Gregory, G. Mannhaupt, K.F. Mayer, M. Munsterkotter, A. Ruepp, M. Spannagl, V. Stumpflen and T. Rattei. 2008. MIPS: analysis and annotation of genome information in 2007. Nucleic Acids Res. 36: D196–D201.

MGI. 2012. MGI-Statistics for the Mouse Genome Informatics database resource. http://www.informatics.jax.org/mgihome/homepages/stats/all_stats.shtml. Accessed: March 2012.

Mitropoulou, C., A.J. Webb, K. Mitropoulos, A.J. Brookes and G.P. Patrinos. 2010. Locus-specific database domain and data content analysis: evolution and content maturation toward clinical use. Hum Mutat 31: 1109–1116.

Mungall, C.J., G.V. Gkoutos, C.L. Smith, M.A. Haendel, S.E. Lewis and M. Ashburner. 2010. Integrating phenotype ontologies across multiple species. Genome Biol. 11: R2.

Mungall, C.J., C. Torniai, G.V. Gkoutos, S.E. Lewis and M.A. Haendel. 2012. Uberon, an integrative multi-species anatomy ontology. Genome Biol. 13: R5.

Nicholas, F.W. 2003. Online Mendelian Inheritance in Animals (OMIA): a comparative knowledgebase of genetic disorders and other familial traits in non-laboratory animals. Nucleic Acids Res. 31: 275–277.

Patrinos, G.P. and A.J. Brookes. 2005. DNA, diseases and databases: disastrously deficient. Trends Genet. 21: 333–338.

PharmGKB. 2012. The Pharmacogenetics and Pharmacogenomics Knowledge Base. http://www.pharmgkb.org/. Accessed: March 2012.

Piano, F., A.J. Schetter, M. Mangone, L. Stein and K.J. Kemphues. 2000. RNAi analysis of genes expressed in the ovary of Caenorhabditis elegans. Curr. Biol. 10: 1619–1622.

Piano, F., A.J. Schetter, D.G. Morton, K.C. Gunsalus, V. Reinke, S.K. Kim and K.J. Kemphues. 2002. Gene clustering based on RNAi phenotypes of ovary-enriched genes in C. elegans. Curr. Biol. 12: 1959–1964.

Prank, K., E. Schulze, O. Eckert, T.W. Nattkemper, M. Bettendorf, C. Maser-Gluth, T.J. Sejnowski, A. Grote, E. Penner, A. von Zur Muhlen and G. Brabant. 2005. Machine learning approaches for phenotype-genotype mapping: predicting heterozygous mutations in the CYP21B gene from steroid profiles. Eur J Endocrinol 153: 301–305.

Reumers, J., L. Conde, I. Medina, S. Maurer-Stroh, J. Van Durme, J. Dopazo, F. Rousseau and J. Schymkowitz. 2008. Joint annotation of coding and non-coding single nucleotide polymorphisms and mutations in the SNPeffect and PupaSuite databases. Nucleic Acids Res. 36: D825–D829.

Reumers, J., S. Maurer-Stroh, J. Schymkowitz and F. Rousseau. 2006. SNPeffect v2.0: a new step in investigating the molecular phenotypic effects of human non-synonymous SNPs. Bioinformatics 22: 2183–2185.

Robinson, P.N. and S. Mundlos. 2010. The human phenotype ontology. Clin Genet. 77: 525–534.

Rodin, A., T.H. Mosley, Jr., A.G. Clark, C.F. Sing and E. Boerwinkle. 2005. Mining genetic epidemiology data with Bayesian networks application to APOE gene variation and plasma lipid levels. J Comput Biol. 12: 1–11.

Rogers, A., I. Antoshechkin, T. Bieri, D. Blasiar, C. Bastiani, P. Canaran, J. Chan, W.J. Chen, P. Davis, J. Fernandes, T.J. Fiedler, M. Han, T.W. Harris, R. Kishore, R. Lee, S. McKay, H.M. Muller, C. Nakamura, P. Ozersky, A. Petcherski, G. Schindelman, E.M. Schwarz, W. Spooner, M.A. Tuli, K. Van Auken, D. Wang, X. Wang, G. Williams, K. Yook, R. Durbin, L.D. Stein, J. Spieth and P.W. Sternberg. 2008. WormBase 2007. Nucleic Acids Res. 36: D612–D617.

Roque, F.S., P.B. Jensen, H. Schmock, M. Dalgaard, M. Andreatta, T. Hansen, K. Soeby, S. Bredkjaer, A. Juul, T. Werge, L.J. Jensen and S. Brunak. 2011. Using electronic patient records to discover disease correlations and stratify patient cohorts. PLoS Comput Biol. 7: e1002141.

Roux-Rouquie, M., M.L. Chauvet, A. Munnich and J. Frezal. 1999. Human genes involved in chromatin remodeling in transcription initiation, and associated diseases: An overview using the GENATLAS database. Mol. Genet. Metab. 67: 261–277.

Rual, J.F., J. Ceron, J. Koreth, T. Hao, A.S. Nicot, T. Hirozane-Kishikawa, J. Vandenhaute, S.H. Orkin, D.E. Hill, S. van den Heuvel and M. Vidal. 2004. Toward improving Caenorhabditis elegans phenome mapping with an ORFeome-based RNAi library. Genome Res. 14: 2162–2168.

Schofield, P.N., G.V. Gkoutos, M. Gruenberger, J.P. Sundberg and J.M. Hancock. 2010. Phenotype ontologies for mouse and man: bridging the semantic gap. Dis. Model Mech 3: 281–289.

Schulz, M.H., S. Kohler, S. Bauer and P.N. Robinson. 2011. Exact score distribution computation for ontological similarity searches. BMC Bioinformatics 12: 441.

Schwarz, E.M., I. Antoshechkin, C. Bastiani, T. Bieri, D. Blasiar, P. Canaran, J. Chan, N. Chen, W.J. Chen, P. Davis, T.J. Fiedler, L. Girard, T.W. Harris, E.E. Kenny, R. Kishore, D. Lawson, R. Lee, H.M. Muller, C. Nakamura, P. Ozersky, A. Petcherski, A. Rogers, W. Spooner, M.A. Tuli, K. Van Auken, D. Wang, R. Durbin, J. Spieth, L.D. Stein and P.W. Sternberg. 2006. WormBase: better software, richer content. Nucleic Acids Res. 34: D475–D478.

Scriver, C.R. 2004. After the genome—the phenome? Journal of Inherited Metabolic Disease 27: 305–317.

Shi, Y. 2003. Mammalian RNAi for the masses. Trends Genet. 19: 9–12.

Simmer, F., C. Moorman, A.M. van der Linden, E. Kuijk, P.V. van den Berghe, R.S. Kamath, A.G. Fraser, J. Ahringer and R.H. Plasterk. 2003. Genome-wide RNAi of C. elegans using the hypersensitive rrf-3 strain reveals novel gene functions. PLoS Biol. 1: E12.

Smith, C.L., C.A. Goldsmith and J.T. Eppig. 2005. The Mammalian Phenotype Ontology as a tool for annotating, analyzing and comparing phenotypic information. Genome Biol. 6: R7.

Sonnichsen, B., L.B. Koski, A. Walsh, P. Marschall, B. Neumann, M. Brehm, A.M. Alleaume, J. Artelt, P. Bettencourt, E. Cassin, M. Hewitson, C. Holz, M. Khan, S. Lazik, C. Martin, B. Nitzsche, M. Ruer, J. Stamford, M. Winzi, R. Heinkel, M. Roder, J. Finell, H. Hantsch, S.J. Jones, M. Jones, F. Piano, K.C. Gunsalus, K. Oegema, P. Gonczy, A. Coulson, A.A. Hyman and C.J. Echeverri. 2005. Full-genome RNAi profiling of early embryogenesis in Caenorhabditis elegans. Nature 434: 462–469.

Sprague, J., L. Bayraktaroglu, Y. Bradford, T. Conlin, N. Dunn, D. Fashena, K. Frazer, M. Haendel, D.G. Howe, J. Knight, P. Mani, S.A. Moxon, C. Pich, S. Ramachandran, K. Schaper, E. Segerdell, X. Shao, A. Singer, P. Song, B. Sprunger, C.E. Van Slyke and M. Westerfield. 2008. The Zebrafish Information Network: the zebrafish model organism database provides expanded support for genotypes and phenotypes. Nucleic Acids Res. 36: D768–D772.

Stephens, P., S. Edkins, H. Davies, C. Greenman, C. Cox, C. Hunter, G. Bignell, J. Teague, R. Smith, C. Stevens, S. O'Meara, A. Parker, P. Tarpey, T. Avis, A. Barthorpe, L. Brackenbury, G. Buck, A. Butler, J. Clements, J. Cole, E. Dicks, K. Edwards, S. Forbes, M. Gorton, K. Gray, K. Halliday, R. Harrison, K. Hills, J. Hinton, D. Jones, V. Kosmidou, R. Laman, R. Lugg, A. Menzies, J. Perry, R. Petty, K. Raine, R. Shepherd, A. Small, H. Solomon, Y. Stephens, C. Tofts, J. Varian, A. Webb, S. West, S. Widaa, A. Yates, F. Brasseur, C.S. Cooper, A.M. Flanagan, A. Green, M. Knowles, S.Y. Leung, L.H. Looijenga, B. Malkowicz, M.A. Pierotti, B. Teh, S.T. Yuen, A.G. Nicholson, S. Lakhani, D.F. Easton, B.L. Weber, M.R. Stratton, P.A. Futreal and R. Wooster. 2005. A screen of the complete protein kinase gene family identifies diverse patterns of somatic mutations in human breast cancer. Nat. Genet. 37: 590–592.

Sutton, W.S. 1902. On the morphology of the chromosome group in Brachystola magna. Biological Bulletin 4: 24–39.

Troyanskaya, O.G., K. Dolinski, A.B. Owen, R.B. Altman and D. Botstein. 2003. A Bayesian framework for combining heterogeneous data sources for gene function prediction (in Saccharomyces cerevisiae). Proc. Natl. Acad. Sci. USA 100: 8348–8353.

Tschermak, E. 1900. Ueber kuenstliche Kreuzung bei Pisum sativum. Berichte der deutschen Botanischen Gesellschaft XVIII: 232–239.

Tuschl, T. 2003. Functional genomics: RNA sets the standard. Nature 421: 220–221.

Tuschl, T. and A. Borkhardt. 2002. Small interfering RNAs: a revolutionary tool for the analysis of gene function and gene therapy. Mol. Interv. 2: 158–167.

Tweedie, S., M. Ashburner, K. Falls, P. Leyland, P. McQuilton, S. Marygold, G. Millburn, D. Osumi-Sutherland, A. Schroeder, R. Seal and H. Zhang. 2009. FlyBase: enhancing Drosophila Gene Ontology annotations. Nucleic Acids Res. 37: D555–D559.

Twigger, S.N., D. Pasko, J. Nie, M. Shimoyama, S. Bromberg, D. Campbell, J. Chen, N. Dela Cruz, C. Fan, C. Foote, G. Harris, B. Hickmann, Y. Ji, W. Jin, D. Li, J. Mathis, N. Nenasheva, R. Nigam, V. Petri, D. Reilly, V. Ruotti, E. Schauberger, K. Seiler, R. Slyper, J. Smith, W. Wang, W. Wu, L. Zhao, A. Zuniga-Meyer, P.J. Tonellato, A.E. Kwitek and H.J. Jacob. 2005. Tools and strategies for physiological genomics—The Rat Genome Database. Physiol Genomics. 23(2): 246–56.

Twigger, S.N., K.D. Pruitt, X.M. Fernandez-Suarez, D. Karolchik, K.C. Worley, D.R. Maglott, G. Brown, G. Weinstock, R.A. Gibbs, J. Kent, E. Birney and H.J. Jacob. 2008. What everybody should know about the rat genome and its online resources. Nat. Genet. 40: 523–527.

van de Peppel, J., N. Kettelarij, H. van Bakel, T.T. Kockelkorn, D. van Leenen and F.C. Holstege. 2005. Mediator expression profiling epistasis reveals a signal transduction pathway with antagonistic submodules and highly specific downstream targets. Mol. Cell 19: 511–522.

van Driel, M.A., J. Bruggeman, G. Vriend, H.G. Brunner and J.A. Leunissen. 2006. A text-mining analysis of the human phenome. Eur J Hum Genet. 14: 535–542.

Van Driessche, N., J. Demsar, E.O. Booth, P. Hill, P. Juvan, B. Zupan, A. Kuspa and G. Shaulsky. 2005. Epistasis analysis with global transcriptional phenotypes. Nat. Genet. 37: 471–477.

Vapnik, V. and O. Chapelle. 2000. Bounds on error expectation for support vector machines. Neural Comput 12: 2013–2036.

Walhout, A.J., J. Reboul, O. Shtanko, N. Bertin, P. Vaglio, H. Ge, H. Lee, L. Doucette-Stamm, K.C. Gunsalus, A.J. Schetter, D.G. Morton, K.J. Kemphues, V. Reinke, S.K. Kim, F. Piano and M. Vidal. 2002. Integrating interactome, phenome, and transcriptome mapping data for the C. elegans germline. Curr. Biol. 12: 1952–1958.

Watson, J.D. and F.H.C. Crick. 1953. A structure for deoxyribose nuclecic acids. Nature 171: 737–738.

Westbrook, T.F., E.S. Martin, M.R. Schlabach, Y. Leng, A.C. Liang, B. Feng, J.J. Zhao, T.M. Roberts, G. Mandel, G.J. Hannon, R.A. Depinho, L. Chin and S.J. Elledge. 2005. A genetic screen for candidate tumor suppressors identifies REST. Cell 121: 837–848.

Wheeler, D.B., A.E. Carpenter and D.M. Sabatini. 2005. Cell microarrays and RNA interference chip away at gene function. Nat. Genet. 37 Suppl: S25–S30.

Wheeler, D.L., T. Barrett, D.A. Benson, S.H. Bryant, K. Canese, V. Chetvernin, D.M. Church, M. Dicuccio, R. Edgar, S. Federhen, M. Feolo, L.Y. Geer, W. Helmberg, Y. Kapustin, O. Khovayko, D. Landsman, D.J. Lipman, T.L. Madden, D.R. Maglott, V. Miller, J. Ostell, K.D. Pruitt, G.D. Schuler, M. Shumway, E. Sequeira, S.T. Sherry, K. Sirotkin, A. Souvorov, G. Starchenko, R.L. Tatusov, T.A. Tatusova, L. Wagner and E. Yaschenko. 2008. Database resources of the National Center for Biotechnology Information. Nucleic Acids Res. 36: D13–D21.

Wilson, R.J., J.L. Goodman and V.B. Strelets. 2008. FlyBase: integration and improvements to query tools. Nucleic Acids Res. 36: D588–D593.

Wood, V., M.A. Harris, M.D. McDowall, K. Rutherford, B.W. Vaughan, D.M. Staines, M. Aslett, A. Lock, J. Bahler, P.J. Kersey and S.G. Oliver. 2012. PomBase: a comprehensive online resource for fission yeast. Nucleic Acids Res. 40: D695–D699.

Yook, K., T.W. Harris, T. Bieri, A. Cabunoc, J. Chan, W.J. Chen, P. Davis, N. de la Cruz, A. Duong, R. Fang, U. Ganesan, C. Grove, K. Howe, S. Kadam, R. Kishore, R. Lee, Y. Li, H.M. Muller, C. Nakamura, B. Nash, P. Ozersky, M. Paulini, D. Raciti, A. Rangarajan, G. Schindelman, X. Shi, E.M. Schwarz, M. Ann Tuli, K. Van Auken, D. Wang, X. Wang, G. Williams, J. Hodgkin, M. Berriman, R. Durbin, P. Kersey, J. Spieth, L. Stein and P.W. Sternberg. 2012. WormBase 2012: more genomes, more data, new website. Nucleic Acids Res. 40: D735–D741.

ZFIN. 2012. The Zebrafish Information Network Data Download. http://zfin.org/zf_info/downloads.html#phenotype. Accessed: March 2012.

Index

1,000 genomes project 1
96-well plates 71, 73

A

ACEDB 238, 242
Adult Phenotypes 74, 80
age-related phenotypes 56
aggression 75
Allantoin 176, 193
Andrew Fire 238
ANN 250
antibiotic 211, 214, 215, 225–227, 231, 232
antibodies 70
antisense mRNA hybridization 70
antisense RNA (asRNA) 212, 213, 225–227
anxiety 66, 75
appearance 239, 243, 253
Arabidopsis 142, 146, 152, 157
Arabidopsis thaliana 111–141
Architecture 112, 115, 120–122, 124, 130, 132
arrayed screen 217, 217
Asian Mouse Phenotyping Consortium 31, 45
assays for human alleles 199
Australian Phenome Bank 30
autofocusing 97
automated microscopes 97
Automatic image analysis 70, 75

B

barcode 211, 220, 221, 229, 230
barley 144, 148, 157, 160
Behaviour 65, 71, 72, 75
BioDBcore 55
bioimage informatics 99
bioinformatics 27, 36, 44
Biolog 208, 210, 222, 225–229, 231
BioMart 243, 53, 57
biotechnological 195
Brachypodium 142, 146
brightfield 71

C

Candida albicans 194
candidate gene selection 89
carbohydrate metabolism 74
cardiovascular disease 73
Carl Linnaeus 14, 15
CASIMIR 44, 45, 57
Cel1 69
Cell Arrays 96, 97
Cell division 111, 119, 120
Cell expansion 111, 119, 123
cell lines 89–91
cells, cell lines 90
Charles Darwin 237
Chemical Screens 73
chemical stress 215
chlorophyll fluorescence 156
Christiane Nüsslein-Volhard 66
chromosomal engineering 209, 211–213
Chromosome 49, 50
Chromosome Substitution Strains 50
Circadian activity 72
classifiers 100, 250
clustering 101
cold tolerance 194
Collaborative Cross (CC) 49, 50
Commercial Mouse Phenotyping Efforts 46
comparative phenomics 249, 250, 252
conditional mutagenesis 28
conditional transgene expression 67
confocal microscope 71
Consortium for Functional Glycomics 49, 57
Controls 34m 37, 41, 52, 54, 56
COSMIC 246
Craig Mello 238
CreZOO database 30, 57
CRISPR/Cas system 29
CRISPR-Cas9 system 68
cross-species 240, 242, 246, 247, 254
cross-spot contamination 96, 97

D

DAS tracks 243
data 1–5
    data Accessibility 3, 4
    data reproducibility 3
    coverage 3, 4
    security 2, 3
Data Access 55
Data capture 44, 51
Data Dissemination 45, 53
Data Integration 160
data mining 34
data openness and exchange 44
Data Quality 45, 52, 54
Data Reproducibility 54
Data Storage 162
Database 126, 127, 130–132
dbGaP 245
Deep phenotyping 10
deletions 28
Design of Targeted Mutations 50
Development 111–141
developmental phenotypes 70
Digital imaging 156, 157
diploid 177, 178, 183–185, 190, 193, 194
disease models 244, 249
Disease Ontology (DO) 15, 16
dissecting microscope 70, 72, 76
distance concentration 101
distance metric 101
Diversity Outbred Mouse Population (DO)
    49, 50
DRSC 240, 245
drug dependency 76
drug stress 209

E

early lethality 75
Electronic Health Records 239, 242, 254
EMBL-EBI 45
Embryonic Development 66, 70, 75, 76, 79,
    80
EMPReSS 36, 37, 40–42, 57
EMPReSS Slim 41
Encode Project 1
Endoreduplication 120
ENSEMBL 243, 247
Entrez Gene 245
ENU 66–69
ENU Mutagenesis 27, 28, 49, 56
ENU mutagenesis projects 27, 28, 56
environmental metadata 152, 161

EQ approach 41, 53
Euclidean distance 101
EUMODIC 3, 34, 36, 41–45, 51–53, 57, 80
Eumorphia 35–37, 40, 41, 44, 57, 242, 248
EUROFAN 251
European Mouse Disease Clinic 36
European Mouse Phenotyping Resource of
    Standardised Screens (EMPReSS) 36,
    37, 40–42, 57
European Mutant Mouse Archive (EMMA)
    30, 31, 57
EuroPhenome 43, 44, 53, 57
evolutionary selection 237
experimental advantages 25
Experimental Design 54
experimental error 4

F

false negative rates 103
false positive rate 103
false positives 196
far-infrared (FIR) imaging 158
Federation of International Mouse
    Resources (FIMRe) 30, 57
Fertility 74, 75
Flowering 111, 112, 115–117, 120, 121
fluorescence sensors 154
fluorescent marker 91, 98
fluorescent protein 87
FlyBase 238, 240, 243, 245
FlyRNAi 240, 245
forward genetics 190, 195
Fry Behaviour 71
Functional assays 70
functional characterization 25

G

Gal4/UAS system 67
galactose 175, 176, 179, 183, 184, 191
GEN2PHEN 242, 246
GenAtlas 240, 245
gene-deletion collection 173
gene deletions 181, 184, 186, 188, 195
gene-gene interactions 5
Gene Ontology 12, 37, 252
gene replacements 28
gene targeting 35
gene trapping 28, 35
genetic background 188
genetic engineering 237, 255
genetic fate mapping 67

Genome engineering 28
GenomeRNAi 240, 247
genome-wide association studies (GWAS) 190, 191
Genotype 8–10, 18, 20, 21
genotype-phenotype repository 254
George Streisinger 66
green area index 155
green fluorescent protein (GFP) 68
Gregor Mendel 237
Growth 112–115, 118–124, 126, 128, 129, 131
growth curves 178–181, 186, 193
GWAS 241, 245

H

haploid 177, 178, 183, 186, 195
hemizygotes 190, 193
HGMD 240, 246
HGVbase 246
hierarchical clustering 101
high-dimensions, dimensions 101
High-Throughput Sequencing 25, 55, 56
hit validation 103
hits 102, 103
human phenome project 1
Human Phenotype Ontology (HPO) 12, 17–20, 254
Human Variome Project 55
humanized yeast 197
Hyper-spectral sensors 155

I

Illumina sequencing 69, 77
image analysis 90, 92, 94, 98, 99, 102, 104
image annotation 99, 100
Imaging 119, 124, 132
IMPRESS 45, 57
in situ staining methods 70
informatics 3
informatics infrastructure 37
Information content 19, 20
Infrafrontier 31, 32, 57
insertional mutagenesis 68
insertions 28
International Classification of Diseases (ICD) 14–16
International Mouse Knockout Consortium (IKMC) 28, 35, 57, 80
International Mouse Phenotyping Consortium (IMPC) 5, 26, 44–46, 50–52, 54, 55, 57, 80

International Mouse Strain Resource (IMSR) 30, 57
Interphenome 44, 45, 57
inversions 28

J

Jackson Laboratory 28, 44, 46, 54, 57
Juvenile Survival 79

K

KASP genotyping 69
kNN 250
knockout library 211, 227
Knockout Mouse Project 30, 35, 57
Knockout Mouse Project (KOMP) repository 30, 35, 41, 50, 57

L

Laboratory mouse 24–26, 36, 50
laboratory zebrafish strain 66
large-scale analysis 254
learning 72, 75, 76
Linkage analysis 190, 191, 193

M

MAGIC 251
Maize 142, 144, 146, 148, 156, 157, 163
Mammalian Phenotype Ontology (MP) 37, 44, 252, 254
mammals 24
manual annotation 87, 98, 100
manual curation 52
mariner transposon 221, 231, 233
MartView 53, 57
medical genetics 238
medical history 9, 10
memory 75
mendelian 237, 238, 240, 241, 245
Mendelian Inheritance in Man 238, 240, 245
Metabolism 66, 73, 74
Metadata 29, 44, 152, 160–162
metagenomic 210, 214, 222, 223, 231, 232
MGD 238, 240
MGI 244
Michael Ashburner 238
Microcultivation 180, 181, 193
Microscopy 118, 119, 123
MIMPP 44
MIPS 240, 251
mitochondrial pheno types 238
MMPC 48, 53, 58

model organism 24, 26, 240, 245, 254, 255
molecular bar code 182
Molecular Phenotyping 55
monogenic 238, 241
Morphology 74, 75, 78, 79
motility mutants 71
Mouse Breeding Logistics 51
mouse clinics 29, 31, 32, 44, 56
mouse embryonic stem cells (mESCs) 28
mouse facilities 29, 32
Mouse genes 25, 45
mouse genome 25, 26, 28, 35, 44, 46, 57
Mouse Genome Database 44
Mouse Genome Informatics 26, 44, 46, 57
Mouse Mutagenesis 27, 58
Mouse Mutant Resource Regional Centers
    (MMRRC) 30
Mouse Phenome Database 33, 34, 44, 46,
    53, 58
Mouse Phenomics Project 32, 34, 50, 54–56,
    242
Mouse Phenotype Database Integration
    Consortium 44
MPD 239, 240, 244
Multi-allelic Phenotype Analysis 77
multifactorial 241
multi-well plates 96
mutagenesis 209, 211–213, 219, 220, 222,
    225, 229, 232
Mutagenesis Methods 67
Mutagenetix 27, 29, 31, 49, 58
Mutant mouse archives (MMAs) 29, 30
Mutant Mouse Production 41

N

Nathan Shock Center of Excellence in the
    Basic Biology of Aging 46, 58
NCBI 245
NDVI 155
Near-infrared 155, 158
negatives 185, 196
Neuromice Consortium 49
Next Generation Sequencing 25
Nile Red 74
nomenclature 27, 33, 34

O

object features, feature vector 99
OBO Foundry 12, 14
off-target effects 88, 89
Oil Red O 73

Online Mendelian Inheritance in Man
    (OMIM) 15, 16, 20
Ontology 128–130
opioid signaling 76
Orphanet 15, 16, 20
overexpression 184, 187, 188, 210, 213, 214,
    227, 228

P

parental effect mutations 75, 79
PATO 53
Paul Berg 237
PED6 73
Permeability 196
personalized medicine 11
PharmGKB 240, 246
PheGenI 245
PhenoBLAST 249
Phenoclusters 249, 252
phenodeviants 27
phenome-interactome network 249
PhenomicDB 240, 243, 246–248, 252
phenomics 1–7, 65–85, 237–262
    aims 1, 5
    definition 2
    origins 1
Phenomizer 18–20
PhenoStat 248
Phenotype 8–12, 16–18, 20, 21
phenotype collection in humans 2
phenotype community 241, 254, 255
phenotype repositories 239, 254
phenotypic abnormalities 10, 18–21
phenotypic abnormality 9, 17, 18
phenotypic profiles 99
phenotypic quality ontology 254
Phenotypic space 149
Phenotyping 111–141
Phenotyping Data Quality 52
Phenotyping Informatics 51
Phenotyping Pipelines 41, 42, 56, 57
Phenotyping Protocols 42, 46, 51
Phenotyping Schemes, Design of
    Phenotyping Scheme 51
phenotyping tests 29, 31, 36, 37, 41–45, 47
phospholipase A_2 73
physical examination 9, 10
Physical Infrastructure Underpinning
    Mouse Phenotyping 29
piggyback 28
Plant 111–141
plant performance 147
plasmids 184

plasticity 148, 150, 161
pleiotropy 186, 187, 191, 195
point mutations 27, 28, 49
pooled screen 211, 219, 220, 222
Population Phenomics 189
population structure 188, 190, 191, 194
precision medicine 11
Prediction of Phenotypes 253
protein expression 56
proteomics 55

Q

QTLs 190, 191

R

Random Mutagenesis 68
recessive mutants 66
reference genome sequence 66
reflectance sensors 154, 155
replicates 103, 104
Retroviruses 68
reverse genetic approach 77
reverse genetics 183, 188, 189, 194, 195
RGD 240, 244
Rhizotrons 153, 157
Rice 142, 144, 146, 148
RNA expression 56
RNA-guided DNA cleavage 29
RNA interference (RNAi) 28, 87, 240, 241,
    243–245, 247–250, 252, 253
RNAi libraries 89
RNAiDB 240, 245
Robustness 149
Rome Agenda 44
Root 112, 114, 122–126, 132
Root imaging 153

S

S score 218, 222
Saccharomyces cerevisiae 172
salt tolerance 156
Scaling 145, 147, 158
Semantic Integration of data 55
Shoot 111–118, 120–122, 125, 126, 132, 133
short hairpin RNAs 28
shuttle box 75
signature-tagged mutagenesis (STM) 220,
    226, 229–231
similarity 101
similarity metrics 252
single nucleotide variations 27
Sleep 66, 72, 73

Sleeping Beauty 28
Smarter Phenotyping 56
SNOMED CT 13–15, 20
SNPeffect 246
Soil 152–154, 156–159
sorghum 144, 148
Sources of Error 4
soybean 144, 158
Spectral measurements 155
Spermatogonial mutagenesis 69
Spores 173, 177, 178, 196
Statistical Analysis of Mouse Phenomics
    Data 43
Statistical Data Analysis 96, 102
stochastic variation in the expression of
    phenotypes 4
supervized machine learning 99, 100
support vector machine 100, 104
Surrogate Genetics 196, 197
SVMs 250
synchrotron micron-scale computer
    tomography (MicroCT) 75
Systematic Phenotyping, Controls 52
Systems Genetics 49, 56

T

target gene inference 89
Targeted Genome Editing 68
Targeted Mutation 28
Targeting-induced local lesions in genomes
    (TILLING) 69
Temperature Sensitivity 78
The Jackson Laboratory 238, 239
Thermal imaging 157
time series 101
time-lapse bright-field microscopy 182
time-lapse imaging 91, 98
T-maze 75
Tools 27, 34, 53
Touch response 71
Trait 8, 9, 17
Transcription activator-like effector
    nucleases (TALENs) 29, 68, 69
Transcriptome Analysis 79
transgenic approaches 67
transgenic plants 147
translocations 28
transposon 183, 184, 188, 194, 209, 211, 216,
    219, 220
Transposon insertional mutagenesis 28
transposons 68
TRIPLES 251
type 2 diabetes 73

U

UMLS 249
UMLS Metathesaurus 13
United Medical Language System 13
unsupervized machine learning 100
US National Mouse Metabolic Phenotyping
    Center (MMPC) 48, 53, 58

V

Valley of Death 148, 149, 154
vectorized phenotypes 249
vegetation indices 154, 155, 157
vertebrate embryo 65
vertebrate genetics 66
Victor A. McKusick 238
Vision 72

W

Wellcome Trust Sanger Institute 65, 66, 76,
    80
Wellcome Trust Sanger Institute Mouse
    Genetics Project 41
wheat 148, 151, 155–157, 159

WormBase 238, 240, 242–245
WormNet 253
WTSI-GMP 41–43

X

XML 44, 51, 52, 243

Y

Yeast 172–207
Yeast Fitness 177
yield 143, 145, 147, 148, 150–153, 155, 163

Z

Z score 222
Zebrafish 65–85
zebrafish forward genetics 66
zebrafish genome-sequencing project 66
Zebrafish Mutation Project (ZMP) 69, 76–79
Zebrafish phenotyping methods 70
ZFIN 239, 240, 243
Zinc finger endonucleases 68

Color Plate Section

Chapter 3

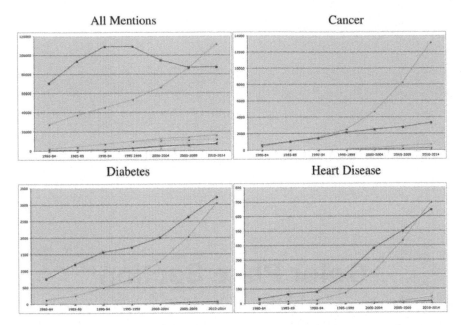

Figure 1. Citations to common model organisms in PubMed by era. The four panels represent the raw numbers of hits to searches using different search terms or combinations thereof to model organisms. X-axis: year range; Y-axis: number of hits (counts for 2010–2014 normalised from counts for 2010–2012). **All Mentions**: searches for mouse (blue), rat (red), *Drosophila* (green), yeast (orange; search for "cerevisiae"), zebrafish (yellow), *C. elegans* (purple; search for "elegans"); **Cancer**: search for keyword "cancer" plus organism term; **Diabetes**: search for keyword "diabetes" plus organism term; **Heart Disease**: search for keyword "Heart Disease" or "Cardiovascular Disease" plus organism term.

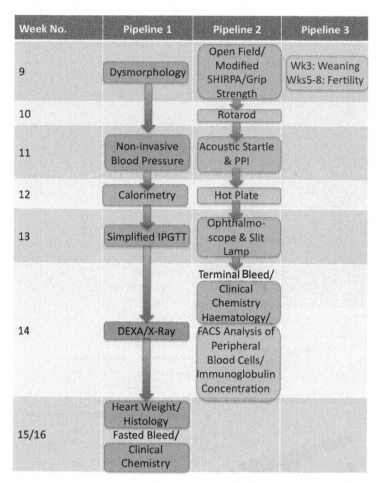

Week No.	Pipeline 1	Pipeline 2	Pipeline 3
9	Dysmorphology	Open Field/ Modified SHIRPA/Grip Strength	Wk3: Weaning Wks5-8: Fertility
10		Rotarod	
11	Non-invasive Blood Pressure	Acoustic Startle & PPI	
12	Calorimetry	Hot Plate	
13	Simplified IPGTT	Ophthalmo-scope & Slit Lamp	
14	DEXA/X-Ray	Terminal Bleed/ Clinical Chemistry Haematology/ FACS Analysis of Peripheral Blood Cells/ Immunoglobulin Concentration	
15/16	Heart Weight/ Histology Fasted Bleed/ Clinical Chemistry		

Figure 2. Overview of the EUMODIC phenotyping pipelines. Phenotyping protocols are arranged by week vertically and by pipeline horizontally. Colouring reflects the class of test carried out: Blue = Morphology & Metabolism; Red = Cardiovascular; Grey = Bone; Green = Neurobehavioural & Sensory; Purple = Haematology & Clinical Chemistry; Orange = Allergy & Immune. Redrawn from http://empress.har.mrc.ac.uk/viewempress/index.php?map=EUMODIC+Pipeline.

Chapter 6

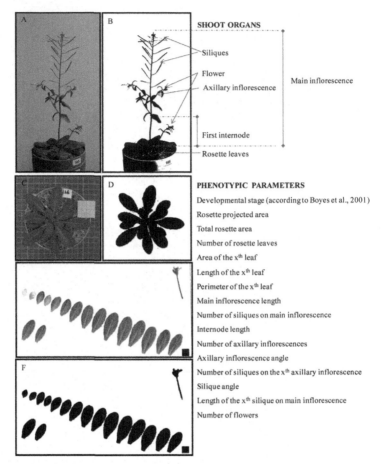

Figure 1. Non-exhaustive list of two-dimensional shoot phenotypic traits extracted from lateral view images (A, B), top view images (C, D) and scans of dissected shoots (E, F). The different shoot organs that are visible on the images are indicated on the left of Figure B and the list of quantitative traits is given below. Depending on the traits, they can be extracted automatically with macros developed on image analysis software, by counting organs manually, or tracing organ length, width or contours.

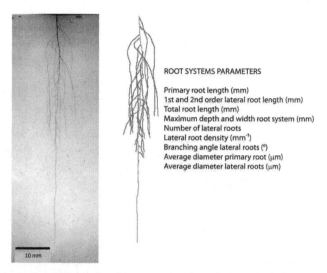

Figure 2. Automated two-dimensional feature extraction of young Arabidopsis root systems is possible in transparent cultivation media.

Chapter 8

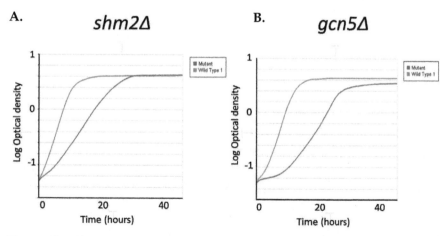

Figure 2. Growth curves of yeast cultures generated by microcultivation. Yeasts were growing in 350 µl of synthetic medium cultures with glucose as carbon and energy source. The optical density was automatically recorded every 20 minutes over two days. (A) Quantitative growth analysis of the *shm2Δ* mutant (red curve) compared to the wild type (grey curve). This mutant displays a clear change in growth rate. (B) Quantitative growth analysis of the *gcn5Δ* mutant (red curve) compared to the wild type (grey curve). This mutant displays clear changes in all three growth variables—growth lag, growth rate and growth efficiency.

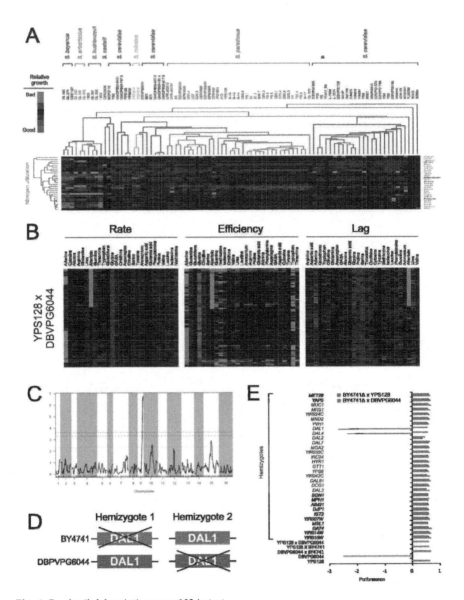

Fig. 4. *For detailed description see p. 193 in text.*

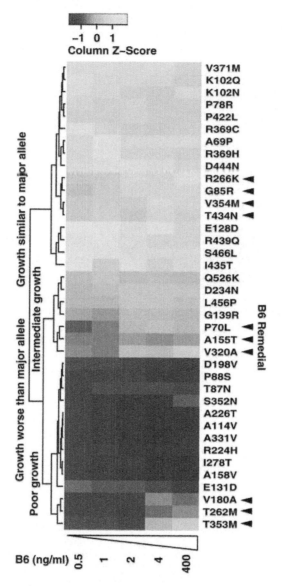

Figure 5. Humanized yeasts expressing different alleles of the enzyme cystathionine-b-synthase show different growth responses. Growth responses in relation to the concentration of vitamine B6, one of the co-factors of the enzyme (Mayfield et al. 2012). Heat maps of growth rates normalized to the growth of the major allele. The column Z-score indicates the mean growth rate (Z-score of 0) and standard deviation (Z-score of 61) of all alleles per column. Arrowheads indicate alleles that respond to cofactor titration more strongly than other alleles in their cluster. Published with permission from The Genetics Society of America.

Chapter 10

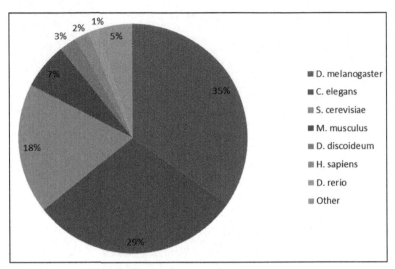

Figure 2. Distribution of phenotype entries in PhenomicDB v. 3.5 by species.